Preventing Occupational Exposures to Bloodborne Pathogens

Articles from Advances in Exposure Prevention, 1994–2003

Janine Jagger, M.P.H., Ph.D., and Jane Perry, M.A., *Editors*

International Healthcare Worker Safety Center

University of Virginia Health System
Charlottesville, Virginia

Publication of this book was made possible by an unrestricted educational grant from

The Safety Institute, Premier Inc.

Published by the International Healthcare Worker Safety Center
Division of Infectious Diseases, Department of Internal Medicine
P.O. Box 800764
University of Virginia
Charlottesville, VA 22908-0764
Phone (434) 924-5159
Fax (434) 982-0821

Designer: Mary Michaela Murray
Printer: CompuDesign & Associates, Inc.

ISBN 0-9655899-1-9

Library of Congress Control Number: 2004102905

TABLE OF CONTENTS

PART I: EPINet Data Reports

Part II: Occupationally Infected Healthcare Workers: Interviews and Personal Accounts

Part III: Other AEP Articles

Part IV: Needle Safety and Needlestick Prevention—Legislation and Policy

Preface

Advances in Exposure Prevention was launched by the International Healthcare Worker Safety Center in 1994 as a forum for rapidly communicating the latest findings from our EPINet multi-hospital sharps injury and blood and body fluid exposure database, and for discussing government policy, legislative developments, and best practices and devices to prevent occupational exposures to bloodborne pathogens. We also intended AEP to be a resource for government policy-makers and legislators, to support initiatives to protect healthcare workers from occupational infection. Over the past ten years AEP has been put to good use in many ways, fulfilling its original objectives. The numerous issues of AEP have been read and circulated, some filed, some scattered to the four winds. Because they represent a coherent body of work, we decided to gather them into a retrospective volume, which we are pleased to present here.

We introduce this volume with our first article published in the *New England Journal of Medicine* in 1988, "Rates of Needle-Stick Injury Caused By Various Devices in a University Hospital," which set us on a national course to safer medical device design. We conclude the volume with an article published in the *Journal* in 2001, "Risks to Health Care Workers in Developing Countries," which set forth our vision for the future: to bring effective protection to healthcare workers in the farthest reaches of the globe.

The cumulative body of work this book represents traces the history of a critical period when risks to healthcare workers had become more lethal than ever, while at the same time preventive technologies were becoming more effective than ever. It tells the story of the coordinated, and ultimately successful, efforts of researchers, government agencies, medical products manufacturers, professional associations, unions and individual healthcare workers to tip the balance in favor of workplace health and safety.

We thank all those who have contributed their ideas, energy and time to AEP—especially our editorial advisory board and our many colleagues who have written articles for AEP. One of AEP's most important roles has been providing a voice for healthcare workers who have experienced the consequences of an occupational blood exposure. Their courage in sharing their experiences in AEP gave names, faces and a heart to this issue, and a sense of urgency to legislators and government policy-makers. They have been a continuing inspiration to us, giving meaning and purpose to our efforts.

We would also like to recognize each hospital that contributed their EPINet sharps injury and blood exposure surveillance data to our multi-hospital research database; their names are listed in the acknowledgements. The database now includes more than 25,000 blood exposure incidents—each one representing a healthcare worker who reported information on his or her exposure. The database as a whole is the source of much of what we know about the risk of blood exposures to healthcare workers; that knowledge rests on the foundation of individual reports diligently provided by thousands of workers. The surveillance coordinators at each participating hospital made a special contribution of their time and effort to compile, maintain, and transmit data to our center. We thank them and we hope that we have been good custodians of their information.

Our international collaborators have brought depth and perspective to our understanding of healthcare worker risks. In particular, we acknowledge our close collaborators at the SIROH group (Italy's national program of surveillance and research on occupational risk of HIV and other bloodborne diseases among healthcare workers), directed by Dr. Giuseppe Ippolito; and also our Japanese colleagues, who developed an EPINet-Japan surveillance database on occupational exposures under the direction of Dr. Satoshi Kimura. We have benefited tremendously from the expertise of these colleagues, and the opportunities for comparative research that they have made possible.

We are particularly proud to document the history of healthcare worker protection in the United States. Nationally, we have accomplished something unique in the time interval represented in this volume. Working together, we have brought about a revolution in the safety of medical device design. Healthcare employers are now required by law to provide to workers sharp devices with engineered sharps injury protection. No other country has achieved the level of safety that has become a standard requirement for U.S. healthcare employers. This is an accomplishment we share. But there remains so much more to do beyond our borders. Let us learn from our experience and set our sights high in order to bring these beneficial changes to every healthcare worker around the globe.

Janine Jagger, M.P.H., Ph.D., Editor-in-Chief, AEP
Jane Perry, M.A., Managing Editor, AEP

Acknowledgements

We want to recognize and thank our colleagues at what was once the Vascular Access unit of Johnson and Johnson—a unit that was first part of a division called Critikon, then Johnson & Johnson Medical, then Ethicon Endo-Surgery. More recently, the vascular access business was sold to Medex, Inc. Their generous financial support for AEP over the last ten years has been unwavering, and crucial to our efforts.

BD and Medisystems, Inc., have also been long-time supporters of AEP and the Center, through bulk subscriptions and educational grants. In addition, Medisystems provided some funding for production of this book, for which we are grateful.

The Safety Institute of Premier, Inc., made possible the printing and distribution of the book through an unrestricted educational grant; we are pleased to acknowledge Premier's generous support.

EPINet Network Hospitals, 1993-2001

The following hospitals participated in the EPINet data-sharing network, coordinated by the International Healthcare Worker Safety Center at the University of Virginia, during the period 1993 to 2001. The occupational exposure data these hospitals provided for the last decade have been the foundation of our research; we are very grateful for their time and efforts, and wish to acknowledge their invaluable contributions.

We thank David Dodge, President and CEO of PHT Services, Ltd. (PHTS), a provider of workers' compensation coverage to South Carolina's healthcare industry, for his leadership role as one of the founders of the EPINet network. Of the 90 hospitals listed below, two-thirds are PHTS member hospitals. We also thank the Occupational Health and Safety staff of the Providence Health System (formerly Sisters of Providence Hospitals) for their dedicated work.

IHWSC Network Hospitals
Florida Hospital Orlando, *Orlando, FL*
Martha Jefferson Hospital, *Charlottesville, VA*
North Broward Hospital District, *Ft. Lauderdale, FL*
Saint Joseph Hospital, *Omaha, NE*
Saint Vincent Health Center, *Erie, PA*
Shands Hospital, *Gainesville, FL*
St. Vincent Indianapolis Hospital, *Indianapolis, IN*
University Hospitals of Cleveland, *Cleveland, OH*
University of Virginia Health System, *Charlottesville, VA*

Providence Health System *(formerly Sisters of Providence)* **Network Hospitals**
Home Health Care, *Anchorage, AK*
Mary Conrad Center, *Anchorage, AK*
Medalia HealthCare LLC Clinics *(Pierce, King and Snohomish counties, WA)*
Mother Joseph Care Center, *Olympia, WA*
Providence Alaska Medical Center, *Anchorage, AK*
Providence Centralia Hospital, *Centralia, WA*
Providence Child Center, *Portland, OR*
Providence Extended Care Center, *Anchorage, AK*
Providence General Hospital and Medical Center, *Everett, WA*
Providence Horizon House, *Anchorage, AK*
Providence Medford Medical Center, *Medford, OR*
Providence Milwaukie Hospital, *Milwaukie, OR*
Providence Mount Saint Vincent, *Seattle, WA*
Providence Newberg Hospital, *Newberg, OR*
Providence Portland Medical Center, *Portland, OR*
Providence Saint Peter Hospital, *Olympia, WA*
Providence Seaside Hospital, *Seaside, OR*
Providence Seattle Medical Center, *Seattle, WA*
Providence St. Vincent Medical Center, *Portland, OR*
Providence Toppenish Hospital, *Toppenish, WA*
Providence Yakima Medical Center, *Yakima, WA*
St. Joseph Medical Center, *Burbank, CA*

PHTS Network Hospitals
Abbeville County Memorial Hospital, *Abbeville, SC*
Allendale/Barnwell Disabilities and Special Needs Board, *Barnwell, SC*
Anderson Area Medical Center, *Anderson, SC*
Bamberg County Memorial Hospital, *Bamberg, SC*
Barnwell County Hospital, *Barnwell, SC*
Beaufort Memorial Hospital, *Beaufort, SC*
Bruce Hospital System, *Florence, SC*
The Byerly Hospital, *Hartsville, SC*
Cannon Memorial Hospital, *Pickens, SC*
Carolinas Hospital System-Lake City, *Lake City, SC*
Charleston Memorial Hospital, *Charleston, SC*
Chester County Hospital, *Chester, SC*
Clarendon Memorial Hospital, *Manning, SC*
Conway Medical Center, *Conway, SC*
Darlington County Disabilities and Special Needs Board, *Hartsville, SC*
Edgefield County Hospital, *Edgefield, SC*
Fairfield Memorial Hospital, *Winnsboro, SC*

Georgetown Memorial Hospital, *Georgetown, SC*
GHS Allen Bennett Memorial Hospital, *Greer, SC*
GHS-Marshall I. Pickens Hospital, *Greenville, SC*
GHS-Roger Huntingdon Nursing Center, *Greenville, SC*
Greenville Memorial Hospital and Medical Campus/Greenville Hospital System, *Greenville, SC*
Greenwood Methodist Home, *Greenwood, SC*
Hampton Regional Medical Center, *Varnville, SC*
Healthcare Center of Wesley Commons (formerly Greenwood Methodist Home), *Greenwood, SC*
Hillcrest Hospital, *Simpsonville, SC*
Hilton Head Hospital, *Hilton Head, SC*
Kershaw County Medical Center, *Camden, SC*
Keystone Substance Abuse Services, *Rock Hill, SC*
Laurens County Hospital, *Clinton, SC*
Lexington Medical Center, *West Columbia, SC*
Lexington Richmond Alcohol and Drug Abuse Council, *Columbia, SC*
Loris Community Hospital, *Loris, SC*
The Lowman Home, Lutheran Homes of South Carolina, *White Rock, SC*
Marion County Medical Center, *Marion, SC*
Mary Black Memorial Hospital, *Spartanburg, SC*
McLeod Regional Medical Center, *Florence, SC*
Medical University of South Carolina, *Charleston, SC*
Methodist Oaks Nursing Home, *Orangeburg, SC*
Mullins County Hospital, *Mullins, SC*
Newberry County Memorial Hospital, *Newberry, SC*
North Greenville Hospital, *Greenville, SC*
Oconee Memorial Hospital, *Seneca, SC*
Palmetto-Richland Memorial Hospital, *Columbia, SC*
The Regional Medical Center of Orangeburg and Calhoun Counties, *Orangeburg, SC*
Roger C. Peace Rehabilitation Hospital, *Greenville, SC*
Saint Eugene Medical Center (McLeod Health), *Dillon, SC*
Saint Francis Hospital, Inc, *Greenville, SC*
Self Regional Healthcare, *Greenwood, SC*
South Carolina Department of Health and Environmental Control, *Columbia, SC*
Springs Memorial Hospital, *Lancaster, SC*
Palmetto Health Richland, *Columbia, SC*
Palmetto Health Baptist, *Easley, SC*
Spartanburg Hospital for Restorative Care, *Spartanburg, SC*
Spartanburg Regional Medical Center, *Spartanburg, SC*
Tuomey Regional Medical Center, *Sumter, SC*
Wallace Thomson Hospital/Union Hospital District, *Union, SC*
Williamsburg Regional Hospital/Carolinas Hospital System, *Kingstree, SC*
Women's Center of Carolinas Hospital System, *Florence, SC*

Note: Some hospitals on this list may have since merged with one or more other hospitals, changed their names, or closed.

Rates of Needle Stick Injury Caused by Various Devices in a University Hospital

Janine Jagger, M.P.H., Ph.D., Ella H. Hunt, R.N., Jessica Brand-Elnaggar, B.A., and Richard D. Pearson, M.D.

Abstract: *We identified characteristics of devices that caused needlestick injuries in a university hospital over a 10-month period. Hospital employees who reported needle sticks were interviewed about the types of devices causing injury and the circumstances of the injuries. Of 326 injuries studied, disposable syringes accounted for 35 percent, intravenous tubing and needle assemblies for 26 percent, prefilled cartridge syringes for 12 percent, winged steel-needle intravenous sets for 7 percent, phlebotomy needles for 5 percent, intravenous catheter stylets for 2 percent, and other devices for 13 percent. When the data were corrected for the number of each type of device purchased, disposable syringes had the lowest rate of needle sticks (6.9 per 100,000 syringes purchased). Devices that required disassembly had rates of injury of up to 5.3 times the rate for disposable syringes.*

One third of the injuries were related to recapping. Competing hazards were often cited as reasons for recapping. They included the risk of disassembling a device with an uncapped, contaminated needle and the difficulty of safely carrying several uncapped items to a disposal box in a single trip. New designs could provide safer methods for covering contaminated needles. Devices should be designed so that the worker's hands remain behind the needle as it is covered, the needle should be covered before disassembly of the device, and the needle should remain covered after disposal. Such improvements could reduce the incentives for recapping needles and lower the risk of needle-stick injuries among health care workers.

The epidemic of the acquired immunodeficiency syndrome (AIDS) has led to intense concern among health care workers about the risks they face in the hospital environment. Needle-stick injuries, in particular, have drawn attention, for despite safety guidelines and employee education, there is little evidence that their incidence is abating.[1-3] Transmission of human immunodeficiency virus is unusual after a needle stick.[4] Nevertheless, infections in health care workers have been attributed to this type of exposure.[5–7] The potential medical and psychological consequences of needle sticks for health care workers and their spouses or sexual partners remain great.

It is difficult to explain the relative complacency that prevailed about needle-stick injuries before the AIDS epidemic, given the serious consequences of hepatitis B virus and other infectious agents transmitted in this manner. The Hepatitis Branch of the Centers for Disease Control has estimated that 200 to 300 health care workers die each year from the direct or indirect consequences of occupationally acquired hepatitis B. As many as 12,000 become infected with the virus (unpublished data). Studies have documented the transmission of at least 20 different pathogens by needle-stick injuries.[8–26]

Overlooked in the debate on this issue has been the perspective of injury epidemiology. The National Academy of Sciences Committee on Trauma Research has concluded that improvements in product design are among the most successful approaches to the prevention of injury.[27] Notably lacking from the literature on needle sticks is information on the types and designs of instruments that injure health care workers.[28] Our study identifies these devices and the mechanisms of the injuries caused by different devices. This information suggests opportunities for improved product design and environmental modifications, thus broadening the number of options for reducing this important occupational hazard.

Methods

A descriptive study of needle-stick injuries was carried out at the University of Virginia Hospitals from January 1 through October 31, 1986. Each employee who reported a needle-stick injury to the employee health department was interviewed by a trained interviewer according to a standard format to determine the nature of the device that caused the injury, the procedure for which it was used, and the setting and mechanism of the injury and to elicit further information about the employee and the outcome of the injury. A detailed description of each incident was recorded. Underreporting of needle-stick injuries was estimated by asking employees how many needle sticks or injuries from sharp objects they had sustained before the current injury and how many were reported to an appropriate hospital authority.

The needle-stick–prevention policy in effect during the study was based on the 1983 guidelines of the Centers for Disease Control.[29] Employees were advised not to recap, bend, or break used needles and to dispose of them immediately in the closest disposal container for sharp objects. Disposal containers were located at the nurses' stations. In addition, during 6 of the 10 months of our study, containers for sharp objects were placed in patients' rooms in 4 of 25 acute care units, as part of a study conducted by the Department of Nursing.[30]

Needle-stick rates for different devices were calculated

by dividing the number of needle sticks attributed to a device by the total quantity of the device purchased by the hospital during the study period. The number of hypodermic needles used to connect intravenous lines to intravenous ports was estimated by subtracting the number of syringes purchased without hypodermic needles attached from the number of unattached hypodermic needles purchased. This yielded the maximal number of hypodermic needles available for connecting intravenous lines and therefore produced a conservative estimate of risk. This report is limited to injuries by hollow-bore needles and incidents in which the device could not be identified but was known to be a needle.

Results

During the study period, 326 needle-stick injuries were reported. Those reporting the injuries were nurses and nursing students (64 percent), laboratory technicians and auxiliary personnel in radiology and respiratory therapy (20 percent), housekeeping personnel (8 percent), attendants (4 percent), and physicians and medical students (3 percent). Past underreporting of needle sticks and injuries from sharp objects averaged 39 percent among the study participants.

The hospital services in which the injuries occurred were acute care units (68 percent), intensive care units (13 percent), laboratories and special-procedure units (11 percent), the emergency department (4 percent), and others (4 percent). None of the 326 incidents resulted in a documented infection or case of infectious disease.

Although it accounted for more injuries than any other device, the disposable syringe had the lowest needle-stick rate, with 6.9 injuries per 100,000 items purchased **(Figure 1)**. Devices that required disassembly had higher rates, ranging up to 36.7 per 100,000 for intravenous tubing and needle assemblies. **Table 1** (next page) shows the types of devices and the mechanisms of injury. Only 17 percent of the incidents occurred during the use of the items. The remainder occurred when workers were preparing the devices for disposal (70 percent) or during or after disposal (13 percent). One third of all the injuries were related to the recapping of devices.

Figure 1. Needlestick Injury Rates per 100,000 Items Purchased, for Six Devices with Needles

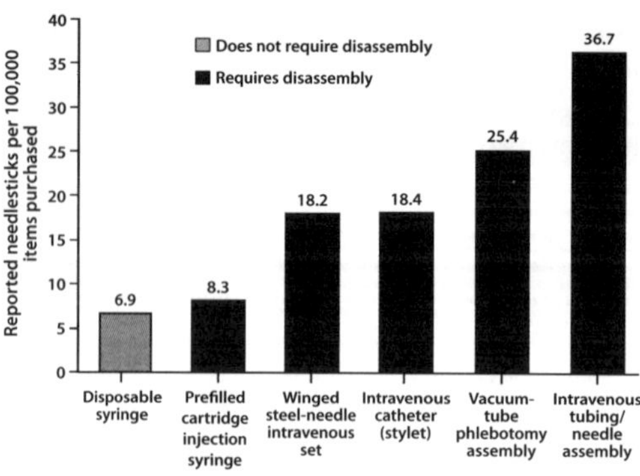

Recapping was the most common mechanism of injury from disposable syringes—a finding that is consistent with other reports. Recapping incidents occurred in three ways. First, the employee missed the cap and the needle stuck the opposing hand; this type of accident underscores the risk of moving a hand in the direction of a contaminated needle. Second, the needle pierced the cap during recapping; in 11 of 13 such incidents, the user was attempting to recap a long needle with a short cap. All such incidents involved 3.8-cm (1.5-inch) needles, suggesting that if needle caps had been uniformly long, the injuries would not have occurred. Third, the cap fell off a recapped needle. This suggests that a hazard remains after a cap is replaced, because of variability in the friction fit securing the cap to the needle hub.

Except for recapping incidents, needle sticks from syringes usually occurred after medical procedures with several steps. Employees attempted to dispose of accumulated debris in a single trip to the trash container to avoid interrupting the procedure. Injuries occurred when workers were picking up the debris or fumbling with it in transit to the disposal box.

Intravenous tubing-needle assemblies are standard hypodermic needles on intravenous tubing. These were the second most common cause of needle sticks and had the highest rate of injury. Only one fourth of incidents involving these devices were related to recapping. Because intravenous lines remain in use for long periods, needle caps are often unavailable when intravenous lines are dismantled. Needle sticks occurred when alternative methods were employed to cover used needles on intravenous needle assemblies, such as introducing them into drip chambers, intravenous ports, or intravenous bags.

Both intentional and inadvertent detachment of intravenous lines also resulted in needle-stick injuries when the needles became concealed in sheets or bedside debris or dangled freely from intravenous poles. Finally, the difficulty of disposing of a needle attached to a length of intravenous tubing was demonstrated by injuries that occurred when employees fumbled with the items on the way to the trash container or were injured by intravenous tubing and needle assemblies protruding from the openings of trash containers.

The prefilled cartridge syringe differs from the disposable syringe in that, after use, the spent cartridge and needle unit must be disengaged from the reusable holder for disposal. Most injuries with this device were directly or indirectly related to disassembly. In all, 82 percent of the needle sticks from these devices occurred during recapping in preparation for disassembly or during disassembly. Prefilled cartridges were most frequently used for perform-

Table 1. Types of Devices and Mechanisms of Injury in 326 Reported Needle Sticks.*

MECHANISM	DISPOSABLE SYRINGE no. (%)	I.V. TUBING/ NEEDLE ASSEMBLY no. (%)	PREFILLED CARTRIDGE SYRINGE no. (%)	WINGED STEEL-NEEDLE I.V. SET no. (%)	VACUUM-TUBE PHLEBOTOMY SET no. (%)	I.V. CATHETER (STYLET) no. (%)	OTHERS no. (%)	TOTAL no. (%)
Before or during use								
Patient jarred item	7 (6.2)	5 (6.0)	1 (2.6)	2 (8.3)	4 (26.7)	0 (0)	0 (0)	19 (5.8)
While withdrawing needle from patient	4 (3.5)	2 (2.4)	1 (2.6)	5 (20.8)	0 (0)	0 (0)	1 (2.3)	13 (4.0)
Item stuck hand holding I.V. port	6 (5.3)	1 (1.2)	1 (2.6)	0 (0)	0 (0)	0 (0)	1 (2.3)	9 (2.8)
Rebound or item slipped off rubber stopper	4 (3.5)	0 (0)	0 (0)	2 (8.3)	0 (0)	0 (0)	0 (0)	6 (1.8)
Other	6 (5.3)	0 (0)	0 (0)	0 (0)	0 (0)	0 (0)	1 (2.3)	7 (2.1)
After use, before disposal								
While recapping (missed cap)	32 (28.3)	11 (13.1)	7 (17.9)	0 (0)	5 (33.3)	1 (12.5)	2 (4.7)	58 (17.8)
While recapping (pierced cap)	13 (11.5)	9 (10.7)	15 (38.5)	1 (4.2)	1 (6.7)	0 (0)	1 (2.3)	40 (12.3)
Contacted item on exposed surface	10 (8.8)	9 (10.7)	2 (5.1)	1 (4.2)	1 (6.7)	3 (37.5)	9 (20.9)	35 (10.7)
Fumbled with item or in transit to trash	8 (7.1)	7 (8.3)	1 (2.6)	5 (20.8)	1 (6.7)	0 (0)	1 (2.3)	23 (7.1)
During disassembly	3 (2.7)	9 (10.7)	7 (17.9)	1 (4.2)	0 (0)	0 (0)	2 (4.7)	22 (6.7)
Item pierced I.V. port, bag, or drip chamber	0 (0)	15 (17.9)	0 (0)	0 (0)	0 (0)	0 (0)	0 (0)	15 (4.6)
Cap fell off after recapping	7 (6.2)	0 (0)	3 (7.7)	0 (0)	0 (0)	0 (0)	0 (0)	10 (3.1)
Hit by detached I.V. line	0 (0)	5 (6.0)	0 (0)	0 (0)	0 (0)	0 (0)	0 (0)	5 (1.5)
Stuck by colleague	1 (0.9)	2 (2.4)	0 (0)	2 (8.3)	0 (0)	0 (0)	0 (0)	5 (1.5)
Other	5 (4.4)	0 (0)	0 (0)	0 (0)	1 (6.7)	3 (37.5)	5 (11.6)	14 (4.3)
During or after disposal								
Item protruded from trash	3 (2.7)	6 (7.1)	0 (0)	3 (12.5)	0 (0)	0 (0)	17 (39.5)	29 (8.9)
While introducing item into disposal box	4 (3.5)	3 (3.6)	1 (2.6)	2 (8.3)	1 (6.7)	1 (12.5)	2 (4.7)	14 (4.3)
Unknown	0 (0)	0 (0)	0 (0)	0 (0)	1 (6.7)	0 (0)	1 (2.3)	2 (0.1)
Total†	**113 (99.9)**	**84 (100.1)**	**39 (100.1)**	**24 (99.9)**	**15 (100.2)**	**8 (100.0)**	**43 (99.9)**	**326 (99.4)**

** I.V. denotes intravenous. "Disposable syringe" also included insulin, tuberculin, and prefilled syringes. "Others" included 18 unattached hypodermic needles, 16 unidentified needles, 3 special-use syringes, 3 spinal needles, 2 Luer adapters, and 1 hypodermic needle mounted on laboratory equipment.*

† Differences from 100 percent are due to rounding.

ing heparin flushes after the intravenous administration of medication, a procedure that produces at least three contaminated needles. It was standard practice to dispose of all contaminated items after completing the procedure. Thus, the incentive to recap the needles was reinforced, because of the apparent risks of transporting several contaminated items to the disposal container.

Winged steel-needle intravenous sets, often referred to as "butterflies," bear some resemblance to intravenous tubing-needle assemblies in that a needle is attached to intravenous tubing. One third of the injuries with these devices occurred when workers fumbled with intravenous tubing on the way to the trash container, when the item protruded from the trash container, or when loops of tubing entangled the needle during disposal. Thirty-seven percent of injuries with these devices occurred during use, while a worker was withdrawing the needle from a patient, when the patient jarred the device, or when the needle rebounded from the rubber stopper of a vacuum tube. The latter situation was characteristic of the removal of a needle from a resistant substance. When the worker pulled hard against the resistance, the needle disengaged suddenly and, in a rebound motion, jabbed the worker's other hand.

Vacuum-tube phlebotomy assemblies accounted for only 5 percent of needle-stick injuries, but the rate per 100,000 items purchased was almost four times as high as for disposable syringes. Injuries most often occurred during recapping. Because the contaminated needle must be unscrewed from a reusable holder, it was common practice either to recap it before disassembly or to secure the needle with clamps during disassembly.

Four injuries were related to attempts to control bleeding from the puncture site when a large-bore phlebotomy needle was removed from the patient. The phlebotomist tried to apply pressure to the puncture site and had only one hand free to place the contaminated device at a safe distance. The patient jarred the device before it could be removed from the bedside, injuring the phlebotomist.

Blood was sometimes drawn into syringes connected to winged intravenous sets and then injected into vacuum tubes through the stoppers. This situation resulted in rebound needle sticks when the needles were withdrawn from the rubber stoppers. The use of the Luer adapter in a phlebotomy holder can preclude this hazard.

Eight injuries involved intravenous catheters; they resulted from difficulties in disposing of the catheter stylet. It was common practice to place the stylet in a container positioned within arm's reach of the employee. Access to the container must be immediate so that the employee can minimize the leakage of blood from the catheter that occurs after the stylet is withdrawn. Two injuries occurred when stylets pierced or fell out of makeshift containers. In three cases stylets were placed on beds or bedside tables because no container was immediately at hand, and employees were later injured when they attempted to pick up the exposed stylets for disposal. Two injuries were sustained during emergency procedures, when employees encountered difficulties in finding a safe place to put the stylets.

Unattached hypodermic needles accounted for 6 percent of needle-stick injuries. The needles were encountered by nurses, housekeeping staff, and technicians, on floors, hidden in bedside debris, in wads of gauze or paper towels, at the bottom of supply trays, or protruding from plastic trash bags. Unidentified needles caused 5 percent of the injuries. In every instance but one, workers were handling trash or were stuck by an item protruding from an overfilled container. Despite safety guidelines, needles continued to be disposed of in inappropriate trash containers, posing a constant risk to housekeeping staff.

Discussion

Design features and handling requirements of devices with needles were important factors in the causation of needle-stick injuries. Previous studies of the occurrence and prevention of needle sticks have also emphasized the dangers of recapping, bending, or breaking used needles and the appropriate design and placement of disposal containers for sharp objects. These factors are addressed in "Guideline for Infection Control in Hospital Personnel," issued by the Centers for Disease Control.[29]

Despite intensified efforts to implement these guidelines, recent studies report disappointing results.[1-3] The lack of success may be explained in part by a failure to deal with the underlying causes of the problem. For example, the recommendation that needles not be recapped is based on the assumption that because recapping accounts for a large proportion of needle-stick injuries, the avoidance of recapping should result in a proportionate decrease in needle-stick rates. However, during interviews, employees spontaneously offered four common reasons for recapping needles: (1) to protect themselves during disassembly of a device with an exposed contaminated needle; (2) to protect themselves from exposed needles when several items had to be carried to a disposal box in a single trip; (3) to store a syringe safely between uses if its contents were to be administered in two or more doses at different times; and (4) to protect others whom the worker had to pass at close quarters on the way to the disposal box.

Rather than disregarding safety guidelines, many employees were perceiving two competing hazards—of recapping and of not recapping the needle—and attempted to choose the lesser. It has yet to be determined whether recapping poses a greater risk than not recapping when a competing hazard is present. Until the incentive for recapping needles is reduced, it is likely that health care workers will continue this hazardous practice.

Our data indicate that there is no single remedy for needle-stick injuries. Each device that causes such injuries must be modified according to its specific use and handling requirements. Several practical recommendations can be made to reduce the hazards of devices with needles.

The optimal solution is to reduce the use of needles by using alternative methods for performing medical procedures or developing novel methods that would require fewer needles or none. Examples of such methods could include a compressed-air injection system as an alternative to the syringe or in-line stopcocks for the incremental administration of drugs through intravenous lines. For instruments that require needles, the optimal solution is to design devices that allow the needle to remain covered during and after use. At a minimum, a fixed barrier should be provided between the hands and the needle after use. The best designs should allow or require the worker's hands to remain behind the needle as it is covered, to preclude the movement of the hands in the direction of used needles, as in recapping. To provide the greatest benefit, the safety feature should be an integral part of the device and not an accessory to be used in combination with a hazardous item.[31] In this way, it is certain to be available precisely when and where it is needed. Moreover, the safety feature should be in effect before disassembly and should remain in effect after disposal, thus protecting the trash handler as well as the user. Finally, safety features should be as simple as possible and should require little or no training to use effectively.

Because our study focused on hollow-bore needles, the hazards encountered in highly specialized areas such as the operating room were not specifically addressed. The surgical setting will require additional study, with specialized approaches to environmental modifications. Furthermore, our study relied on passive surveillance for the identification of cases, and it is probable that the actual number of needle sticks was higher than reported. Hamory's survey of hospital employees revealed a 40 percent rate of underreporting of needle-stick injuries occurring in the preceding three months and a 75 percent rate of underreporting of those occurring in the preceding year.[32] The 39 percent rate of underreporting for all previous injuries due to sharp objects among participants in this study may be an underestimate, since those who reported current injuries may have been more likely than others to report injuries in the past. Comparisons of injury rates for specific devices were based on the assumption that underreporting does not differ substantially among types of hollow-bore devices.

The replacement of hazardous devices with safer designs will not take place immediately but will require sustained effort on the part of both product manufacturers and the health care community. The current climate of concern has resulted in a demand for a safer work environment; this may be viewed as a historic opportunity to accelerate the transfer of new technology into the work place. A timely response to this opportunity can bring about substantial and lasting improvements in an area in which progress is long overdue.

Acknowledgements: We are indebted to the employees of the University of Virginia Hospitals who willingly contributed their time to describe the circumstances of their injuries; to Dr. John T. Ashley, executive director of the University of Virginia Hospitals, for technical and financial support; to Dr. John A. Jane, chairman of the Department of Neurosurgery; to the staffs of the Employee Health Department, the hospital storeroom, and the hospital pharmacy for providing records; and to Ruth J. Johnson for assistance with the collection of the data and the preparation of the manuscript.

Supported in part by a grant from the University of Virginia Hospitals.

References

1. Ribner BS, Landry MN, Gholson GL, Linden LA. Impact of a rigid, puncture resistant container system upon needlestick injuries. *Infect Control.* 1987; 8:636.
2. Krasinski K, LaCouture R, Holzman RS. Effect of changing needle disposal systems on needle puncture injuries. *Infect Control.* 1987;8:59–62.
3. Edmond M, Khakoo R, McTaggart B, Solomon R. Effect of bedside needle disposal units on needle recapping frequency and needlestick injury. *Infect Control Hosp Epidemiol.* 1988; 9:1146.
4. McCray E, Cooperative Needlestick Surveillance Group. Occupational risk of the acquired immunodeficiency syndrome among health care workers. *N Engl J Med.* 1986; 314:112732.
5. Needlestick transmission of HTLVIII from a patient infected in Africa. *Lancet.* 1984; 2:13767.
6. Stricof RL, Morse DL. HTLVIII/LAV seroconversion following a deep intramuscular needlestick injury. *N Engl J Med.* 1986; 314:1115.
7. Oksenhendler E, Harzic M, Le Roux JM, Rabian C, Clauvel JP. HIV infection with seroconversion after a superficial needlestick injury to the finger. *N Engl J Med.* 1986; 315:582.
8. Meyers JD, Dienstag JL, Purcell RH, Thomas ED, Holmes KK. Parenterally transmitted nonA, nonB hepatitis: an epidemic reassessed. *Ann Intern Med.* 1977; 87:579.
9. Cannon NJ Jr, Walker SP, Dismukes WE. Malaria acquired by accidental needle puncture. *JAMA.* 1972; 222:1425.
10. Chacko CW. Accidental human infection in the laboratory with the Nichols rabbitadapted virulent strain of Treponema pallidum. *Bull WHO.* 1966; 35:80910.
11. Baldwin AH, McCallum F, Doull JA. A case of pharyngeal diphtheria probably due to autoinfection from a diphtheric lesion of the thumb. *JAMA.* 1923; 80:1375.
12. Hambrick GW Jr, Cox RP, Senior JR. Primary herpes simplex infection of fingers of medical personnel. *Arch Dermatol.* 1962; 85:5839.
13. Su WPD, Muller SA. Herpes zoster: case report of possible accidental inoculation. *Arch Dermatol.* 1976; 112:17556.
14. Emond RTD, Evans B, Bowen ETW, Lloyd G. A case of Ebola virus infection. *Br Med J.* 1977; 2:5414.
15. Jacobson IT, Burke JP, Conti MT. Injuries of hospital employees from needles and sharp objects. *Infect Control.* 1983; 4:1002.
16. Sexton DJ, Gallis HA, McRae JR, Cate TR. Possible needleassociated Rocky Mountain spotted fever. *N Engl J Med.* 1975; 292:645.
17. Sahn SA, Pierson DJ. Primary cutaneous inoculation drug-resistant tuberculosis. *Am J Med.* 1974; 57:6768.
18. Beverley JKA, Skipper E, Marshall SC. Acquired toxoplasmosis: with a report of a case of laboratory infection. *Br Med J.* 1955; 1:5778.
19. Chappler RR, Hoke AW, Borchardt KA. Primary inoculation with Myco bacterium marinum. *Arch Dermatol.* 1977; 113:380.
20. Collins CH, Kennedy DA. Microbiological hazards of occupational needlestick and "sharps" injuries. *J Appl Bacteriol.* 1987; 62:385402.
21. Evans N. A clinical report of a case of blastomycosis of the skin from accidental inoculation. *JAMA.* 1903; 40:17725.
22. Glaser JB, Garden A. Inoculation of cryptococcosis without transmission of the acquired immunodeficiency syndrome. *N Engl J Med.* 1985; 313: 266.
23. Hill A. Accidental infection of man with mycoplasma caviae. *Br Med J.* 1971; 2:7112.
24. Joffe B, Diamond MT. Brucellosis due to selfinoculation. *Ann Intern Med.* 1966; 65:5645.
25. Sarasin G, Tucker DN, Arean VM. Accidental laboratory infection caused by *Leptospira icterohaemorrhagiae. Am J Clin Pathol.* 1963; 40:14650.
26. Gugel EA, Sanders ME. Needlestick transmission of human colonic adeno-carcinoma. *N Engl J Med.* 1986; 315:1487.
27. Committee on Trauma Research. Injury in America: A Continuing Public Health Problem. Washington, D.C.: National Academy Press, 1985.
28. Jagger J, Pearson RD. A view from the cutting edge. *Infect Control.* 1987; 8:512.
29. Williams WW. Guideline for infection control in hospital personnel. *Infect Control*. 1983;4:Suppl 4:32649.
30. Sanborn C, Luttrell N, Hoffmann K. Creating a safer environment for health care workers: implementing a pointofuse sharps disposal system. *Nurs Adm Q.* 1988; 12:2431.
31. Jagger J, Pearson RD, Brand JJ. Avoiding the hazards of sharp instruments. *Lancet.* 1986; 1:1274.
32. Hamory BH. Underreporting of needlestick injuries in a university hospital. *Am J Infect Control.* 1983; 1:1747.

PART I

EPINet Reports

About EPINet

In 1991, Janine Jagger and colleagues at the International Healthcare Worker Safety Center at the University of Virginia developed the EPINet surveillance system in order to provide healthcare facilities with a standardized program for tracking needlestick injuries and blood and body fluid exposures among employees. Because prevention is a primary aim of EPINet, it provides important detail on the performance of medical devices. The widespread dissemination of EPINet over the last decade to over 1500 hospitals in the United States, and many others overseas, resulted in a massive increase in data on the causes of needlesticks and blood exposures.

In 1992, the Center established a voluntary data-sharing network among some of the hospitals in the U.S. using EPINet. These facilities send their exposure data to the Center on an annual basis, and the data are merged into an aggregate database. The number of facilities participating in the network varies from year to year, but a cumulative total of 84 hospitals have contributed data over the last 10 years. It is the longest-standing database of healthcare workers' at-risk exposures to blood and body fluids in the U.S.

The EPINet database is the foundation of the Center's research. In the articles from AEP that follow, EPINet data were analyzed from a variety of perspectives in order to understand exposure risk patterns for specific job categories, clinical areas and devices, and develop strategies for prevention.

Report on Blood Drawing: Risky Procedures, Risky Devices, Risky Job

Janine Jagger, M.P.H., Ph.D.

Vol. 1, no.1, 1994

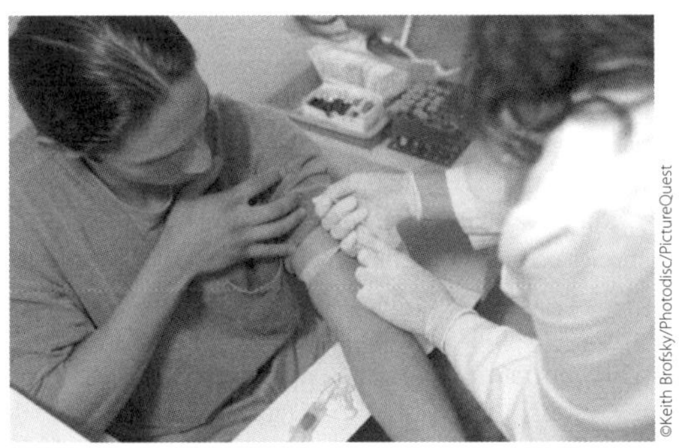

Introduction

The United States has the largest number of recognized cases of occupationally acquired HIV in the world. Although no surveillance system in the U.S. or elsewhere identifies all, or even most, occupationally infected healthcare workers, the number of recognized cases alone is cause for serious concern. A review of international literature through June 1993 identified 176 documented or probable cases of occupationally acquired HIV infection worldwide[1,2]; two-thirds of those were in the United States. One use that can be made of these unfortunately large, and growing, numbers is to identify transmission patterns, which can lead in turn to focused and effective prevention measures.

The U.S. Centers for Disease Control and Prevention (CDC) identified 123 documented or possible cases of occupationally acquired HIV in the United States through December 1993.[3] Two professional categories, nurses and clinical laboratory workers—primarily phlebotomists—ranked first among HIV-infected healthcare workers, accounting for 24% each of the 123 reported cases. Physicians ranked next, accounting for 12% of cases. The high number of cases among nurses is consistent with the fact that they comprise the largest group of healthcare workers in the U.S., numbering 2.2 million (source: American Nurses Association). Similarly, the number of HIV-infected physicians is consistent with the smaller number of physicians—approximately 670,000—in the U.S. workforce (source: American Medical Association). The number of infected phlebotomists, however, is disproportionate to their numbers in the workforce: there are fewer than 100,000 phlebotomists employed in the U.S., according to the National Phlebotomy Association and the American Society of Phlebotomy Technicians, less than 1/20th the number of nurses.

A notable difference between nurses and phlebotomists is that the latter consistently perform blood drawing procedures, while nurses perform a wider variety of procedures. When a phlebotomist sustains a needlestick, the device causing injury is most likely to be a blood-filled needle. In contrast, a nurse may be stuck by a needle used for an intramuscular injection or intravenous infusion, which would not be blood-filled, and less often by a needle used for blood drawing.

Needles used for different purposes appear to carry different risks for transmitting HIV.[4] The same may be true for other bloodborne pathogens, such as hepatitis B and hepatitis C, but there is insufficient surveillance data to confirm this. In a report by the CDC on the exposure circumstances of workers with occupationally acquired HIV infection, those who were exposed by needlestick had all been stuck by hollow bore, blood-filled needles. Furthermore, phlebotomy was the procedure most frequently associated with HIV exposures.[5] Similar conclusions have been drawn from data reported in other countries.[1,2]

This evidence suggests that (a) blood drawing presents a high risk of exposure to bloodborne pathogens and that (b) the risk profiles of different professional groups are in part linked to the frequency with which they perform blood drawing procedures. These are compelling reasons to focus prevention efforts on the devices, procedures, and professional groups involved in blood drawing.

Two reports follow. The first describes the patterns of percutaneous injuries associated with blood drawing procedures in a national network of hospitals. The second presents the results of a survey conducted at a national phlebotomy conference, which looked at blood exposure patterns and risks among phlebotomists.

I. Characteristics of Percutaneous Injuries Associated with Blood Drawing in a National Network of 58 Hospitals

Methods

Fifty-eight hospitals which voluntarily participated in three data-sharing networks contributed one year of data for this report. All the hospitals use the Exposoure Prevention Information Network (EPINet) program for tracking percutaneous injuries in their institutions.[6] Network A includes nine hospitals located in six states in the eastern half of the United States; they report their data to the University of Virginia. Network B consists of 50 hospitals in South Carolina that report their data to the Palmetto Hospital Trust Needlestick Prevention Demonstration Project in

Columbia, South Carolina. Network C includes 11 Sisters of Providence hospitals in the Pacific Northwest that report their data to Johnson & Higgins of Washington, Inc., in Seattle, Washington. The hospitals represent a cross-section of institutions in diverse geographic locations; of the hospitals included in this report, 26 had an average daily census of less than 100 beds, 16 had from 100 to 299 occupied beds, and 16 had 300 or more occupied beds. Seventeen were teaching hospitals.

Data included all percutaneous injuries reported by healthcare workers to the employee health department or similar designated authority in their institution. Data collection began in September 1992. Each participating facility provided one year of data on disk to investigators; of the 70 hospitals in the three networks, 58 with complete data at the time of this report were included.

Results

There was an overall total of 3,829 percutaneous injuries in the merged database, and a cumulative total of 11,978 occupied beds in the 58 hospitals. Needlestick incidents were selected in which the device associated with the injury was (1) a syringe used for drawing venous or arterial blood, (2) a winged steel needle (butterfly) used for blood drawing, or (3) a vacuum tube phlebotomy needle. Four hundred and seventy-one cases met these criteria—12.3% of all injuries from the 58 hospitals. **Table 1.1** shows the job categories of workers reporting needlesticks from blood drawing needles. Nurses and phlebotomists together accounted for two-thirds of cases. The remaining cases were reported primarily by respiratory therapists (a job category that does not exist in many countries), who often draw blood for arterial blood gas analysis, and by clinical laboratory technicians and physicians (mainly residents).

Table 1.1 Job Classification of healthcare workers Reporting Needlesticks Associated with Blood Drawing (58 Hospitals, 1 Year)

Job	Number	Percent
Nurse	157	33.3
Plebotomist	150	31.8
Respiratory Therapist	43	9.1
Clinical Lab Worker	40	8.5
Physician*	39	8.3
Attendant (non-surgical)	14	3.0
Other	28	5.9
TOTAL	471	100 %

*30/39 reported needlesticks were to residents

Table 1.2 Place of Occurrence of Needlesticks Associated with Blood Drawing (58 Hospitals, 1 Year)

Location	Number	Percent
Patient Room	251	53.3
Emergency Department	66	14.0
Intensive/Critical Care	39	8.3
OutpatientFacility	24	5.1
Clinical Laboratory	21	4.5
Operating Room	13	2.8
Venipuncture	12	2.5
Procedure Room	9	1.9
Other	36	7.6

Table 1.2 shows where these incidents occurred. Most injuries occurred in patient rooms; the next most frequent locations were emergency departments and intensive or critical care units. Less frequent locations were outpatient clinics, operating rooms, and clinical laboratories.

Figures 1.1 through 1.4 show the mechanism of needlestick injuries for four major needle devices used for blood drawing. This breakdown of injuries shows a profile of when the injuries occurred during the use/disposal cycle; each device has a unique profile, reflecting the design characteristics of the device and the handling requirements for performing specific procedures. For instance, recapping injuries are more frequently associated with syringes used for drawing arterial blood **(Figure 1.1)** than for similar syringes used for drawing venous blood **(Figure 1.2).** This may be due to the need to remove needles from arterial blood gas syringes before delivering the filled syringes to the laboratory. On the other hand, injuries that occur when withdrawing a needle from the stopper of a tube are more frequent with syringes used for drawing venous blood because the blood is often injected into tubes before delivery to the laboratory.

The risks associated with winged steel (butterfly) needles are largely related to problems in transporting the devices safely to the disposal containers and difficulties in pushing the needles through the openings of containers **(Figure 1.3).** The coiled tubing attached to the needles makes the devices awkward to handle.

There were 184 injuries from vacuum tube phlebotomy needles **(Figure 1.4).** The circumstances of the injuries were similar to those for winged steel needles—that is, they were caused by problems in transporting needles to disposal containers and in introducing needles into the containers.

Hollow bore needles were not the only devices associated with injuries incurred during blood drawing procedures. Lancets used for fingersticks and heelsticks, glass capillary tubes, and glass vacuum tubes containing blood samples also caused injuries. Most numerous among these were 108 injuries caused by fingerstick and heelstick lancets that lacked an automatic retracting safety feature. Fifty-four percent of the injuries occurred when disengaging a disposable lancet from a reusable holder or when handling

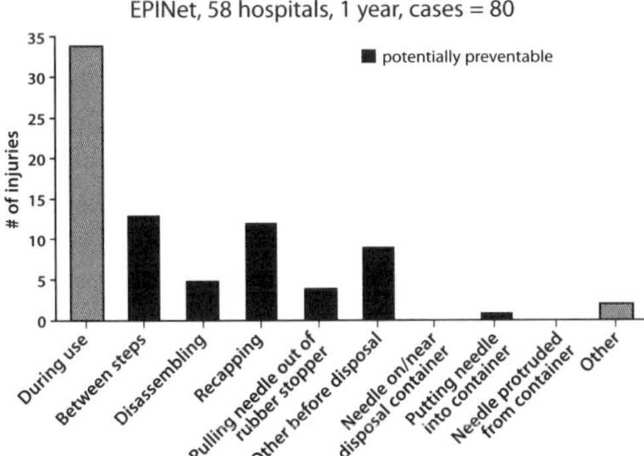

Figure 1.1 Mechanism of Injury from Syringes Used for Drawing Arterial Blood

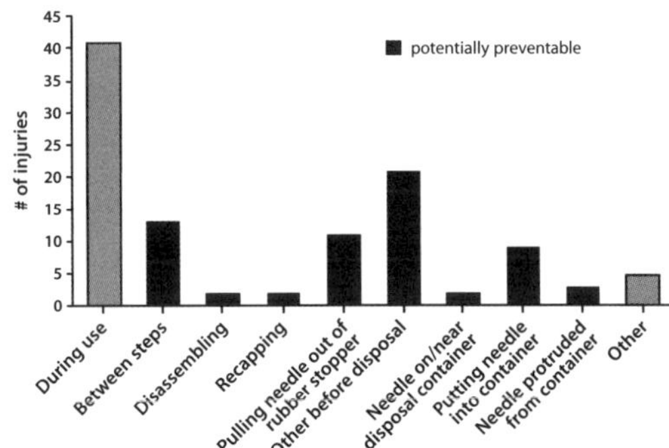

Figure 1.2 Mechanism of Injury from Syringes Used for Drawing Venous Blood

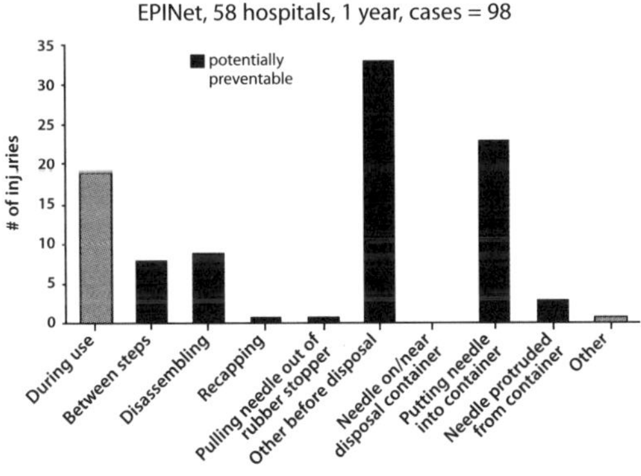

Figure 1.3 Mechanism of Injury from Winged Steel Needles Used for Drawing Blood

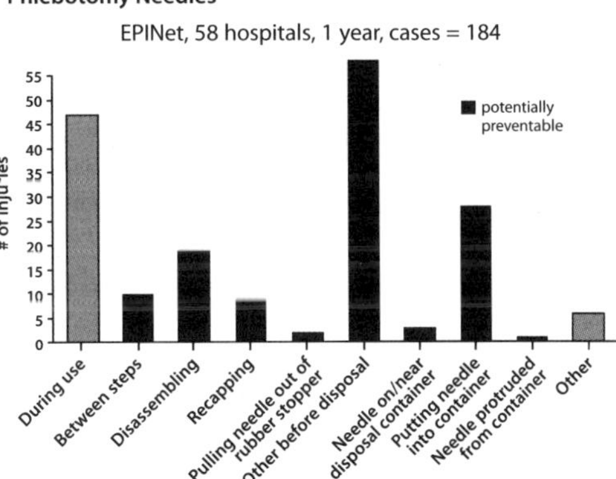

Figure 1.4 Mechanism of Injury from Vacuum Tube Phlebotomy Needles

International Healthcare Worker Safety Center, University of Virginia

the used lancet prior to disposal. These injuries reflect the difficulty of handling such small devices which, if no protective shield is provided, require the fingers to remain close to the exposed point as the device is disassembled and discarded.

Eighteen injuries were caused by glass capillary tubes used to contain small-volume blood samples following fingersticks or heelsticks. Injuries occurred during all phases of handling. Force must be applied when pushing the tubes into putty to close off one end, and again the tubes are subject to significant force when they are centrifuged for hematocrit determination. Under such stresses the fragile glass easily fractures. The devices are blood-filled and have the potential to produce sizable lacerations with significant inoculation of blood.

Nine injuries were caused by broken glass vacuum tubes. Although there were relatively few such injuries when compared with the 184 from phlebotomy needles used to draw blood into the tubes, vacuum tube injuries are particularly serious because of the large amounts of blood that can be introduced into the lacerations. Breakage often occurred at the top of the tubes, when the tight-fitting stoppers were removed either to introduce or withdraw blood samples.

Preventable Injuries

The EPINet system is designed to identify the proportion of injuries that are potentially preventable through improvements in the design of sharp medical devices. Injuries are considered preventable if an alternative already exists that can eliminate the sharp device or unsafe feature. For instance, all injuries from needles used to inject fluid into or withdraw fluid from intravenous access ports are preventable, because needleless equipment can be used. All injuries caused by the breakage of glass devices for which nonbreakable alternatives are available can also be prevented.

Some, but not all, needles can be eliminated. Many devices have needles that are used to pierce skin or tissue; these are necessary needles. Needlesticks that occur *after use* or *between uses* of a necessary needle, however, are potentially preventable, when a safety feature that shields the hand from the needle can be put into place. The percentage of needlesticks that occur *during* the performance of the procedure, when the needle must be exposed for use, is not included in the preventable fraction for that device.

The concept of the "preventable fraction" is intended to project a target that may be achieved by the implementation of feasible measures. It is not intended to imply that there is a limit to the potential reductions that can be achieved. It is possible that some safety devices on the market may already exceed the estimated preventable fraction in a given device category. It is also possible that in the future additional prevention strategies, such as procedure changes or new technology, may increase the preventable fraction or even eliminate injuries in a specific device category altogether.

Figures 1.1 through 1.4 highlight the preventable fraction of needlesticks for each device. They were: 58% for syringes used for drawing venous blood; 55% for syringes used for drawing arterial blood; 80% for winged steel (butterfly) needles; and 71% for vacuum tube phlebotomy needles. Across all four devices, 67% (316/471) of injuries fell into the potentially preventable fraction.

Seventy-four percent of injuries from fingerstick or heelstick lancets fell into the preventable fraction. Nearly *all* injuries, however, caused by such lancets are preventable, because there are safety lancets presently on the market with a built-in spring action that automatically withdraws the sharp point into a shielded position immediately after being discharged. The sharp lancet retracts so quickly that it is highly improbable that a healthcare worker could sustain a puncture injury after performing a fingerstick or heelstick procedure. The percentage of preventable injuries, therefore, is likely to be significantly greater than 74%. The safest retracting lancets are those that do not permit a contaminated lancet to be inadvertently discharged a second time.

An additional advantage of the safety lancets is that they reduce the potential for patient-to-patient cross-contamination, because when the used lancet is disposed of, all parts of the device that have come into contact with the patient are automatically discarded with the lancet. The problem of cross-contamination from spring-loaded lancets was linked to an outbreak of hepatitis B and was subsequently the subject of a 1990 FDA Safety Alert.[7,8]

All injuries (100%) from glass capillary tubes used for hematocrit determination are preventable, because the use of these devices is unnecessary—unbreakable plastic capillary tubes are available, as well as a system for hematocrit determination that does not require the use of capillary tubes at all. These injuries could be eliminated tomorrow with the appropriate selection of products.

It is difficult to estimate the proportion of preventable injuries from broken vacuum tubes. A plastic vacuum tube, which would be an appropriate alternative to glass, is not yet available; the plastic would have to meet specifications both for breakage resistance and for impermeability to air, which is necessary for a tube to retain its vacuum over a long shelf life. There is, however, a redesigned vacuum tube stopper for glass vacuum tubes that is intended to reduce stresses on the top edge of the tube and thereby lower the risk of breakage if the stopper is removed. The potential effect of redesigned stoppers on reducing injuries from broken vacuum tubes cannot be estimated at present.

As a final point, the handling requirements of blood drawing equipment must be taken into consideration when implementing prevention programs or evaluating safer devices. After withdrawing a phlebotomy needle from a patient, the healthcare worker must apply pressure to stem bleeding at the puncture site. This leaves only one hand free to handle the exposed needle. The healthcare worker must either dispose of the needle with one hand or leave it on a nearby surface until he or she is free to move away from the patient. This situation points to the need for an appropriate disposal container within arm's reach of the patient. It also emphasizes the advantage of having a safety blood drawing device that requires only one hand to activate. If two hands are needed to activate a safety feature after the needle has been removed from the patient, the healthcare worker is in the same situation as with a conventional device, and cannot activate the needle protection feature until both hands are free.

II. Survey of Needlesticks and Other Blood Exposures Among Phlebotomists

"This survey documents the hazards we see everyday in various sites. It is evident that these hazards could be prevented if devices that have proven to be safer were used in every facility where phlebotomy is being done. For instance, employers could make improvements tomorrow if they replaced glass capillary tubes with plastic tubes, or purchased self-shielding needles, or made proper needle disposal readily available. In addition to providing safer technologies in workplaces, employers should also provide adequate training. These simple steps would go a long way to reducing the hazards for phlebotomists."

—Diane Crawford, President and CEO,
National Phlebotomy Association, Inc.

In most countries there is no specific professional group dedicated to blood drawing. Usually nurses, but also physicians and other healthcare workers, draw blood as one of their many responsibilities. In North America, however, the specialized job category of phlebotomist has been established—although, of course, other healthcare professionals draw blood as well. Because phlebotomists as a group are at high risk for occupationally acquired bloodborne pathogens, it is important to understand the risk patterns

Table 2.1 Frequency of Risk Factors Among Phlebotomists During Past 12 Months (cases = 140)

Uses two-handed recapping of blood collection needles	7%
Does not routinely wear gloves	7%
Sometimes cuts top off index finger glove	16%
Hepatitis B vaccine started but incomplete	19%
Has never received hepatits B vaccine	20%
Uses needles to draw blood from IV, arterial or central lines	24%
Empoyer did not give training to prevent blood exposure	25%
Changes needles to inoculate blood culture medium	28%
Uses glass capillary tubes	36%
Injects blood through stoppers into blood collection tubes	63%

Table 2.2 Percentage of Phlebotomists Sustaining Blood Contact to Skin or Mucosa during Past 12 Months (cases=140)

While applying pressure to a puncture site	16%
While disposing of a blood-filled specimen container	15%
While uncapping a blood-filled vacuum/speciman tube	9%
When a vacuum tube stopper blew off a vacuum tube	5%

Table 2.3 Devices Causing Injury to Phlebotomists

Device	Number	Percent
Vacuum tube phlebotomy needle	13	42.0
Winged steel needle (butterfly) attached to vacuum tube set	10	32.3
Winged steel needle (butterfly) attached to syringe	1	3.2
Disposable syringe	2	6.4
Lancet	2	6.4
Broken capillary tube	1	3.2
Broken vacuum tube	1	3.2
Undetermined	1	3.2
TOTAL	31	100.0

and exposure mechanisms associated with their work. Such an understanding may help reduce their risk as well as that of other healthcare workers who draw blood.

Methods

On July 30, 1994, a survey was conducted of phlebotomists attending the annual conference of the National Phlebotomy Association in New York City. The survey instrument was designed by the International Healthcare Worker Safety Center at the University of Virginia, in collaboration with the Service Employees International Union. Of approximately 200 registrants, 152 phlebotomists responded to the survey; 12 of the responses were excluded because the respondents either were currently unemployed, were students, or provided incomplete information. A total of 140 respondents were included in the final analyses.

Results

Fifty-four percent of the respondents worked in hospitals, 16% were employed by private laboratories, 11% worked in free-standing clinics, 5% worked as private contractors, 4% were employed by health maintenance organizations, and 10% were employed in other settings. The average work week was 37.3 hours, and the average number of blood drawing procedures performed in a typical eight-hour work day was 32.

Respondents were asked questions relating to risk factors for blood exposures and percutaneous injuries during the previous 12 months. **Table 2.1** shows the frequency of those risk factors. Overall, there was a high frequency of unsafe procedures and use of unnecessarily hazardous equipment. Two-thirds of the respondents injected blood through stoppers into blood collection tubes, a hazard which can be eliminated by using equipment that draws blood directly into vacuum tubes, or by using a large-volume syringe with a safety shield that can be locked in place, allowing blood tubes to be inserted into the shield so that the needle is covered as blood is injected.

Over one-third of the respondents used glass capillary tubes frequently or occasionally; this is an unnecessary risk since unbreakable alternatives exist. Respondents also indicated that the use of unnecessary needles is commonplace: 24% used needles to draw blood from intravenous, arterial, or central lines, although needleless systems allow blood to be drawn directly from lines into specimen containers; and 28% changed needles for injecting blood into blood culture medium, even though extra needle manipulation increases the risk of needlestick and has not been shown to reduce bacterial contamination of blood samples.[9]

Twenty percent of respondents had not been vaccinated against hepatitis B, and another 19% had started but not completed the vaccine series—a serious shortcoming in a group at such high risk for exposure to bloodborne pathogens. Twenty-five percent of employers had never provided safety training to the phlebotomists on the prevention of bloodborne pathogen transmission.

Table 2.2 shows the percent of phlebotomists sustaining blood contact to skin or mucosa during the previous year under a variety of circumstances. Thirty-four percent of all respondents experienced one or more of these incidents; blood contact occurred most frequently when disposing of a blood-filled specimen container or while applying pressure to a puncture site. The problem of blood exposures due to stoppers popping off of vacuum tubes emphasizes the hazard of injecting blood into vacuum tubes, which can create a positive pressure inside the tube, thus dislodging the stopper.

Thirty-one respondents reported a total of 47 percutaneous injuries during the previous year. When adjusted for an average 40-hour work week, the average injury rate was .36 injuries per full-time phlebotomist per year. The injury rate per procedure was 12.5 injuries per 100,000 blood-drawing procedures.

The 31 respondents who experienced at least one percutaneous injury in the previous year were asked to provide details about their most recent injury. The locations of the injuries were as follows: patient room, 45.2%; venipuncture facility, 19.4%; clinical laboratory, 9.7%; outpatient office, 6.4%; emergency department, 3.2%; procedure room, 3.2%; other settings, 12.9%. The procedures performed at the time of injury included: drawing venous blood, 74.2%; fingersticks, 6.4%; blood culture, 3.2%; other or no response, 16.2%. Devices causing the injuries are shown in **Table 2.3** *(page 7)*.

Four of the 31 injuries occurred before use of the device. Of the remaining 27 injuries, 17 (68%) fell into the potentially preventable fraction—that is, they could potentially have been prevented by eliminating an unnecessarily hazardous product (such as glass capillary tubes) or by utilizing devices with integrated safety features that shield the hands from used needles.

Comparisons were made between phlebotomists who had sustained injuries during the previous year and those who had not, in order to identify possible risk factors for percutaneous injuries. An important difference between the two groups was found in one area: *phlebotomists who had had percutaneous injuries during the previous year worked in facilities that provided fewer safety devices than phlebotomists who had had no injuries.* This pattern was consistent across every category of safety device listed. **Figure 2.1** shows the percent of facilities providing safety devices in each device category, comparing phlebotomists who had injuries during the past year to those who did not. The differences between the two groups of phlebotomists were statistically significant in two device categories—"needle shield" and "lancet" ($p<.05$ Fisher's exact test, χ^2, respectively). The enclosed poster provides complete descriptions of the specific safety devices identified on the survey.

Figure 2.1 Percent of Facilities Using Safety Devices, for: Workers Reporting Needlesticks in Past Year (31 cases) vs. Workers Reporting No Needlesticks in Past Year (109 cases)

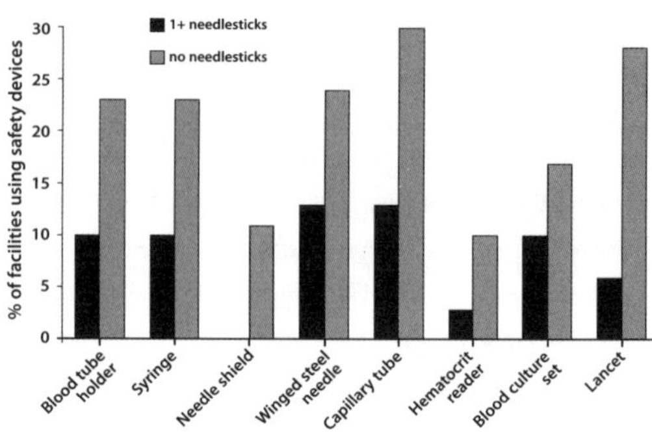

International Healthcare Worker Safety Center, University of Virginia

Recommendations

These findings suggest that phlebotomists are routinely exposed to risks that are preventable *today.* A number of steps should be taken to minimize the risks they face:

- Every effort must be made to insure that all susceptible phlebotomists are fully vaccinated against hepatitis B.
- All unnecessary hazardous devices should be eliminated from use, including glass capillary tubes and needles used for drawing blood from intravenous, arterial, and central lines. In addition, the practice of changing needles for blood culture phlebotomy should be abandoned.
- The practice of injecting blood through a stopper into a vacuum tube with an exposed needle should be stopped. Methods of drawing blood directly into vacuum tubes or other specimen containers should be employed instead; alternatively, syringes with a needle shield locked in place over the needle may reduce risk of needlestick and contain blood splatter from dislodged tube stoppers when injecting blood into tubes.
- Blood drawing devices with integrated safety features designed to prevent percutaneous injuries should be rapidly implemented and closely monitored for user and patient safety and for reliability of laboratory values.
- Automatically retracting fingerstick or heelstick lancets should be used in place of manual or non-retracting spring-loaded lancets.
- All facilities should provide puncture-resistant disposal containers within arm's reach of the phlebotomist for blood drawing procedures.
- All facilities should comply with the Occupational Safety and Health Administration's bloodborne pathogens standard and provide annual in-service training for phlebotomists on appropriate methods for reducing occupational transmission of bloodborne pathogens.[10] Hazardous practices still employed by phlebotomists such as two-handed recapping and cutting off the tip of the index finger of gloves should be explicitly prohibited.

These prevention strategies are important not only for phlebotomists but for all healthcare workers who perform blood drawing procedures. Because of the disproportionately high risk of infection from bloodborne pathogens among healthcare workers who perform such procedures, and the potential for substantially reducing this risk, we should strive to implement all possible prevention measures as quickly as possible.

References

1. Ippolito G, Puro V, De Carli G. Infezione professionale da HIV in operatori sanitari: descrizione dei casi con seroconversione documentata segnalati al 30 giugno 1993. *Giornale Italiano dell'AIDS.* 1993;4:63–75.
2. Ippolito G, Puro V, De Carli G. Infezione professionale da HIV in operatori sanitari: 2 - descrizione dei casi senza seroconversione documentata segnalati al 30 giugno 1993. *Giornale Italiano dell'AIDS.* 1993; 4:186–193.
3. Centers for Disease Control and Prevention. *HIV/AIDS Surv Rep.* 1994;5:19.
4. Berry, AJ. Are some types of needles more likely to transmit HIV to healthcare workers? *Am J Infect Control.* 1993;21:216–218.
5. Metler R, Ciesielski C, Marcus R, Ward J. Exposure circumstances of workers with occupationally acquired HIV infection. Presented at Frontline Healthcare Workers: A National Conference on Prevention of Device-Mediated Bloodborne Infections. Washington, D.C., August 18, 1992.
6. Jagger J, Cohen M, Blackwell B. EPINet: A tool for surveillance of blood exposures in health care settings. In: Charney W, ed. *Essentials of Modern Hospital Safety* (vol. 3). Boca Raton, FL: Lewis Publishers/CRC Press, 1994; 223–239.
7. Food and Drug Administration, Center for Devices and Radiological Health. FDA safety alert: hepatitis B transmission via spring-loaded lancet devices. August 28, 1990.
8. Douvin C, Simon D, Zinelabidine H, et al. An outbreak of hepatitis B in an endocrinology unit traced to a capillary blood sampling device. *N Engl J Med.* 1990;4:57–58.
9. Leisure MK, Moore DM, Schwartzman JD, et al. Changing the needle when inoculating blood cultures: a no benefit and high risk procedure. *JAMA.* 1990;264:2111–2112.
10. Occupational Safety and Health Administration: 29 CFR Part 1910.1030: Occupational exposure to bloodborne pathogens. *Fed Regist.* 1991;56(235). 64004–65182.

Blood and Body Fluid Exposures to Skin and Mucous Membranes

Janine Jagger, M.P.H., Ph.D. and Melanie Balon, B.S.

Vol. 1, no. 2, 1995

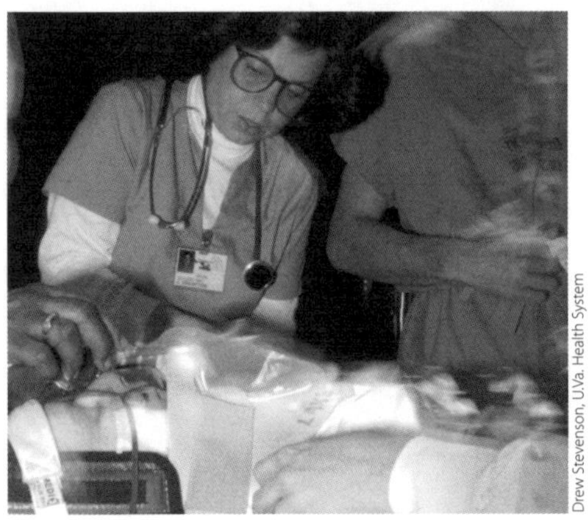
Drew Stevenson, U.Va. Health System

Introduction

Percutaneous injuries, needlesticks in particular, are generally recognized as the most common mechanism of bloodborne pathogen transmission among healthcare workers. However, contact of workers' skin and mucous membrane with infected blood from patients has also been associated with documented seroconversions to bloodborne pathogens, although with lower frequency than percutaneous injuries.[1-3] Only 4 out of 61 documented cases of occupational HIV infection reported internationally were linked to skin or mucous membrane contact with blood.[4] There are several possible explanations for these relatively small numbers, including a lower transmission rate following mucocutaneous exposures. But another factor may be the difficulty in attributing an infection to a specific skin or mucous membrane exposure.

Blood and body fluid (BBF) exposures are so common in many clinical settings, such as emergency departments and obstetrics, that healthcare workers are not likely to report an event they view as a routine occurrence. Thus, some occupational infections resulting from routine and repeated contact with BBF are probably not recognized or documented as such. Cases of occupational HIV infection have been reported in which there was a history of contact with HIV-contaminated (or potentially contaminated) blood or lab samples, but in which no percutaneous injury or other specific exposure was recalled.[5] Some studies have shown higher prevalence rates of bloodborne pathogens in healthcare workers employed in exposure-intensive settings than in control groups or in the general population, providing indirect evidence that transmission may occur despite the lack of documented seroconversions.[6-10] These reports illustrate the difficulty of linking exposure events to infections, and may in part explain why relatively few documented cases of bloodborne pathogen transmission via mucocutaneous route are found in the medical literature.

The healthcare worker's first line of defense against exposure to BBF is personal protective equipment (PPE), including gloves, liquid-resistant gowns, face masks, face shields, goggles, head covers, shoe covers, and other more specialized garments. In the United States, recommendations for the appropriate use of PPE are described in the policy of universal precautions, first published by the Centers for Disease Control and Prevention in 1987 and enacted as a mandatory national standard in December 1991.[5,11] The effectiveness of PPE depends on whether the design of the protective garments matches the distribution of BBF contact, whether the garment material is adequately liquid resistant, and whether the appropriate garments are worn when needed. When BBF exposures occur, they reflect either non-compliance with PPE requirements, a failure of protective garments to provide an adequate barrier, or unanticipated circumstances that the healthcare worker was unable to prepare for—a common situation in the complex environment of a hospital.

Skin or mucous membrane contact with patients' BBF is frequent in most health care institutions. In a survey of workers in a hospital's clinical laboratories—a highly regulated environment—respondents reported an average of 4.6 BBF contacts per year. Only 2% reported their most recent BBF contact to their employee health department, a much higher underreporting rate than is generally noted for percutaneous injuries.[12] A similar survey conducted among personnel in an emergency department—a less predictable and more body fluid-intensive environment—revealed an average of 56.5 BBF contacts per person per year. Only 4% reported their most recent BBF contact to employee health.[13] Not all incidents, however, were considered at-risk exposures because the vast majority of BBF contacts in both the clinical labs and the emergency department were to intact skin only (95% and 88%, respectively). Of BBF contacts to non-intact skin or mucous membranes among emergency department personnel, 38% were reported to employee health. In the same study, 39% of emergency department personnel reported their most recent percutaneous injury to employee health. This comparison suggests that underreporting rates for BBF contacts to non-intact skin or mucous membranes may be similar to underreporting rates for percutaneous injuries.

Data originating from employee health records represent only a small fraction of actual BBF contacts occurring in clinical areas, and are more likely to include incidents that involve BBF contact with non-intact skin or mucous membranes. When interpreting these data, potential reporting biases must be kept in mind. Nevertheless, employee health records provide a valuable source of information for defining the characteristics of body fluid contact and the circumstances under which they occur. These data are also useful for monitoring compliance with current policies, and are especially important for identifying prevention opportunities and emerging exposure risks.

Findings in a Nine-Hospital Network

Nine hospitals which voluntarily participate in a data-sharing network contributed data for this report; they are located in six states in the eastern half of the U.S. All hospitals use the EPINet system for tracking both percutaneous injuries and blood and body fluid exposures in their institutions; they report their data quarterly to the University of Virginia. The cumulative total average daily census for the nine hospitals is 4,886 occupied beds. The present study includes descriptions of BBF exposures reported to the employee health department of each institution during a 24-month period from September 1992 through August 1994.

There were 1,150 BBF reports during that two-year interval, which accounted for 24% of all reports (including both sharp object injuries and BBF contacts). The average annual reported BBF exposure rate was 12 BBF exposures per 100 hospital beds. Healthcare workers were permitted to file exposure reports regardless of the seriousness of the incident. Therefore, the data include a mix of high-risk incidents (i.e., blood contact with non-intact skin), and lower-risk incidents (i.e., saliva contact with intact skin).

Job categories of workers reporting BBF contacts

Nurses reported 52% of all incidents, far more than workers in any other job category (see **Figure 1**). This reflects the large number of nurses employed, as well as the frequent patient contact required of nurses. Attendants and physicians (mainly residents) reported 9% of incidents each. Other job categories, including laboratory technicians, respiratory therapists, phlebotomists, intravenous nurses, and nursing and medical students, together accounted for 16% of cases.

Of interest were personnel whose job classification was listed as "other". Thirteen percent of incidents fell into this category. Seven percent were healthcare workers to whom universal precautions clearly apply, such as paramedics, perfusionists, home health aides, housekeeping and laundry workers, and radiology students. However, the remaining cases (6.3%) involved personnel who might not anticipate exposure to patients' body fluids, and are not likely to be prepared for body fluid contact. The most numerous in this category were physical or occupational therapists (20 cases), and security personnel (20 cases). There were 9 incidents reported by clerical personnel, and 8 incidents reported by mental health workers. One or two incidents each were reported by a wide variety of other personnel not providing direct patient care, including two audiologists, two athletic trainers, an electrician, a chaplain, a child care director, a food service worker, an information systems technician, a systems analyst, a psychologist, a teacher's aide, a teacher, a translator, and a volunteer. These incidents suggest that personnel who do not provide patient care are sometimes at risk of exposure to patients' BBF. Some categories of personnel, especially physical and occupational therapists and security guards, should receive routine in-service training in universal precautions, and be instructed to anticipate situations in which BBF contact may occur.

Figure 1. Job Catagories of Workers Reporting Blood and Body Fluid Exposures

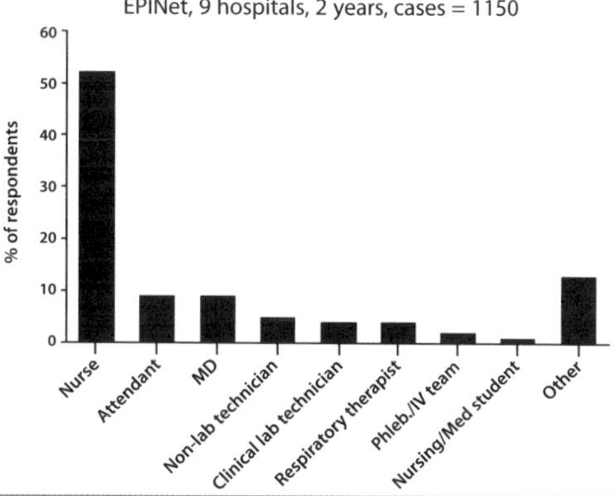

International Healthcare Worker Safety Center, University of Virginia

Location of BBF contacts

Forty-five percent of reported exposures occurred in patient rooms, a substantially higher percentage than in any other area of the hospital (see **Figure 2**). This may reflect

Figure 2. Location of Blood and Body Fluid Contact

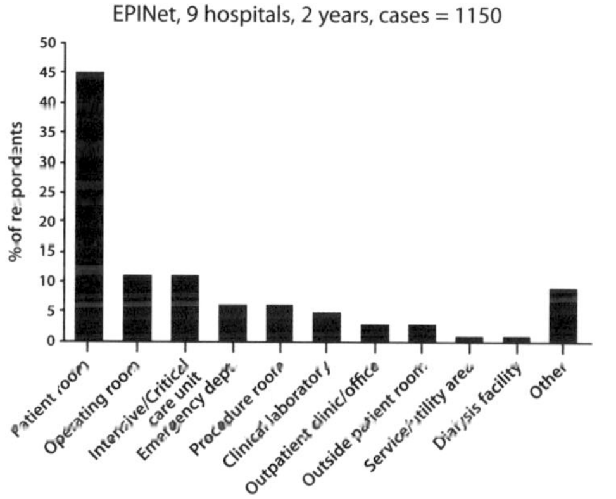

EPINet, 9 hospitals, 2 years, cases = 1150

% of respondents: 0, 5, 10, 15, 20, 25, 30, 35, 40, 45, 50

Patient room, Operating room, Intensive/Critical care unit, Emergency dept., Procedure room, Clinical laboratory, Outpatient clinic/office, Outside patient room, Service/utility area, Dialysis facility, Other

International Healthcare Worker Safety Center, University of Virginia

the high proportion of care that is provided in patient rooms. The operating room accounted for 11% of all reports. Although blood contact is relatively frequent in the blood-intensive environment of surgery, reporting an exposure to employee health is often considered inconvenient because personnel are not free to file a report until after the surgical procedure is complete. The remaining incidents were widely distributed among intensive and emergency care areas, diagnostic and treatment procedure areas, and clinical laboratories where blood and body fluid specimens are continually handled.

Ninety-seven incidents (9%) were described as occurring in "other areas" of the hospital. Many of these incidents occurred in a non-treatment area, such as a cafeteria, bathroom, gym, elevator, pediatric activity room, pharmacy, equipment repair room, autopsy/pathology lab, or research lab. Twenty-four cases occurred outside of hospitals, including six incidents in emergency transport helicopters. Other incidents occurred in patient homes (during home care), in parking lots, on the sidewalk, at hospital fair grounds, and at accident scenes. These cases illustrate the complex and varied interactions between hospital personnel and patients.

Body fluids involved in exposures

Fifty-nine percent of incidents involved blood or blood products (see **Figure 3**). A wide variety of other fluids were identified in the remaining 41% of cases. Urine was the second most frequent body fluid reported (8%), followed by vomit, saliva, and sputum (5% each). Gastric fluids, feces, peritoneal fluid, pleural fluid, and amniotic fluid together accounted for 7% of reports. Fifteen percent of reports did not fall into any of the provided categories and involved fluids such as abdominal irrigation fluids, bile, abscess and cyst drainage fluids, seminal fluid, endotracheal secretions, nasal secretions, vaginal secretions, wound discharge, and inoculated culture medium.

Items worn at the time of exposure

In 87% of all cases, BBF exposure was to an unprotected body area, indicating that in the majority of cases, an appropriate protective garment was not worn at the time of contact (see **Figure 4**). In 10% of cases, BBF exposure occurred at the gap between garments, often at the wrist between a glove and sleeve. In 8% of cases BBF soaked through clothing, and in 3% of cases BBF penetrated protective garments.

Gloves were worn at the time of BBF exposure in two-thirds of reported incidents, and were therefore the most commonly worn protective garment. However, hands were the second most frequent body area (after the face) exposed to blood and body fluids. **Table 1** shows the frequency of BBF contact with hands in workers who were wearing gloves at the time of exposure as compared to those who were not. The difference was statistically significant (c2=157, P<.0001). These data demonstrate the protective benefit that gloves provide against BBF exposure to hands, but they also show that gloves do not always provide total protection for the hands.

Eyeglasses (non-protective) were worn by 15% of workers at the time of BBF exposure, but protective eyewear, including goggles or faceshields, was worn at the time of exposure in only 5% of cases. **Table 2** shows the relative frequency of BBF exposure to eyes in workers with different levels of eye protection at the time of exposure. There was an apparent gradient showing a decreasing frequency in eye exposures with increasing levels of eye protection. The difference in frequency of eye exposures among those wearing goggles or faceshields versus those wearing no eyewear was statistically significant (c2=10.4, p=.001). This comparison strongly supports the benefit of goggles and faceshields as protective eyewear. There was also a lower frequency of eye exposures among those wearing eyeglasses in comparison to those with no eyewear (c2=2.6, p=.03). However, the difference in frequency of eye exposures among those wearing

Figure 3. Body Fluids Involved in Exposures

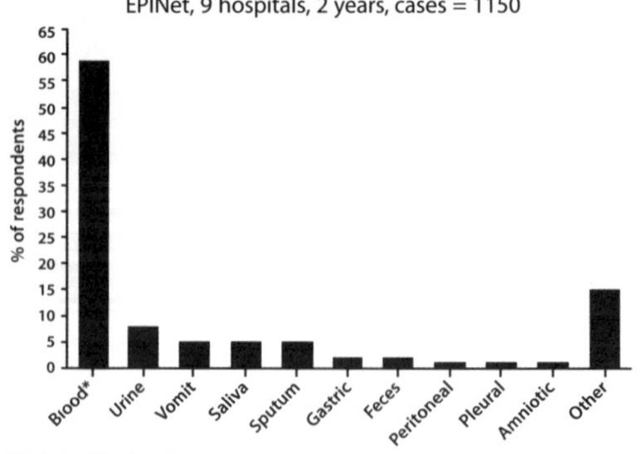

*Includes blood products

International Healthcare Worker Safety Center, University of Virginia

Figure 4. Items Worn at the Time of Exposure

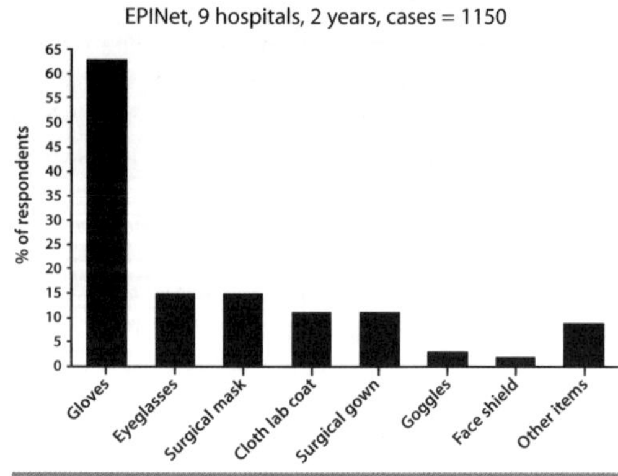

International Healthcare Worker Safety Center, University of Virginia

Table 1. Percent of Workers with Hand Exposures:
Comparing workers wearing gloves to workers not wearing gloves at the time of exposure

	# in group	# with hand exposures	% with hand exposures
Did not wear gloves	420	166	40%
Wore gloves	723	63	9%

Table 2. Percent of Workers with Eye Exposures:
Comparing workers with and without eyewear at the time of exposure

	# in group	# with eye exposures	% with eye exposures
No eyewear	918	485	53%
Wore eyeglasses	173	76	44%
Wore goggles or faceshield	58	18	31%

Table 3. Percent of Workers with Torso or Arm Exposure:
Comparing workers with and without cover garments at the time of exposure

	# in group	# with torso/arm exposures	% with torso/arm exposures
No cover garment	881	145	16%
Wore cloth lab coat	132	21	16%
Wore surgical/protective gown	128	11	9%

eyeglasses versus those wearing goggles or faceshields was somewhat less than that required for statistical significance ($c2=3.0$, $p=.08$).

Despite their relative effectiveness, even goggles and faceshields do not provide absolute protection for the eyes, since 31% of healthcare workers wearing them at the time of exposure nevertheless sustained BBF contact to the eyes. This finding suggests the need to reevaluate the design of protective eyewear to more effectively shield the eyes from exposure to BBF.

The infrequent use of cover garments for the torso and arms was notable. Only 11% of healthcare workers were wearing protective or surgical gowns at the time of BBF contact; another 11% of workers wore cloth lab coats. Table 3 shows that healthcare workers wearing a surgical-type fluid-resistant gown at the time of exposure sustained the fewest torso or arm exposures. The difference in frequency of torso or arm exposures among workers wearing surgical-type gowns and those wearing no cover garment was statistically significant ($c2=5.3$, $p=.02$).

Of special note was the finding that cloth lab coats provided no protective benefit since the same proportion of torso or arm exposures were reported in those who wore cloth lab coats as in those who wore no protective garment.

The design of traditional lab coats is mismatched to the bodily distribution of BBF exposures. Lab coats are front-opening, and are frequently worn open; even when buttoned, the V-neck design leaves the upper torso unprotected. The open cuffs also leave forearms and wrists—common areas of BBF exposure—unprotected. Traditional lab coats are made of loose-weave cotton or cotton-polyester blends, fabrics that are not designed to be resistant to liquid penetration. Cotton is actually an absorbent material that wicks liquid through fabric, potentially increasing the amount of body fluid coming into direct contact with a healthcare worker's skin.

Workers were asked to describe the quantity of blood or body fluid that came into contact with their skin or mucous membranes. In most cases (86%), the quantity was less than 5cc of fluid. In 10% of cases the reported BBF quantity was from 5cc to 50cc. In only 3% of cases was BBF quantity greater than 50cc. The quantity of BBF (especially blood) coming into contact with a healthcare worker, however, is not necessarily correlated with the risk of pathogen transmission. A small quantity of infected blood on non-intact skin or conjunctiva, for example, may have a higher risk of pathogen transmission than a larger quantity of blood on intact skin.

Type of skin or mucous membrane exposed

Healthcare workers indicated the type of skin or mucous membrane exposed to BBF **(see Figure 5).** In some cases more than one type per exposure was reported. BBF contact with intact skin occurred in 50% of reported incidents, and with non-intact skin in 16% of incidents. Mucosa of the mouth was involved in 13% of incidents and mucosa of the nose in 6% of incidents.

Of note was the high percentage of incidents (50%) which involved BBF exposures to the eyes. This finding suggests a bias in the type of incidents that healthcare workers report to employee health; an illustration of this potential bias is shown in Figure 6. A survey was conducted of all

Figure 5. Type of Skin or Mucous Membrane Exposed
EPINet, 9 hospitals, 2 years, cases=1150

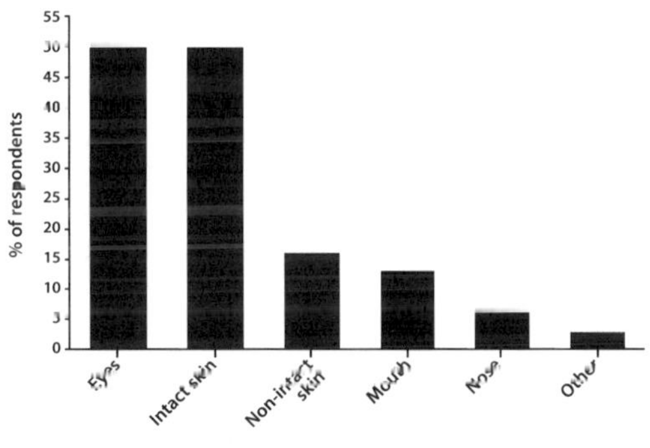

International Healthcare Worker Safety Center, University of Virginia

Figure 6. **Blood & Body Fluid Contact Locations in Emergency Department Staff: A Comparison of Two Reporting Methods**

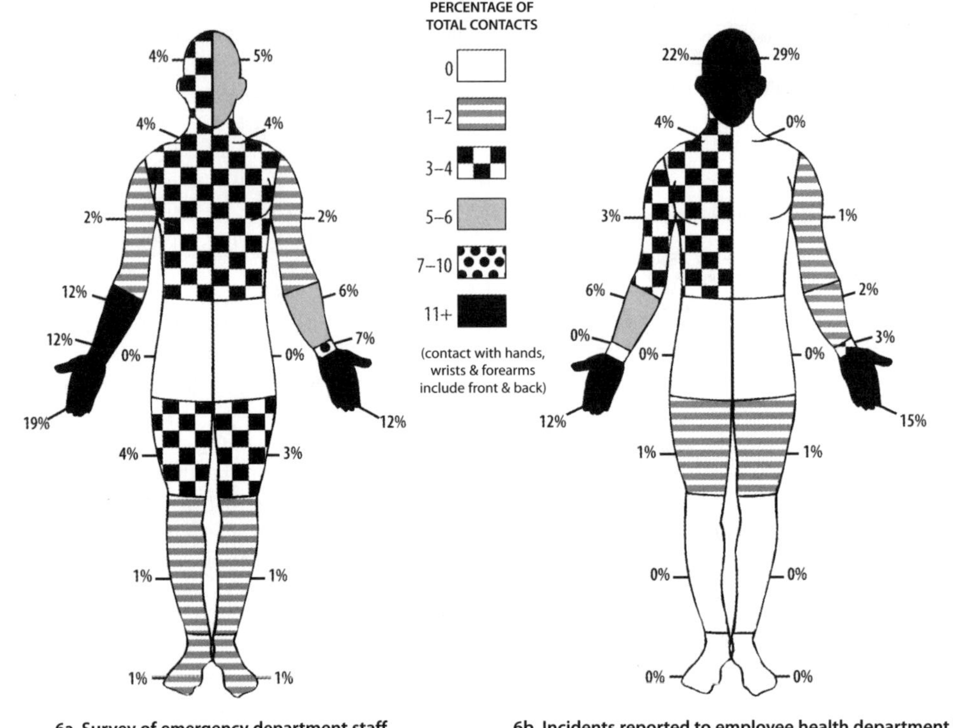

6a. Survey of emergency department staff (most recent incident recalled), n=68 incidents

6b. Incidents reported to employee health department by emergency department staff, n=65 incidents

International Healthcare Worker Safety Center, University of Virginia

healthcare workers in an emergency department of a university hospital (one of the nine participating EPINet hospitals).[13] Seventy-eight respondents were asked to describe the most recent BBF exposure they recalled regardless of the severity of the exposure. The bodily distribution of BBF exposures they described in 68 exposure incidents is shown in **Figure 6** (left side). For purposes of comparison, the BBF exposures (65) reported by emergency department personnel in the nine-hospital database was selected and the bodily distribution of BBF contacts was plotted (**Figure 6,** right side). While the head and face were involved in only 9% of the BBF incidents most recently recalled in the survey, 51% of incidents reported to employee health involved BBF contact with the head and face. One possible explanation for this bias is that exposures to the face are often alarming or repulsive, intruding upon the senses of sight, smell, and even taste. It is likely that shocking events, which would include body fluid exposures to the face, have a greater probability of being reported.

One mechanism of facial exposures was identified that has not been previously described in the literature. A review of 222 detailed descriptions of facial exposures from one of the nine EPINet hospitals showed that 14 different incidents (6%) involved patients spitting into the faces of healthcare workers. Most of these incidents occurred when restraining combative patients; protective faceshields, therefore, should be considered appropriate protective equipment in such situations.

Mechanism of exposure

In 49% of all cases reported to employee health, the BBF exposure was the result of direct patient contact **(Figure 7)**. An unexpected finding, however, was that in most of the remaining cases (nearly 50%), a medical device or product

Figure 7. **Mechanism of Blood and Body Fluid Exposures**

EPINet, 9 hospitals, 2 years, cases=1150

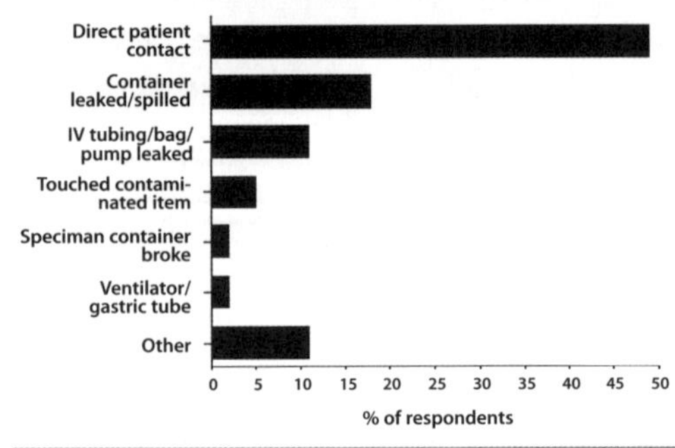

International Healthcare Worker Safety Center, University of Virginia

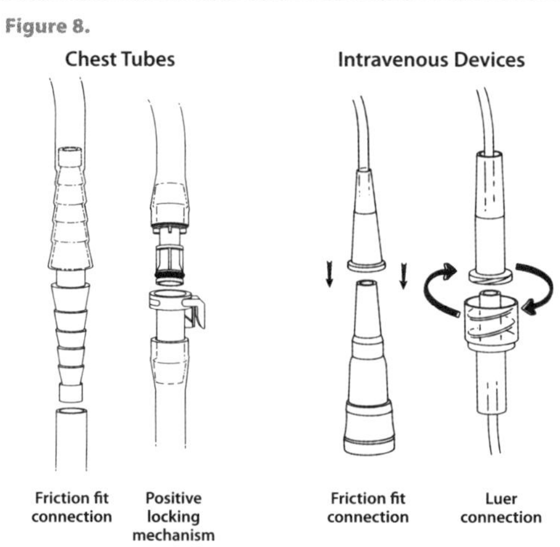

Figure 8.

served as a vehicle of exposure. In 18% of cases a specimen container or other type of body fluid container, such as a suction canister, leaked or spilled. In 2% of cases a specimen container such as a vacuum tube broke. In 5% of cases healthcare workers touched contaminated items such as soiled drapes, laboratory equipment, or surgical instruments. Incidents in the "other" category involved splashes, squirts, or sprays with a variety of devices and items, including syringes, chemstrips, drainage bags, wound dressings, and wash basins; these incidents occurred while irrigating, washing, injecting, disconnecting equipment components, or emptying body fluid containers.

Eleven percent of cases involved tubes, bags or pumps mainly associated with intravenous, arterial, or central lines, equipment used to pump blood under pressure, and urinary drainage devices. Exposures occurred when vascular lines were removed, when junctions in the tubing circuits separated, when lines were flushed under pressure, and when stopcock ports were inadvertently turned to open position.

Two percent of exposures involved elastic tubing, such as is used for gastric and endotracheal tubes. Exposures often occurred during extubation when residual fluid in the rubbery tubing was flung in unpredictable directions. Other exposures occurred when disconnecting tubing segments; as the tubing was pulled apart, it snapped and sprayed droplets of fluid.

One product feature, the friction fit connection, was associated with several of the reported incidents. A friction fit is a connection in which two compatible parts are simply pushed together with enough force to secure the junction by friction. **Figure 8** shows a friction fit junction on chest tubes and on intravenous devices. Friction fit connections are found on many medical devices, and present a common hazard to device users as well as to patients. The security of the junction depends on whether there is a tight fit between the parts. Friction fit connections are prone to inadvertently giving way under certain circumstances, such as when pressure builds up inside the joined tubing segments or when tubing is jostled or manipulated.

During disassembly tubing segments joined by a friction fit must be pulled apart by force, which can result in a rebound motion of the hands and splattering of the fluid contents of the tubes. Alternatives to friction fit connections include the Luer lock, where the two parts are screwed together, and a positive lock, where the connecting parts snap together and are released by pressing a lever. Both of these locking mechanisms provide a junction that is more secure and reliable than a friction fit, and prevent rebound of the hands during disconnection.

Conclusions

Healthcare workers are still far from the ideal situation of being able to accurately anticipate and adequately prepare for exposure situations. At present, universal precautions recommendations are non-specific and discretionary, and provide little guidance for appropriate precautions under particular conditions. The future challenge will be to increase the efficacy of universal precautions by more accurately targeting precautions to specific risk situations. The data presented here do not address all the remaining questions on this issue, but they indicate several areas where improvements are needed, and provide the basis for the following recommendations:

- Increase the availability and use of appropriate personal protective equipment.
  (a) Increase the use of gloves, because of the frequency of hand exposures and the documented transmission of bloodborne pathogens following hand contact with infected blood.
  (b) Increase the use of faceshields and goggles, because of the frequency of exposures to the face and the documented transmission of bloodborne

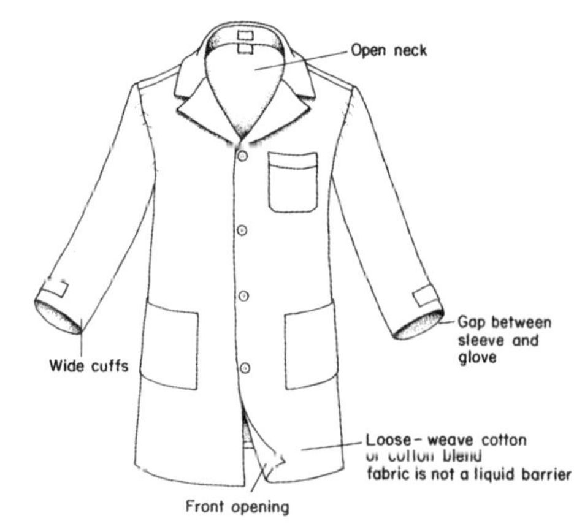

Figure 9. The Common Lab Coat: Not a Protective Barrier

pathogens following BBF exposures to the eyes, nose, and mouth. Face protection should be used when there is the potential for splashing or spraying of body fluids, when restraining combative patients, when in proximity to equipment that contains or pumps body fluid under pressure, and when performing (or near someone performing) intubation or extubation procedures.

(c) The use of loose-weave cotton or cotton-blend cloth lab coats in areas where contact with patient body fluids is possible should be discontinued (see **Figure 9**).

(d) The use of cover garments made of liquid-resistant material and providing a continuous barrier in the front (i.e., not front-opening or v-neck) should be encouraged. Cover garments should have long sleeves and snug cuffs that overlap with gloves in order to provide adequate protection for the arms.

- Identify and improve medical devices and products that act as vehicles of blood and body fluid exposures in healthcare settings.

  (a) Specimen and other body fluid containers should have closures that provide a tight, positive-locking seal, and should be resistant to breakage.

  (b) Equipment that pumps blood under pressure should have positive-locking junctions between connecting components, and should have pressure sensors linked to an alarm or pump cut-off, or an equivalent safeguard to prevent high-pressure ruptures of tubing.

  (c) Reporting systems for healthcare worker exposures to blood and body fluids should explicitly identify products involved in exposures, in order to assist with future product selection, more effectively communicate with product manufacturers, and improve understanding of product design features that best protect healthcare workers.
- Reduce to a minimum the amount of patient blood and body fluids handled by healthcare workers. The quantities of blood and urine collected for laboratory tests, for example, often exceed the amounts required for testing.
- Expand the standardized surveillance of healthcare worker exposures to blood and body fluids, in order to further define the circumstances and mechanisms of exposures, determine the frequency and risk of occupational infection transmission under different exposure conditions, and document the efficacy of specific prevention interventions.

References

1. Centers for Disease Control. Update: Human immunodeficiency virus infections in healthcare workers exposed to blood of infected patients. *MMWR.* 1987;36:285–9.
2. Ippolito G, Puro V, De Carli G, and the Italian Study Group on Occupational Risk of HIV Infection. The risk of occupational human immunodeficiency virus infection in healthcare workers: Italian multicenter study. *Arch Intern Med.* 1993;153:1451–8.
3. Sartori M, La Terra G, Aglietta M, et al. Transmission of hepatitis C via blood splash into conjunctiva. *Scand J Infect Dis.* 1993;25:270–271.
4. Ippolito G, Puro V, De Carli G. Infezione professionale da HIV in operatori sanitori: descrizione dei casi con seroconversione documentata segnalati al 30 giugno 1993. *Giornale Italiano dell'AIDS.* 1993;4:63–75.
5. Department of Labor, Occupational Safety and Health Administration. 29 CFR Part 1920.1030, Occupational exposure to bloodborne pathogens, final rule. *Fed Regist.* December 6, 1991;56:64004–64182.
6. Levy BS, Harris BC, Smith JL, et al. Hepatitis B in ward and clinical laboratory employees of a general hospital. *Am J Epidemiol.* 1977;106:330–5.
7. Wruble LD, Masi AT, Levinson MJ, et al. Hepatitis-B surface antigen (HB Ag) and antibody (Anti-HB) prevalence among laboratory and nonlaboratory hospital personnel. *South Med J.* 1977;70:1075–1079.
8. Pattison CP, Boyer KM, Maynard JE, et al. Epidemic hepatitis in a clinical laboratory. *JAMA.* 1974;230:854–857.
9. Alter MJ, Gerety RJ, Smallwood LA, et al. Sporadic non-A, non-B hepatitis: frequency and epidemiology in an urban U.S. population. *J Infect Dis.* 1982;145:886–893.
10. Klein RS, Freeman K, Taylor PE, et al. Occupational risk for hepatitis C virus infection among New York City dentists. *Lancet.* 1991;338:1539–1542.
11. Centers for Disease Control. Recommendations for prevention of HIV transmission in health care settings. *MMWR.* 1987;36:Suppl 2S.
12. Jagger J, Detmer DE, Cohen ML, Scarr PR, Pearson RD. Reducing blood and body fluid exposures among laboratory workers: meeting the OSHA standard. *Clin Lab Manage Rev.* 1992;6:415–424.
13. Jagger J, Powers RD, Day JS, Detmer DE, Blackwell B, Pearson RD. Epidemiology and prevention of blood and body fluid exposures among emergency department personnel. *J Emerg Med.* 1994;12:753–765.

The nine hospitals contributing data to this report were:
Florida Hospital, Orlando, Florida
Martha Jefferson Hospital, Charlottesville, Virginia
North Broward Hospital, Ft. Lauderdale, Florida
St. Joseph Hospital, Omaha, Nebraska
St. Vincent Hospital, Erie, Pennsylvania
St. Vincent Hospital, Indianapolis, Indiana
Shands Hospital, Gainesville, Florida
University Hospitals of Cleveland, Cleveland, Ohio
University of Virginia Hospital, Charlottesville, Virginia

Suture Needle and Scalpel Blade Injuries: Frequent But Underreported

Janine Jagger, M.P.H., Ph.D. and Melanie Balon, B.S.

Vol. 1, no. 3, 1995

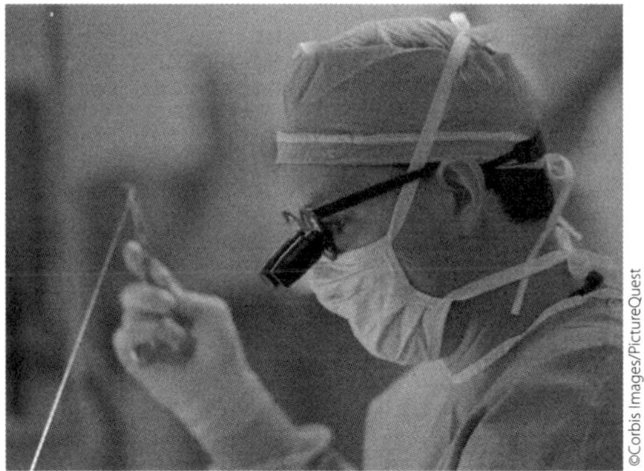

Introduction

Most reports on percutaneous injuries among healthcare workers focus on hollow-bore needle devices such as syringes, blood-drawing devices, and vascular access devices. One reason is that the majority of reported percutaneous injuries in healthcare settings are caused by hollow-bore needles. In 58 U.S. hospitals participating in an EPINet data-sharing group from September 1992 through August 1993, hollow-bore needles accounted for 67% (2,565/3,829) of all reported percutaneous injuries.[1] Furthermore, hollow-bore needles have been most often associated with occupational transmission of bloodborne pathogens, especially HIV.[2]

However, injuries that occur in the surgical setting, where non-hollow-bore needles and sharps are predominantly used, are far less likely to be reported to a hospital authority than those that occur elsewhere in the hospital. In one prospective study, as few as 4% of percutaneous injuries or blood exposures to mucous membranes or non-intact skin that were observed in the operating room were subsequently reported to the employee health department of the hospital.[3] This finding highlights the fact that, while sharp instruments used in surgery are a frequent cause of percutaneous injuries, injuries from these devices are not well documented because data have been less accessible than for other devices and clinical settings.

Suture needles and scalpel blades are the two most common non-hollow-bore devices causing percutaneous injuries in hospitals, together accounting for 17% of all reported injuries in one 58-hospital study.[1] Since most of these injuries occurred in the operating room, where only a fraction of injuries are reported, their true frequency is much higher. Despite the high number of injuries, few instances of occupational infection have been directly linked to suture needles and scalpel blades. This may be because the high frequency of undocumented injuries makes it difficult to link an infection to a specific exposure. Additionally, pathogen transmission rates may be related to the viral titer in the blood of the source patient and the amount of inoculum introduced into the wound. The quantity of blood inoculum introduced by a suture (solid) needle has been shown to be significantly less than that introduced by a blood-filled hollow-bore needle.[4]

One case of HIV transmission following a scalpel blade injury has been documented in Italy [personal communication, Dr. G. Ippolito]. There are at least four cases of possible occupational HIV transmission in the literature, in which multiple percutaneous injuries and blood exposures were recalled while working in surgical settings in high HIV-prevalence areas.[5] The four cases do not provide direct evidence of HIV transmission via suture needles or scalpel blades, but the possibility that these devices may have been transmission vehicles or caused a wound that permitted viral penetration in one or more of these cases must be considered.

The localized transmission of tuberculosis to the hands of pathologists who sustained scalpel injuries during the performance of autopsies was common enough early in the 20th century to be referred to as "prosector's finger."[6] The problem has not been eliminated, and can be compounded by the potentially grave consequences of infection with a resistant strain of tuberculosis.[7]

A further concern is that suture needles and scalpels come into contact with the open wounds of patients, creating the additional risk of healthcare-worker-to-patient pathogen transmission when a percutaneous injury occurs. To date, no cases of HIV transmission from operating room personnel to patients have been reported, despite large-scale look-back studies testing the serology of patients whose surgeons were HIV positive.[8] On the other hand, numerous reports document the transmission of HBV from operating room personnel to patients.[9] The specific exposure mechanisms in these cases were not determined, although inadequate barrier precautions were cited as a contributing factor in several cases.

In order for a patient to be exposed to the blood of a healthcare worker, the worker must first sustain an injury from a sharp instrument, and then the contaminated instrument must recontact the patient, or the worker must bleed directly into the patient's wound. It is necessary to determine the probability of this sequence of events in order to estimate the risk of bloodborne pathogen transmission from healthcare worker to patient. Two observational studies reported the proportion of percutaneous injuries to surgical personnel in which an instrument contaminated with the worker's blood recontacted a patient. The studies found that patient recontact with the sharp instrument occurred after 20% and 24% of injuries respectively.[10,11]

Although the mechanism of injuries from suture needles and scalpel blades is similar whether the injuries occur in the operating room, emergency department, pathology, or elsewhere in the hospital, published reports based on observational methods primarily describe injuries in the operating room, and the number of recorded events in these studies is too small to evaluate device-specific patterns of occurrence.[10-12] This report will analyze the patterns of occurrence of a large number of suture needle and scalpel blade injuries reported in a nine-hospital data-sharing network in order to enhance understanding of healthcare worker risk when handling these sharp instruments, and to identify the most promising opportunities for minimizing the risk of pathogen transmission to healthcare workers and to patients.

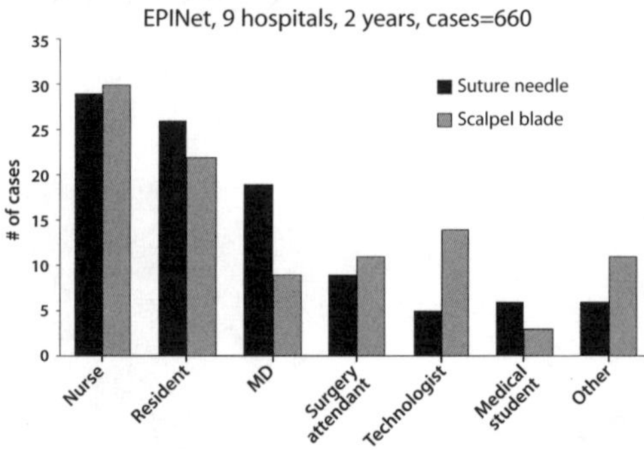

Figure 1. Job Categories of Workers Reporting Injuries with Suture Needles and Scalpel Blades

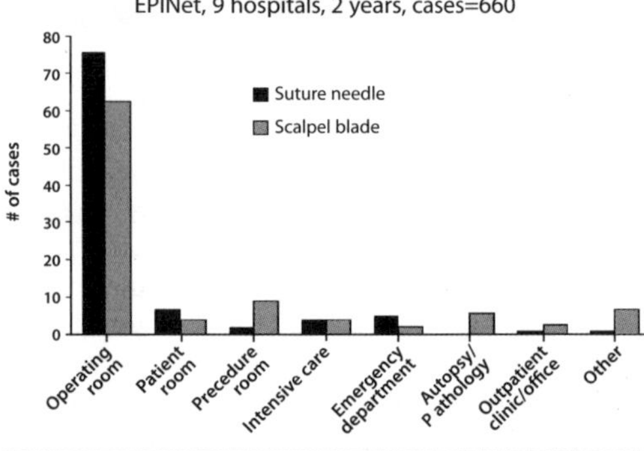

Figure 2. Location of Workers Reporting Injuries with Suture Needle and Scalpel Blades

Findings

Nine hospitals which voluntarily participate in a data-sharing network contributed data for this report [see list at end of article]. They are located in six states in the eastern half of the U.S. All the hospitals use the Exposure Prevention Information Network (EPINet) system for tracking both percutaneous injuries and blood and body fluid exposures in their institutions, and they report their data quarterly to the International Healthcare Worker Safety Center. The cumulative average daily census for the nine hospitals is 4,886 occupied beds. Data in this report include sharp-object injuries only and were collected during a two-year period from September 1992 through August 1994. A total of 3,666 percutaneous injuries were reported in the nine hospitals during two years, including 389 (10.6%) suture needle injuries and 271 (7.4%) scalpel blade injuries. The following results will be restricted to these 660 cases.

Job categories and location of injuries

The distribution of injury occurrence is similar for suture needles and scalpel blades **(Figures 1 and 2).** In contrast to patterns noted with most other devices, physicians (including attending physicians and residents) outnumber nurses in frequency of injury with both devices. Forty-five percent of injuries from suture needles were to attending or resident physicians, and 29% were to nurses. For scalpel blades, 31% of injuries were to physicians, and 30% were to nurses. Technologists (mainly surgical) sustained 14% of scalpel blade injuries while surgery attendants sustained 11% of scalpel injuries.

The majority of injuries from these devices occurred in the operating room: 76% for suture needle injuries and 63% for scalpel blade injuries. Outside the operating room, scalpel blade injuries occurred mainly in pathology/autopsy and in procedure rooms such as angiography and cardiac catheterization labs. Suture needle injuries outside the operating room occurred in patient rooms and emergency departments; however, despite the frequency of wound-suturing in the emergency department, only 5% of the injuries occurred there.

Mechanism of injuries

More than half (54%) of reported suture needle injuries occurred while the device was in use, that is, during suturing **(Figure 3)**. This profile is in direct contrast to that for most hollow-bore devices, where injuries occur primarily after use and during disposal. Of interest was the comparison between users and non-users of suture needles. Of 213 injuries that involved the original user of the device, 68% were self-inflicted during suturing. Of 172 injuries to non-users (scrub nurses or assistants) of the device causing injury, 36% of injuries occurred during suturing and were inflicted by the person using the needle. This indicates that the hands of assistants are often close to needles during

Figure 3. **Mechanism of Suture Needle Injuries**
EPINet, 9 hospitals, 2 years, cases=389

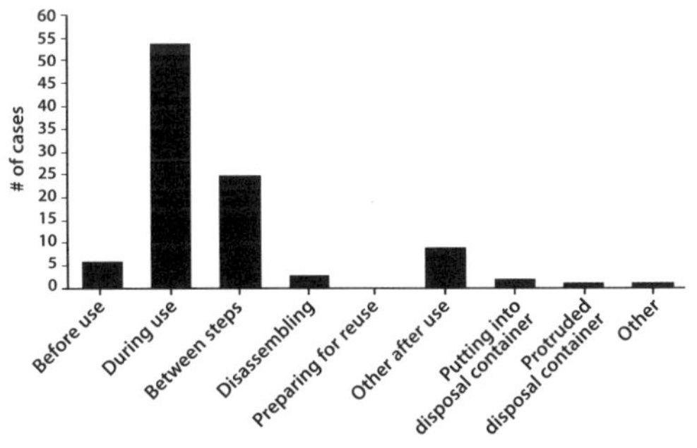

International Healthcare Worker Safety Center, University of Virginia

Figure 4. **Mechanism of Scalpel Blade Injuries**
EPINet, 9 hospitals, 2 years, cases=271

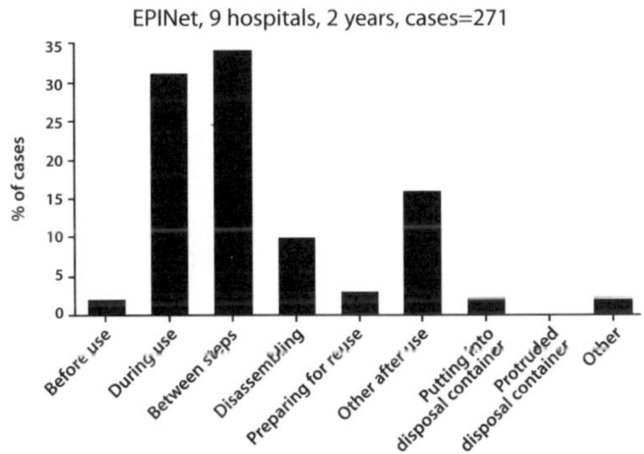

International Healthcare Worker Safety Center, University of Virginia

Figure 5. **Mechanism of Injuries from Suture Needles and Scalpel Blades: Nurses vs. Physicians**
EPINet, 9 hospitals, 2 years, cases=449

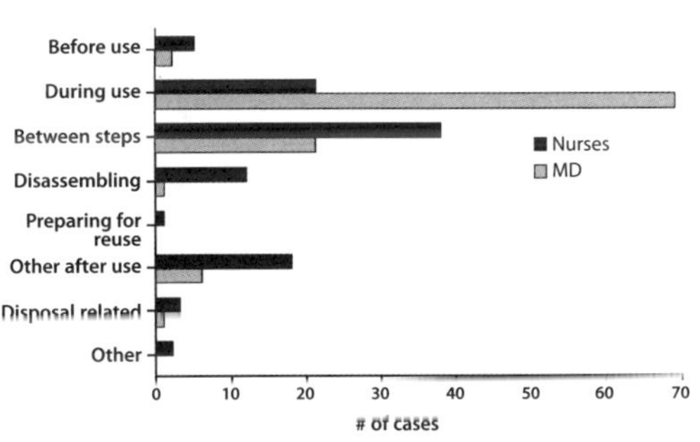

International Healthcare Worker Safety Center, University of Virginia

suturing and are therefore highly vulnerable to injury. Twenty-five percent of all suture needle injuries occurred between steps, mainly during passing of the device. Only 12% occurred after use or in disposal. These findings highlight the need for suture needle protection not only after use but during suturing and passing as well.

Table 1. **Percent of Injuries from Suture Needles and Scalpel Blades Inflicted by Another Worker** (by job category)

	# in group	# injuries inflicted by other worker	% injuries inflicted by other worker
Nurses	193	85	44%
Resident MD's	160	34	21%
Attending MD's	96	16	17%

For scalpels, a smaller proportion of injuries (34%) occurred during use of the device (that is, while cutting) compared to suture needles (see **Figure 4**). This is closer to the injury profile for hollow-bore needles. Unique to scalpel blades, however, is that more injuries occurred between steps of a procedure, such as while passing, than any other phase of use. Scalpel blade injuries were also unique in that, in the majority of cases (61%), the person injured was not the user of the device but rather an assistant. Of 105 injuries to actual users of the device, 45% were self-inflicted. Of 164 injuries to non-users, 22% were inflicted by the users during cutting. Ten percent of all scalpel blade injuries occurred during disassembly of blades from reusable handles.

A comparison of the mechanism of injury for physicians versus nurses (**Figure 5**) reflects a pattern of interaction between the two groups. Sixty-nine percent of physicians sustained injuries while cutting or suturing; this compares to 21% for nurses, and many of these injuries were inflicted by another person using the device. Only 22% of injuries to physicians occurred during passing or disassembling, compared to 50% for nurses.

If the injured worker was not the original user of the device that caused the injury, and the injury occurred during cutting, suturing, or between steps, such as while passing, the injury was determined to be inflicted by another worker. **Table 1** compares the percentage of injuries to nurses, resident physicians, and attending physicians that were inflicted by another worker.

The difference between attending and resident physicians in percentage of incidents inflicted by others was not statistically significant (χ^2=.8, p=.37). The difference between all physicians and nurses, however, was statistically significant (χ^2=31, p<.0001), with 44% of injuries to nurses related to the actions of others. This finding demonstrates the crucial role of interactions between physicians and their assistants in injuries from suture needles and scalpel blades, and highlights the need to involve all professional groups in efforts to reduce injuries from these devices.

Distribution of injuries to the hands

Figures 6 and 7 show the location of injuries to the hands from suture needles and scalpel blades, comparing injuries that occur during use with those not occurring during use for both devices. For both devices together, 39% of injuries were to the right hand, and 53% were to the left hand. Although handedness was not reported in this study, the finding is consistent with a pattern previously reported in which the non-dominant hand sustained the majority of injuries from surgical instruments.[10] However, factors other than handedness also play a role in the distribution of these injuries. In particular, when the injuries occurred during the use of the instrument, while cutting or suturing, the left hand was most frequently injured. But when injuries occurred during passing, disassembly, or disposal, the discrepancy in injury frequency between the two hands was reduced; in the case of scalpel blades the pattern was reversed, with a higher frequency of injuries to the right hand. This reversed pattern may be related to hand-to-hand passing, in which the dominant hand is used for receiving instruments.

Nineteen scalpel blade injuries were not to the hands. Most of these injuries were to the forearms and upper arms, and often occurred when workers inadvertently rested arms on or brushed against the instrument stand where scalpels were placed.

Severity of injuries

On average, scalpel blades are likely to cause more severe injuries than suture needles **(Table 2)**. Fifty-two percent of suture needle injuries were classified as superficial (little or no bleeding), while only 29% of scalpel blade injuries were superficial, the remaining 71% being moderate (skin punctured, some bleeding) or severe (profuse bleeding). The dif-

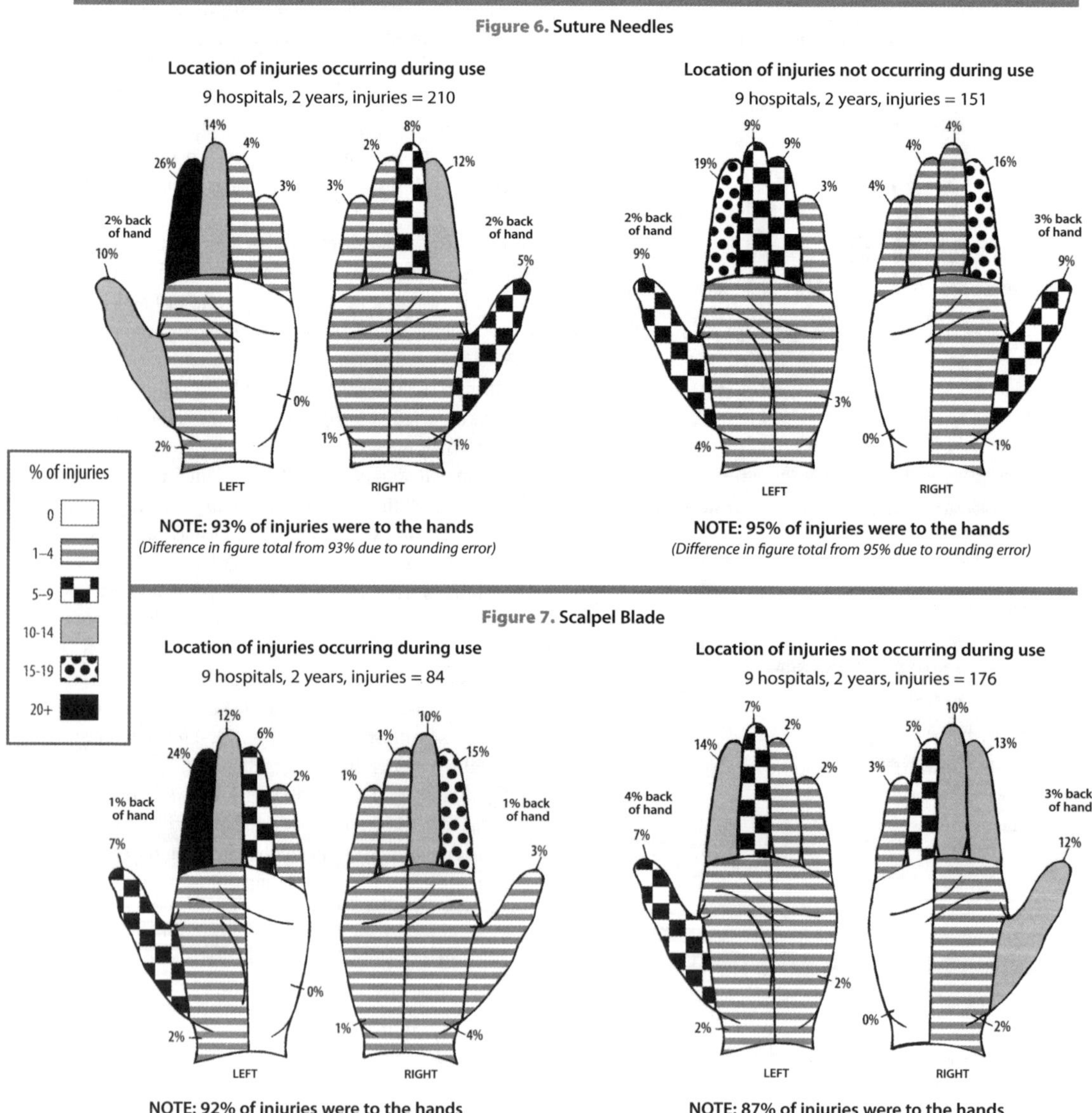

Figure 6. Suture Needles

Figure 7. Scalpel Blade

Table 2. Comparative Severity of Injuries from Suture Needle and Scalpel Blades

	Suture Needles # (%)	Scalpel Blades # (%)
Superficial	202 (52%)	79 (29%)
Moderate	177 (46%)	156 (58%)
Severe	6 (2%)	35 (13%)

ference in injury severity between the two devices is statistically significant (c2=57.25, p<.0001).

When an injury results in bleeding and the healthcare worker's hands are in or near the surgical site—which is the case when injuries occur during use of suture needles or scalpel blades—there is a potential risk of healthcare worker-to-patient pathogen transmission. Scalpel blades present a special risk because the injuries they cause are more likely to result in significant bleeding. Another factor to be considered is that personnel other than the physicians using the instruments, such as nurses and other assistants, can sustain injuries while in contact with the surgical site. Previous studies which estimated healthcare worker-to-patient pathogen transmission risks have focused primarily on surgeons.[13,14] However, any personnel in contact with the surgical site who sustain a percutaneous injury have the potential for transmitting bloodborne pathogens to patients. Of injuries sustained by physicians and nurses during suturing and cutting, 19% were to nurses, although most were not the original users of the devices.

Conclusions

Although suture needles and scalpels are among the devices that most commonly cause percutaneous injuries to healthcare workers, few data have been available describing injury patterns because of extremely low injury reporting rates among surgical personnel, the group most likely to be injured by these devices.

There is currently a lack of direct evidence linking bloodborne pathogen transmission via suture needles or scalpel blades with patient-to-worker or worker-to-patient bloodborne pathogen transmission, yet these transmission routes remain biologically plausible. There is likely a lower probability of patient-to-worker transmission via suture needle injuries than for injuries from blood-filled hollow-bore needles, because they involve a smaller blood inoculum. It is also possible, however, that the lack of transmission evidence may be due to the failure to document suture needle and scalpel injuries.

Prevention of these injuries remains important because healthcare workers still must face the consequences of possible infection, including post-exposure follow-up, emotional stress, and, if the source patient is the carrier of a bloodborne pathogen, changes in personal behavior until transmission can be ruled out. These injuries also have financial consequences for the institutions where they occur, which are further compounded if an infection results from an exposure.

There are many opportunities for changes in devices, procedures, and barrier equipment for reducing injury risk from suture needles and scalpel blades. One approach is to reduce the use of these devices by utilizing them only when necessary. For example, using stapling devices or adhesive products for wound closure whenever possible would reduce the use of suture needles.

The integration of safety features into the designs of suture needles and scalpels is another approach. Scalpels with blade shields that can be placed in protected position during passing and after use are already available. If these designs perform as they are intended to, a large proportion of injuries from scalpels (as much as 69%) could potentially be prevented, since so many injuries occur when the scalpel blade could be covered (i.e., during passing or disposal). It is important, however, to document compliance in activating the safety feature and prevention efficacy. Devices that allow the non-manual release of scalpel blades from reusable handles provide a method for reducing injury risk during disassembly of scalpels, potentially addressing 10% of scalpel blade injuries; such devices include specially designed scalpel handles that release scalpel blades, and accessory devices that mechanically remove blades from reusable handles. The use of disposable scalpels, which do not require blade removal, also have the potential to prevent disassembly-related injuries.

Blunt suture needles have recently become available that are sharp enough to penetrate internal tissues such as muscle and fascia but are not sharp enough, under normal conditions, to cause percutaneous injuries to healthcare workers. The potential for injury prevention depends on the proportion of tissue-suturing that can be performed with blunt needles. In response to a survey, general surgeons at the University of Virginia estimated that 33% of sutures were placed in muscle or fascia, while resident surgeons estimated 37%.[15] Although blunt suture needles cannot be used for all suturing, their potential applicability is nevertheless substantial, at least in the surgical setting, since suture needles cause more injuries in the operating room than any other device.

Changes in procedures, in particular techniques for passing instruments, have been implemented in many institutions in an effort to reduce surgical instrument injuries. Twenty-five percent of suture needle injuries and 34% of scalpel blade injuries in this study occurred between steps of a procedure, which includes passing, although the percent associated with hand-to-hand passing cannot be determined from these data. Hands-free transfer is a technique that is intended to minimize collisions of hands with sharp instruments by designating a neutral zone where instruments can be placed and picked up. The neutral zone may be a small basin or other appropriate receptacle, or a specially designed surgical drape with a recessed area for instrument placement. This technique is becoming more widespread, although controlled studies are needed to fully validate it as an effective injury-reduction method. The high proportion of injuries that occur during interaction of device users and assistants indicates the importance of cooperation among different surgical personnel when implementing new procedures or techniques, such as hands-free instrument transfer.

Barrier products for the prevention of punctures and cuts to the hands are now available, such as cut-resistant gloves and puncture-resistant adhesive pads. Since these products are intended for hand protection, the issues of flexibility and touch sensitivity are important to users. The products are often recommended for use on hand locations that are most likely to sustain sharps injuries; for example, since the non-dominant is most likely to be injured, a cut-resistant glove worn on that hand would likely have a higher efficacy than if it were worn only on the dominant hand, where it might interfere with touch sensitivity or dexterity. Selective use of puncture- and cut-resistant hand protection should be carefully considered, however, because the distribution of injury locations on hands is markedly different for device users as opposed to those who are assisting users. The inappropriate placement of such barriers can greatly compromise their potential efficacy.

Finally, standardized surveillance systems that document the devices and products involved in blood exposures, and which track follow-up serology of exposed healthcare workers, need to be widely used so that device- and mechanism-specific transmission rates can be determined, not only for HIV but for HBV and HCV as well. Comprehensive surveillance systems incorporating these elements have been implemented recently in Italy, Canada, and Australia. Data from such systems will result in a better understanding of pathogen transmission as it relates to different classes of medical devices; the data will also aid in the development of more effective prevention measures for minimizing the occupational transmission of bloodborne pathogens.

References

1. Jagger J, Blackwell B, Fowler M, Carter K, Funderburk S, Bradshaw E, Swapp J. Percutaneous injury surveillance in a 58-hospital network. Tenth international conference on AIDS, Yokohama, Japan, 8/9/94.
2. Ippolito G, Puro V, De Carli G. Infezione professionale da HIV in peratori sanitari: descrizione dei casi con seroconversione documentata segnalati al 30 giugno 1993. *Giornale Italiano dell'AIDS.* 1993;4:63–75.
3. Lynch P, White MC. Perioperative blood contact and exposures: a comparison of incident reports and focused studies. *Am J Infect Control.* 1993;21:357–363.
4. Howard R, Bennet NT. Quantity of blood inoculation in a needlestick injury from suture needles. *J Am Col Surg.* 1994;178:107–110.
5. Ippolito G, Puro V, De Carli G. Infezione professionale da HIV in operatori sanitari: II descrizione dei casi senza seroconversione documentata segnalati al 30 giugno 1993. *Giornale Italiano dell'AIDS.* 1993;4:186–193.
6. Collins C, Kennedy DA. Microbiological hazards of occupational needlestick and "sharps" injuries: a review. *J Appl Bacteriol.* 1987;62:385–402.
7. Sahn SA, Pierson DJ. Primary cutaneous inoculation drug-resistant tuberculosis. *Am J Med.* 1974;57:676–678.
8. Centers for Disease Control: Investigations of patients who have been treated by HIV-infected healthcare workers. *MMWR.* 1992;41:344–346.
9. Centers for Disease Control. Recommendations for preventing transmission of human immunodeficiency virus and hepatitis B virus to patients during exposure-prone invasive procedures. *MMWR.* 1991;40 (RR-8):1–6.
10. Tokars JI, Bell DM, Culver DH, et al. Percutaneous injuries during surgical procedures. *JAMA.* 1992;267:2899–2904.
11. Gerberding JL, Littell C, Tarkington A, et al. Risk of exposure of surgical personnel to patients' blood during surgery at San Francisco General Hospital. *N Engl J Med.* 1990;322:1788–1793.
12. Panlilio AL, Deretha RF, Edwards JR, et al. Blood contacts during surgical procedures. *JAMA.* 1991;265:1533–1537.
13. Nease RF, Owens DK. Estimating the risk posed to patients of HIV-infected surgeons and dentists performing invasive procedures. *Med Decis Making.* 1991;11:325.
14. Schulman KA, McDonald RC, Lynn LA, et al. Screening surgeons for HIV infections: assessment of a potential public health program. *Infect Control Hosp Epidemiol.* 1994;15:147–155.
15. Jagger J, Detmer DE, Blackwell B, Litos M, Pearson RD. Percutaneous injuries among operating room personnel. Conference on Preventing Bloodborne Pathogen Transmission in Surgery and Obstetrics. American College of Surgeons, Centers for Disease Control and Prevention, Atlanta, Georgia, 2/14/94.

The nine hospitals contributing data to this report were:
Florida Hospital, Orlando, Florida
Martha Jefferson Hospital, Charlottesville, Virginia
North Broward Hospital, Ft. Lauderdale, Florida
St. Joseph Hospital, Omaha, Nebraska
St. Vincent Hospital, Erie, Pennsylvania
St. Vincent Hospital, Indianapolis, Indiana
Shands Hospital, Gainesville, Florida
University Hospitals of Cleveland, Cleveland, Ohio
University of Virginia Hospital, Charlottesville, Virginia

Disposal-Related Sharp-Object Injuries

Janine Jagger, M.P.H., Ph.D. and Melanie B. Bentley, B.S.

Vol. 1, no. 5, 1995

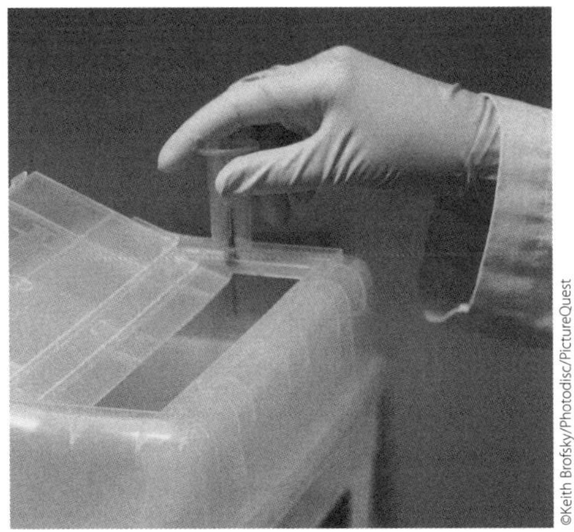

©Keith Brofsky/Photodisc/PictureQuest

Introduction

For as long as the risk of percutaneous injuries among healthcare workers has been an issue of major concern, the safe disposal of used needles and sharp devices has been recognized as an important factor in minimizing injury risk. Official recommendations for the safe disposal of contaminated sharp devices were among the prevention measures promoted by the Centers for Disease Control (CDC) in 1983.[1] In addition to other measures, the 1983 document specified that sharp medical devices should be discarded in puncture-resistant disposal containers placed near the point of use. These guidelines became an enforceable national workplace standard in December 1991 when the Occupational Safety and Health Administration (OSHA) enacted the final bloodborne pathogens standard.[2]

Although there is little documentation of the most common disposal systems and practices in effect in hospitals prior to the promotion of universal precautions guidelines, it is nevertheless apparent that many changes and improvements have occurred, particularly since 1987 when OSHA first announced the preliminary bloodborne pathogens standard. It is now common to find sharps disposal containers located in each patient room in clinical units, rather than at central locations such as nurses' stations. Flimsy disposal containers have for the most part been removed from the U.S. market, and manufacturers have recently developed containers with enhanced features, including greater puncture resistance, one-way openings, and mechanisms to prevent overfilling. However, because of the lack of standard data collection methods for occupational blood exposures before 1987, the impact of recent improvements has not been well documented.

EPINet data used for this analysis began to be collected in September 1992, after the implementation of OSHA-mandated standards. Despite recent advances, it is clear that disposal-related injuries continue to occur in significant numbers and that many problems remain. This analysis is intended to describe the current characteristics of disposal-related percutaneous injuries in a network of hospitals, and to provide a baseline for assessing future progress in addressing specific disposal issues identified by this and other research.

Methods

Nine hospitals participating in the EPINet data-sharing network coordinated by the International Healthcare Worker Safety Center provided two years of data on percutaneous injuries reported to each hospital's employee health department. The nine hospitals had a mean average daily census of 4,550, and reported a total of 3,666 percutaneous injuries during the two-year period.

The descriptive categories on the EPINet report form that pertain to mechanism of injury were reviewed in order to determine which categories could be classified as "disposal-related." Recapping injuries were excluded from the analysis, although they are to some degree relevant to disposal. While not included in the analyses, recapping injuries will be addressed in the discussion section. Other injuries that occurred after use but before disposal of sharp items were considered to be disposal-related and were included in the analysis. The rationale for including these cases is that they are disposal "failures": the likelihood of contaminated sharp devices being left on beds, tables, or floors where they can cause pre-disposal injuries may be related to the inaccessibility of appropriate disposal facilities at the time of their use. Other included categories were injuries occurring from devices left on or near a disposal container; from introducing a sharp item into a container; from a device protruding from the opening of a disposal container; from a sharp item that pierced the side of a disposal container; and from a sharp item placed in an inappropriate trash container. The characteristics of injuries occurring under these circumstances were evaluated in relation to the job categories of injured workers, the location of injuries, the devices causing injuries, the original purpose of devices causing injuries, and whether the source patient was identifiable. These analyses can help in defining the current characteristics of disposal-related percutaneous injuries and in setting priorities for their prevention.

Results

Of 3,666 injuries, 1,335 injuries (36.4%) were classified as disposal-related based on the criteria described above. This figure does not include injuries that occurred while recapping, disassembling a sharp device, or between steps of a multi-step procedure. The following analyses refer only to disposal-related injuries using 1,335 (100%) cases as the denominator, unless otherwise stated.

Figure 1 shows mechanism of injury for 1,335 cases of disposal-related injuries. Of these, 910 (68.2%) occurred after use but before disposal, indicating a failure to properly dispose of a sharp item that was left on a table, bed, or floor, or an intervening situation which prevented safe disposal, such as fumbling with a sharp item on the way to the disposal container. This category also includes 20 cases (1.5%) that occurred from sharp items left on or near a disposal container. Because there were few such incidents and they were similar to cases in which devices were left in other inappropriate places, these 20 cases were included in the 910 injuries occurring after use but before disposal.

Figure 1. Mechanism of Disposal-Related Sharp-Object Injuries

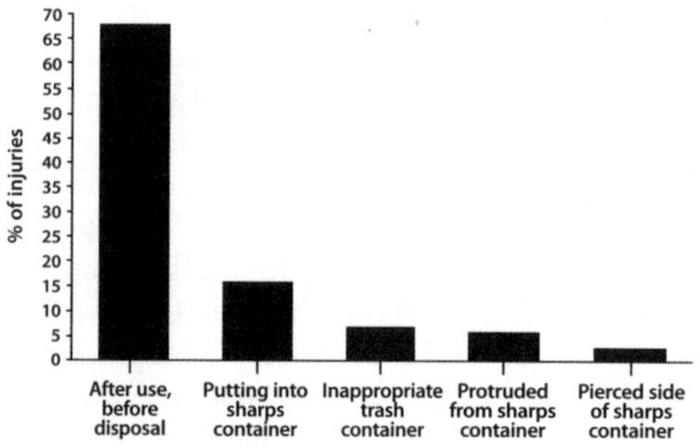

The remaining 425 (31.8%) injuries were directly related to the act of disposal or to a device that had already been disposed of. In 213 cases (16.0%), healthcare workers were stuck by a sharp device they were holding as they introduced it into a disposal container. In 97 cases (7.3%), workers were stuck by devices that were disposed of in inappropriate, non-puncture-resistant trash containers, such as waste baskets in patients' rooms. Many of these injuries occurred when a needle pierced a plastic bag while trash was being removed from a patient's room or other clinical area. All of these injuries were caused by inappropriate disposal practices on the part of the original user of the sharp device.

Eighty-one injuries (6.1%) occurred when workers were stuck by devices protruding from the openings of sharps disposal containers. These injuries can be attributed mainly to overfilled containers, and highlight the importance of an effective hospital policy on regular and frequent waste removal. Thirty-four injuries (2.5%) occurred when needles pierced the sides of sharps disposal containers, indicating that the containers were not adequately puncture-resistant. Twenty of the 34 injuries occurred in one hospital where a cluster outbreak of needlesticks was attributed to one type of disposal container, which was subsequently modified by the manufacturer to increase puncture resistance.

Figure 2 shows the job categories of workers sustaining disposal-related injuries. Nurses constituted the largest group, with slightly more than 50% of cases. Housekeepers and laundry workers were the second largest group with 10.3% of cases. The remaining cases were broadly distributed across a number of job categories, with each having fewer than 10% of cases.

Of interest was the comparison of mechanism of injury for the top two job categories, nurses and housekeepers/laundry workers, as shown in **Figure 3**. This figure compares the percentage of all workers that were either nurses

Figure 2. Job Categories of Workers Reporting Disposal-Related Sharp-Object Injuries

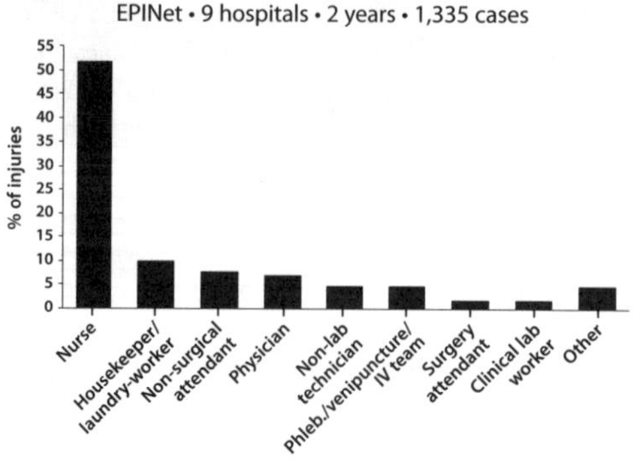

Figure 3. Comparison of Injury Mechanism for Nurses vs. Housekeepers/Laundry Workers

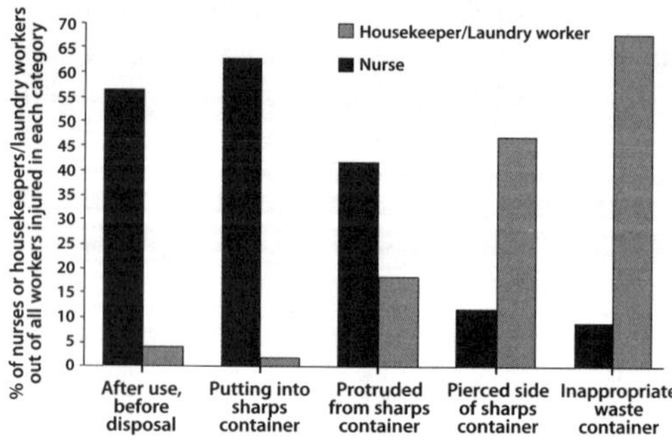

Figure 4. Unusual Body Locations of Injuries from Needles in Inappropriate Trash Containers and Needles Piercing the Sides of Disposal Containers

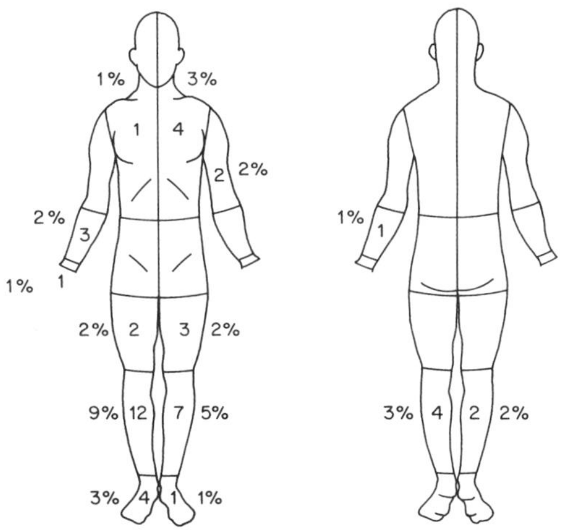

Figure 5. Location of Workers Reporting Disposal-Related Sharp-Object Injuries

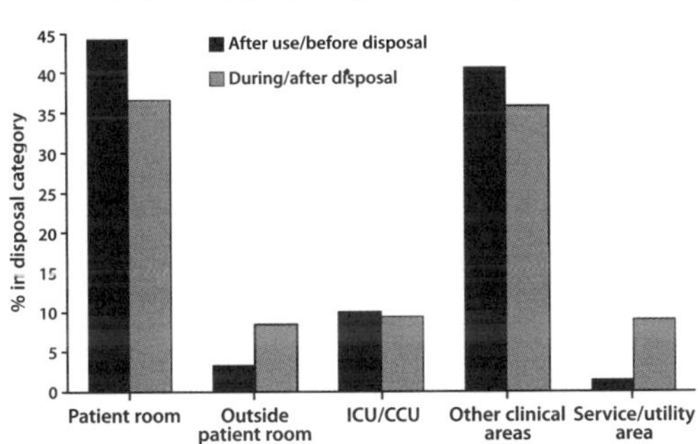

Figure 6. Cases in Which the Source Patient Identity was Known

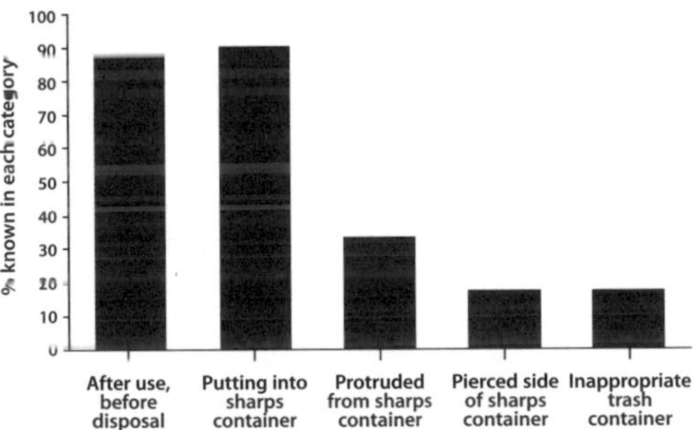

or housekeepers/laundry workers (mainly housekeepers) in each disposal categories. Injuries to nurses occurred primarily after use and before disposal of devices, and during the act of disposal. However, housekeepers' injuries usually occurred after devices had been disposed of. This comparison suggests that the actions of one group affects the risk of injury for other groups. For example, if disposal containers are not removed promptly by housekeepers when full, this increases the risk of other personnel being injured from devices protruding from openings of the containers. Conversely, if nurses (or other patient care personnel) dispose of sharp devices in patient waste baskets, this increases the risk of housekeepers being injured from needles sticking through plastic trash bags. However, injuries caused by needles piercing the sides of sharp containers indicate a design flaw in the containers—that is, inadequate puncture resistance.

Of interest was the body distribution of injuries that were caused by needles piercing containers and devices disposed of in inappropriate trash containers, as shown in **Figure 4**. An unusually high percentage of injuries were to locations other than hands (37%). Although injuries were widely distributed in different body locations, the highest number of non-hand injuries were from needles that pierced plastic trash bags and grazed the calves of housekeepers as they carried the bags at their sides.

Figure 5 shows the general locations where disposal-related injuries occurred. Injuries were separated into two categories: those occurring after use and before disposal (68.2%), and those occurring during or after disposal (31.8%). These two categories are presented separately for the purpose of comparison with previous studies that did not consider injuries which occurred after use and before disposal to be disposal-related. This figure shows the percentage distribution of each group across hospital locations. The majority of percutaneous injuries occur in areas where the sharp devices are used—in patient rooms, critical care units, and other clinical areas. A minority of disposal-related injuries occur outside clinical areas where sharps waste is transported and manipulated, such as corridors, elevators, and utility areas, and those injuries are more likely to happen during and especially after the act of disposal. Although service workers sustain most of the injuries that occur in utility areas, they nevertheless sustain more injuries in clinical areas (57.2%) than they do in utility and non-clinical areas (42.8%).

Figure 6 shows the percentage of cases in which the source patient identity was known in relation to different mechanisms of injury. Source patient identity was known in more than 80% of cases for injuries that occurred after use, before disposal, and during the act of disposal. But for injuries that occurred after disposal, when devices were already in a disposal container or other trash, source patient identity was known for only one-third or fewer cases. The consequence of this discrepancy is that housekeepers are

Figure 7. Original Purposes of Devices Causing Disposal-Related Sharp-Object Injuries

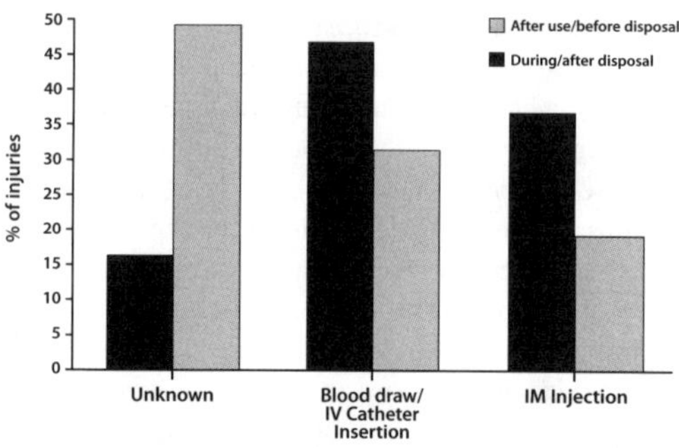

Figure 8. Devices Causing Disposal-Related Sharp-Object Injuries

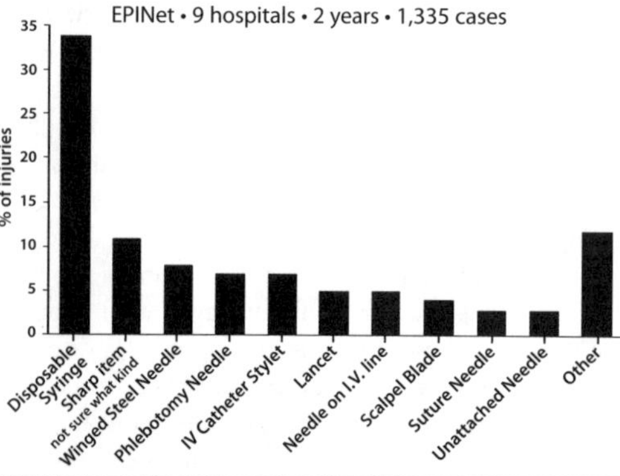

disproportionately affected by the additional testing and uncertainty that occurs when source patient status cannot be verified.

Figure 7 shows the original purpose of devices causing disposal-related injuries for the three categories most commonly reported. The percentage distributions are presented separately for injuries occurring after use/before disposal, and for injuries occurring during/after disposal. This figure highlights the large proportion of disposal-related injuries from needles used for blood-drawing or intravenous catheter insertion, which carry the highest risk of bloodborne pathogen transmission. On the other hand, injuries from devices whose original purpose is unknown may or may not carry real risk, but by default are considered high-risk since transmission risk cannot be ruled out. Injuries from needles used for intramuscular injections are lower risk for bloodborne pathogen transmission than those from phlebotomy needles, but they are not risk-free and, therefore, follow-up does not differ from other injuries at present.

The devices associated with disposal-related injuries can be seen in **Figure 8**. The distribution of devices is similar to that for non-disposal-related injuries. Very few devices—only 10.0%—could not be identified, despite the large number of cases in which the purpose of the device was unknown, as seen in Figure 7.

Discussion

Although disposal-related percutaneous injuries constitute a significant proportion of reported sharps injuries overall, a precise definition of "disposal-related" has not been established. Injuries that occur while introducing a sharp device into a sharps container, and those that are caused by devices protruding from sharps or other trash containers, are clearly disposal-related. Other injuries, however, such as those associated with recapping, may be directly affected by the placement and quantity of disposal containers in a facility. One of the reasons healthcare workers give for recapping needles is to protect themselves from an exposed needle as they bring it to the disposal container.[3] When containers are placed within arm's reach or a short distance from the point of use, the incentive to recap may decline. A possible indicator, therefore, of an improved disposal system that includes conveniently placed containers may be a reduction in injuries from recapping. There is indirect evidence consistent with this hypothesis. One study of needlestick injuries conducted in 1986 reported that 23% of injuries from hollow-bore needles were caused by recapping. The present study, using the same data collection definitions, showed that for the same hospital during 1992-1994 only 5.0% of injuries from hollow-bore needles were caused by recapping (6.8% for all 9 hospitals). Although it is not possible to determine to what degree these reductions were due to improved disposal systems implemented after 1986, as opposed to training and education to discourage recapping, they nevertheless suggest that changes in disposal systems influence the disposal practices of healthcare workers. Similarly, injuries occurring after use and before disposal of sharp devices, such as those from devices left on tables or in beds, may be due in part to the lack of convenient sharps containers where the device was used. The availability and convenience of disposal containers in different areas of the hospital should continue to be evaluated and improved.

These data highlight the continued need for a consistently communicated protocol on handling and disposing of used sharps. Injuries caused by devices in inappropriate trash containers or protruding from sharps containers can be prevented if workers adhere to waste disposal recommendations. Hospitals need an effective policy for monitoring sharps containers so that they are replaced on a consistent basis before becoming full. Healthcare workers need to understand the consequences to others if they place used needles and sharps in inappropriate trash containers.

Product design also plays an important role in the causation and prevention of disposal-related injuries. Injuries that occur when needles pierce the sides of disposal containers can be eliminated by requiring all sharps containers to have high puncture resistance; this is a readily achievable goal. Although the design of disposal containers has greatly improved in recent years, those who make product decisions for hospitals have no basis for comparing the puncture resistance of different containers. To date there is no standard test for judging puncture resistance and no industry-wide agreement on product labelling to indicate the level of puncture resistance for different containers. This remains an important need.

The design of sharp medical devices also has a bearing on the risk of sustaining disposal-related injuries. Devices with safety features that shield or blunt needles after use will protect not only the user but the waste-handler as well. As safety-engineered devices become more widespread, the frequency of disposal-related injuries should decline. It is important, however, to note the differences in safety features of various devices. If a safety feature is "passive" or automatic (i.e., does not require user activation), then the protective feature should prevent all disposal-related injuries from that device. If, however, a safety feature requires activation by the user, then activation rates need to be monitored to ensure that the safety feature is providing the intended benefits for users and waste-handlers alike.

Finally, disposal-related injuries impose a disproportionate burden of risk on housekeepers. Most of their injuries are associated with waste-handling. Because the identity of the source patient cannot be established in a large percentage of disposal-related injuries, pathogen transmission must be considered a possibility in many instances where no risk exists. The lifestyle changes recommended until infection can be ruled out may have a major impact on the lives of the affected individuals. Continued efforts—including monitoring of sharps disposal policies and practices, and continued improvements in the safety of disposal containers and sharp medical devices—are needed to better protect all healthcare workers, but especially those at the receiving end of the medical device waste stream.

References

1. Williams WW. Guideline for infection control in hospital personnel. *Infect Control.* 1983;4(4 Suppl): 326–49.
2. Department of Labor, Occupational Safety and Health Administration. 29 CFR Part 1920.1030, Occupational exposure to bloodborne pathogens, final rule. *Fed Regist.* December 6, 1991;56:64004–64182.
3. Jagger J, Hunt EH, Brand-Elnaggar J, Pearson RD. Rates of needlestick injury caused by various devices in a university hospital. *N Engl J Med.* 1988;319:284–288.

The nine hospitals contributing data to this report were:

Florida Hospital, Orlando, Florida
Martha Jefferson Hospital, Charlottesville, Virginia
North Broward Hospital, Ft. Lauderdale, Florida
St. Joseph Hospital, Omaha, Nebraska
St. Vincent Hospital, Erie, Pennsylvania
St. Vincent Hospital, Indianapolis, Indiana
Shands Hospital, Gainesville, Florida
University Hospitals of Cleveland, Cleveland, Ohio
University of Virginia Hospital, Charlottesville, Virginia

EPINet Data Report:

ABG Syringes Associated with More Injuries During Procedure

Vol. 2, no. 1, 1995

EPINet data provides indirect support for the clinical observation that arterial blood drawing causes more patient discomfort than other needle procedures such as intramuscular or subcutaneous (IM/SQ) injections. A higher proportion of healthcare worker injuries would be expected during the performance of more painful procedures as a result of unexpected patient movements related to pain response. Injuries, however, occurring after the procedure but before disposal, during disposal or after disposal are not considered indicative of patient responses or discomfort. We compared the mechanism of injuries caused by arterial blood gas (ABG) syringes to those caused by syringes used for IM/SQ procedures. The data were obtained from a 64-hospital EPINet database during a two-year period beginning in September 1992.

Figure 1 shows that only 22% of injuries caused by syringes used for IM/SQ injections occurred during the procedure (injection), whereas Figure 2 shows that 47% of injuries caused by ABG syringes occurred during the procedure (arterial blood drawing). This difference was statistically significant ($x^2 = 31, p < .0001$). Clearly, there is a proportionately higher risk of healthcare worker injury during arterial blood drawing than during IM/SQ injection. Factors other than pain may also contribute to this difference, since arterial blood drawing is often a more difficult and complex procedure than IM/SQ injection. The degree to which patient discomfort accounts for the difference seen in this comparison cannot be precisely determined. However, the open-ended descriptions of incidents with ABG syringes showed that injured healthcare workers noted patient movements as a contributing factor in more than half of the injuries that occurred during arterial blood drawing.

Measures to reduce patient discomfort should be considered as one potential strategy for minimizing healthcare worker risk of needlestick during arterial blood drawing. Because needlesticks from blood drawing needles in general are among the most likely to transmit bloodborne pathogens, all potential risk-reducing measures with these needles should be actively investigated.

Figure 1. Mechanism of Injuries for Syringes Used for IM/SQ Injections

EPINet • 64 hospitals • 2 years • 1,103 injuries

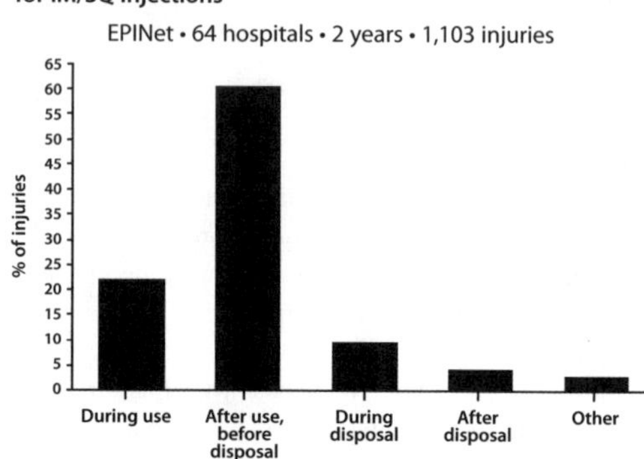

International Healthcare Worker Safety Center, University of Virginia

Figure 2. Mechanism of Blood Gas Syringe Injuries

EPINet • 64 hospitals • 2 years • 107 injuries

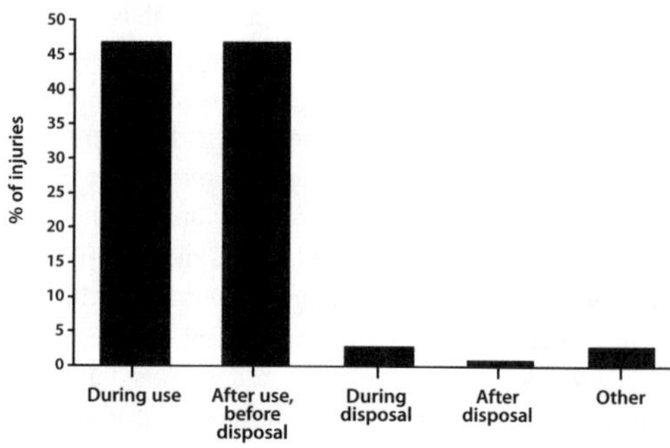

International Healthcare Worker Safety Center, University of Virginia

Clinical Laboratories: Reducing Exposures to Bloodborne Pathogens

Janine Jagger, M.P.H., Ph.D. and Melanie Bentley, B.S.

Vol. 2, no. 3, 1996

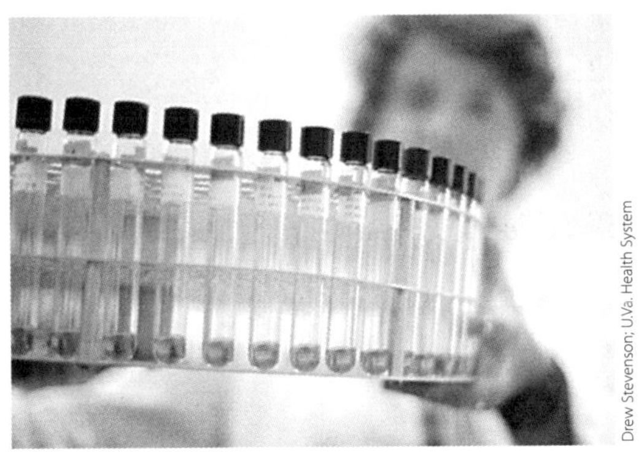

Drew Stevenson; U.Va. Health System

Introduction

In the past decade, occupational risks to laboratory workers have become more serious than ever with the advent of the AIDS epidemic. One of the most disturbing statistics on the occupational transmission of HIV is the large proportion of clinical laboratory workers among the documented and probable cases of occupationally transmitted HIV in the United States. Clinical laboratory workers accounted for 20% (30/151) of all cases reported through December 1995, ranking second only to nurses despite the fact that nurses are far more numerous in the health care work force than clinical laboratory workers.[1] One explanation for this is that most of the cases attributed to clinical laboratory workers involved phlebotomists who sustained needlesticks from blood-drawing needles. It has become clear that the types of exposures that are most likely to result in HIV transmission are those that involve the direct inoculation of significant quantities of blood.[2]

In addition to HIV, numerous studies have documented an impressive array of pathogens transmitted to clinical laboratory personnel, foremost among them hepatitis B.[3-8] In one study, hepatitis B virus (HBV) seropositivity rates were highest among lab workers who routinely got blood on their clothes, those who frequently performed blood gas analysis, and those working on multi-channel autoanalyzers.[9] These observations suggest that transmission of some bloodborne pathogens can also be associated with frequent contact and skin contamination. The potential for exposure to laboratory specimens contaminated with bloodborne pathogens remains high. One study in a hospital chemistry laboratory found that 6% of serum or plasma specimens received were HBV-contaminated, and 3% were HIV-contaminated.[10]

Nevertheless, during the past ten years great strides have been made to reduce the risk of laboratory worker exposures to, and infections by, bloodborne pathogens. The two most important landmarks in prevention have been the availability of an effective hepatitis B vaccine, and the implementation of the Occupational Safety and Health Administration (OSHA) bloodborne pathogens standard, which was enacted in its final form in December of 1991.[11] The Standard requires employers to provide the hepatitis B vaccine at no cost to employees, to provide an adequate supply of appropriate personal protective equipment, and to clean, maintain, and/or replace this equipment regularly. Puncture-resistant, leak-proof disposal containers must be provided in convenient locations, and must be replaced before becoming overfilled. If an employee is exposed to blood or body fluids, the employer must provide post-exposure follow-up, including employee and source patient testing for bloodborne pathogens, and records of reported exposures must be maintained. Finally, the employer must provide training in universal precautions and safe work practices, and explain the facility's obligations under the standard to its employees.

Although it is widely believed that these measures have resulted in significant improvements, there is little documentation of the effects of the standard on the frequency of exposures and infections in the clinical laboratory environment. This is because there were no standardized exposure surveillance programs in place before the implementation of these measures that would have made before-and-after comparisons possible. Another reason is that the number of exposure incidents reported from clinical laboratories in a single institution is likely to be too small to draw general conclusions about exposure and risk patterns. One approach for obtaining enough data to draw meaningful conclusions is to combine data from a number of institutions, provided they all use a standard surveillance system. The present analysis, including data from 70 hospitals, was carried out to determine exposure patterns in clinical laboratory settings in order to identify high-risk equipment and activities, and to accurately focus prevention initiatives in the areas of greatest need and opportunity.

Methods

All percutaneous injuries or mucocutaneous exposures to blood or body fluids that occurred in clinical laboratory settings and were reported to the 70 participating EPINet[12] hospitals during a two-year period, beginning in September 1992, were included in this analysis (306 cases). In order to identify the unique characteristics of the clinical laboratory environment that distinguish it from other areas of the hospital, these cases were compared to exposures occurring in patient rooms (3,468 cases) during the same period.

Incidents involving clinical laboratory personnel outside of the clinical laboratory environment and incidents that occurred with non-contaminated items were not included in this analysis.

Results

More than 80% of the 306 exposures occurring in clinical labs were to lab technicians, technologists or phlebotomists. A small proportion were to nurses, physicians, and housekeepers. **Figure 1** shows the types of devices causing sharp-object injuries in clinical labs. Needles caused the greatest number of injuries, accounting for 38% of all injuries. Of needle injuries, 51% were caused by blood-drawing needles in clinical labs, while in patient rooms only 23% of needle injuries were caused by blood-drawing needles ($\chi^2 = 32.8$, $p < .0001$). This is significant because blood-drawing needles are among the devices most commonly associated with the transmission of blood-borne pathogens. Another unique finding was that 22% of needle injuries in clinical labs were associated with needles that were used as tools, while only 3% of needle injuries in patient rooms were attributed to needles used as tools ($\chi^2 = 85.5$, $p < .0001$). It is a common practice in clinical labs to use needles and syringes for purposes other than what they were designed for. The lack of safer equipment specifically designed for laboratory applications often leaves lab workers no alternative but to use available equipment that puts them at unnecessary risk of injury.

Of particular interest was the finding that injuries from glass items (specimen tubes, capillary tubes, pipettes, and slides) accounted for 39% of injuries overall, as opposed to only 0.3% of injuries in patient rooms ($\chi^2 = 918.9$, $p < .0001$). These are especially serious injuries because glass items are often containers for blood, so that when a laceration occurs there is potential for a large inoculum of blood to be introduced into the wound. **Figure 2** shows how glass injuries occured. The greatest number of incidents occurred during use of the glass item, which means that glass items frequently broke as they were being handled. There are many alternative plastic products available that are safer than conventional glass equipment. Vacuum tubes, specimen tubes, and capillary tubes, the devices most commonly used as blood receptacles, are now available in plastic and are highly resistant to breakage. Plastic pipettes are now more common in most labs than glass ones. The use of glass pipettes should be strictly limited to procedures for which plastic cannot be used.

There were 76 mucocutaneous exposures reported in clinical labs during the two-year interval. In 74% of these incidents the body fluid involved was blood, while in patient rooms only 54% of mucocutaneous exposures involved blood ($\chi^2 = 10.9$, $p < .001$). However, the quantity of blood or body fluid involved in exposures was smaller in clinical lab settings, where only 1% of exposures exceeded 5cc of biological fluid, whereas in patient rooms 12% of exposures

Figure 1. Devices Causing Sharp-Object Injuries in Clinical Laboratories

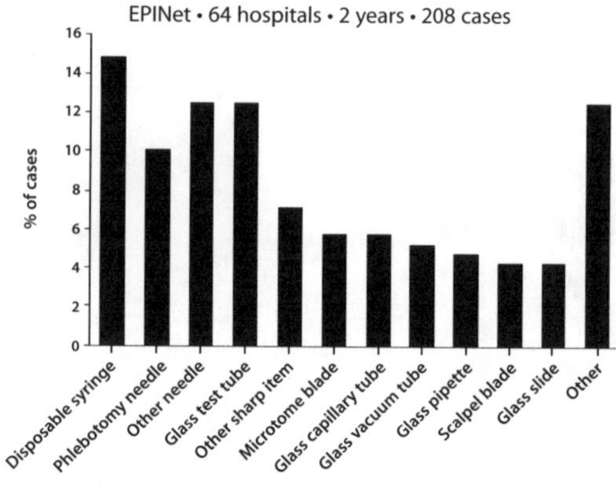

Figure 2. Mechanism of Glass Injuries in Clinical Laboratories

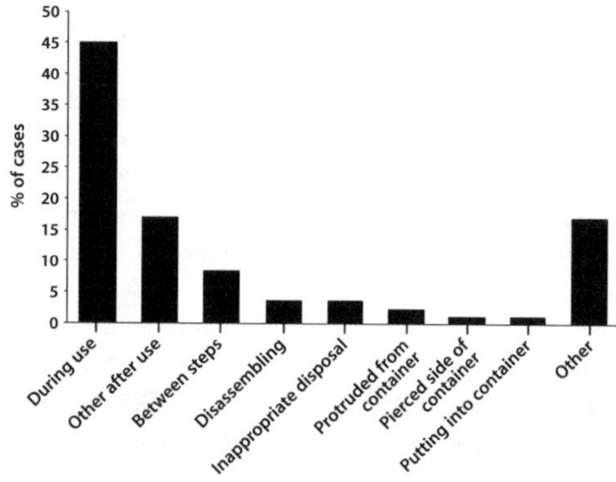

Figure 3. Mechanism of Blood and Body Fluid Exposures in Clinical Laboratories

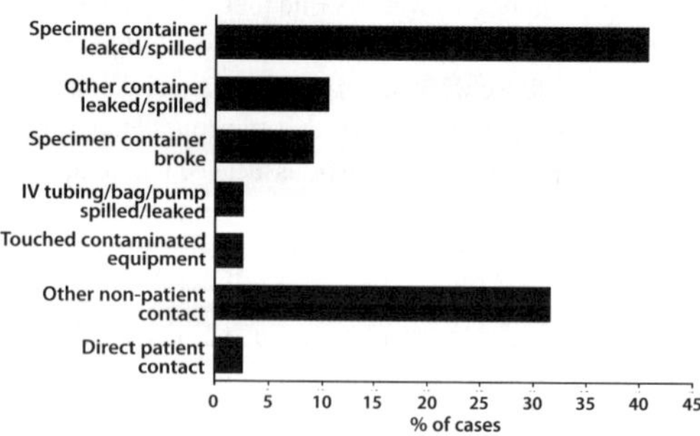

did ($\chi^2 = 8.4, p<.01$). Smaller amounts of biological fluids pose less risk to healthcare workers. In clinical labs, the amount of biological fluid in a single exposure usually does not exceed the capacity of a specimen container, which is relatively small. However, risk could be even further reduced by limiting the amount of blood or body fluids collected from patients to the minimum required. Often several cc's are collected, when only drops are needed for testing.

A notable characteristic of blood and body fluid exposures in clinical labs was the high percentage of cases in which the exposure was due to specimen containers leaking or breaking, or other product-related failures, as shown in **Figure 3.** In clinical labs, 66% of exposures were due to equipment failure, and 31% were product-related (where a product served as the primary vehicle of exposure); only 3% were the result of direct patient contact. By contrast, in patient rooms 52% of exposures were the result of direct patient contact, 31% were due to equipment failure and 16% were product-related. The design and integrity of specimen containers is an issue requiring more attention. In addition to leakage and breakage, the need to uncap specimen tubes to access contents creates a significant risk. Methods of specimen access and transfer that do not require uncapping of specimen tubes would constitute an important advance in safety.

Another finding of interest is shown in **Figure 4.** Clinical lab personnel more frequently wore personal protective garments at the time of mucocutaneous exposures than healthcare workers exposed in patients' rooms. In particular, 78% of lab workers were wearing gloves at the time of exposure, as opposed to 64% of healthcare workers exposed in patients' rooms ($\chi^2 = 5.9, p<.05$). Sixty-seven percent of lab workers were wearing cloth lab coats or gowns at the time of their exposures, while only 15% of workers in patients' rooms were wearing them ($\chi^2 = 123.2, p<.0001$). Clinical lab personnel appear to be more consistent in their use of personal protective equipment than other healthcare workers. This may be explained in part by the controlled environment of the clinical lab in contrast to patient care, where contact with blood or body fluids can occur without warning.

On the other hand, previous studies we have conducted provide evidence that cloth lab coats give only an illusion of protection without reducing the risk of skin contact with biological fluids.[13,14] The cloth lab coat remains the most common garment for covering the torso and arms of clinical lab personnel. If its purpose is to serve as a work uniform, or to prevent the soiling of clothes, the cloth lab coat can meet these goals. If its purpose is to prevent skin contact with biological fluids, the cloth lab coat is irrelevant. At a minimum, fluid-resistant gowns or coats should be available in locations where there is a risk of splashing or spraying of biological fluids, although it is not clear that all lab workers would need this degree of protection at all times.

There is another reason, however, to explain why lab coats and gloves did not prevent mucocutaneous exposures. We found that 74% of reported mucocutaneous exposures were to the eyes, nose, mouth, or other areas of the face. There is surely a reporting phenomenon contributing to this finding. We have noted similar selective reporting patterns among emergency department workers.[14,15] A blood exposure to the eyes, mouth or other area of the face is a shocking experience and more likely to be reported than blood contact with arms or hands. Face exposures are relatively infrequent, but it is clear that lab personnel rarely wear protective eyewear or faceshields. These findings point to the need to identify situations and procedures that may result in splashing or spraying in which face exposure is a risk, and to selectively implement the use of protective face and eyewear under those circumstances.

Figure 4. Personal Protective Equipment Worn at the Time of Exposure: Clinical Laboratories vs. Patient Rooms

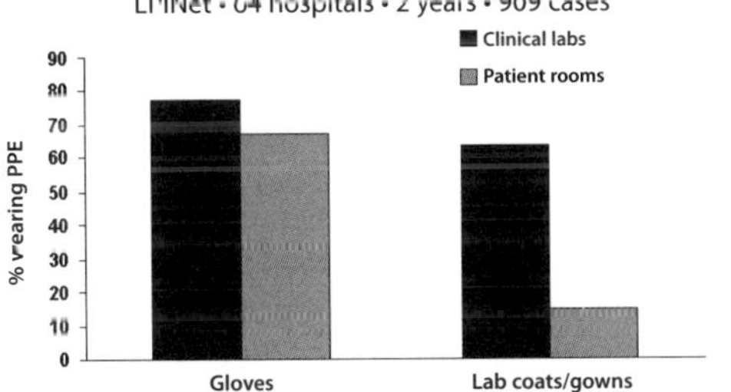

International Healthcare Worker Safety Center, University of Virginia

Conclusions

The most serious risk of exposure to bloodborne pathogens among clinical lab workers is from needles used to draw or transfer blood samples, and from glass specimen tubes and capillary tubes used for collecting and storing blood samples. Plastic specimen and capillary tubes should be substituted for glass tubes wherever possible. For blood drawing, protective devices that shield or blunt phlebotomy and syringe needles after their use should be implemented, not only among clinical lab personnel but hospital-wide. The use of needles and syringes as lab tools, which place lab workers at unnecessary risk, should be eliminated as much as possible. The use of the cloth lab coat as a fluid barrier should be eliminated and replaced, in at-risk locations and for at-risk procedures, by fluid-resistant gowns. The use of protective face and eyewear should be implemented under circumstances in which face exposure is a risk. Manufacturers should improve the designs of specimen containers to minimize risk of leakage, breakage, and spillage, and should introduce features that would allow the transfer of fluid contents without the need to open containers. Quantities of blood and body fluid samples collected from patients should be limited to the minimum required for testing. Finally, every effort should be made to achieve a 100% hepatitis B vaccination rate among clinical laboratory personnel.

References

1. Centers for Disease Control and Prevention. *HIV/AIDS Surv Rep.* 1995;7(2):21.
2. Centers for Disease Control and Prevention. Case-control study of HIV seroconversion in healthcare workers after percutaneous exposure to HIV-infected blood (France, United Kingdom, and United States, January 1988-August 1994). *MMWR.* 1995;44:929–933.
3. Collins CH, Kennedy DA. Microbiological hazards of occupational needlestick and "sharps" injuries. *J Appl Bacteriol.* 1987;62:385–402.
4. Grist NR, Emslie J. Infections in British clinical laboratories. *J Clin Pathol.* 1985;38:721–5.
5. Jacobson JT, Orlob RB, Clayton JL. Infections acquired in clinical laboratories in Utah. *J Clin Microbiol.* 1985;21:486–9.
6. Vesley D, Hartman HM. Laboratory-acquired infections and injuries in clinical laboratories: a 1986 survey. *Am J Public Health.* 1988;78:1212–5.
7. Levy BS, Harris BC, Smith JL, et al. Hepatitis B in ward and clinical laboratory employees of a general hospital. *Am J Epidemiol.* 1977;106:330–335.
8. Wruble LD, Masi AT, Levinson MJ, et al. Hepatitis-B surface antigen (HB Ag) and antibody (anti-HB) prevalence among laboratory and nonlaboratory hospital personnel. *South Med J.* 1977;70:1075–1079.
9. Pattison CP, Boyer KM, Maynard JE, et al. Epidemic hepatitis in a clinical laboratory. *JAMA.* 1974; 230:854–857.
10. Handsfield HH, Cummings MJ, Swenson PD. Prevalence of antibody to human immunodeficiency virus and hepatitis B surface antigen in blood samples submitted to a hospital laboratory. *JAMA.* 1987;258:3395–3397.
11. U.S. Occupational Safety and Health Administration. 29 CFR Part 1920.1030, Occupational exposure to bloodborne pathogens, final rule. *Fed Regist.* 1991; 56:64004–64182.
12. Jagger J, Cohen M, Blackwell B. EPINet: A tool for surveillance of blood exposures in health care settings. In: Charney W, ed. *Essentials of Modern Hospital Safety* (vol. 3). Boca Raton, FL: Lewis Publishers/CRC Press, 1994; 223–239.
13. Jagger J, Detmer DE, Cohen ML, et al. Reducing blood and body fluid exposures among clinical laboratory workers: meeting the OSHA standard. *Clin Lab Manage Rev.* 1992;6:416–424.
14. Jagger J, Balon M. EPINet Report: Blood and body fluid exposures to skin and mucuous membranes. *Adv Exposure Prev.* 1995;1(2):1–9.
15. Jagger J, Powers R, Detmer DE, et al. Epidemiology and prevention of blood and body fluid exposure among emergency department staff. *J Emerg Med.* 1994;12:753–765.

Blood and Body Fluid Exposures to Healthcare Workers' Eyes While Wearing Faceshields or Goggles

Melanie Bentley, B.S.

Vol. 2, no. 4, 1996

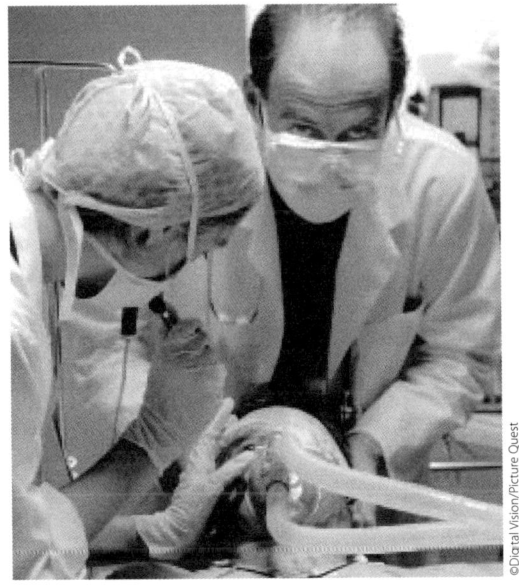

In a 1995 issue of AEP (vol. 1, no. 2), an analysis of exposure data from nine hospitals participating in an EPINet data-sharing network strongly supported the benefit of wearing goggles or faceshields as a method of preventing eye exposures to blood and body fluids. However, these data also revealed the surprising finding that goggles and faceshields do not always provide total protection. In a group of healthcare workers reporting blood or body fluid exposures (n=1,150), among those who were wearing either goggles or faceshields at the time of their exposures (n=58), 31% (n=18) still experienced eye exposures despite wearing eye protection. In order to gain insight into the circumstances under which goggles and faceshields fail to protect eyes from blood and body fluid exposures, descriptions of a number of these events have been provided by seven of the participating EPINet hospitals. These descriptions suggest that such failures often occur when blood and body fluids are ejected or squirt under pressure, when goggles or faceshields either slip or leave unprotected gaps, or when they provide no seal above the eyes, allowing blood or body fluids to run down the forehead and into the eyes. Many of these incidents occurred in labor and delivery and operating rooms. Improved designs for protective eye and facewear may provide greater protection for the highest-risk circumstances.

- A scrub nurse in an operating room was assisting in an exploratory laparotomy. The surgeon was handling the small bowel, while facing the surgical field. The nurse felt something splatter into her right eye. Forty-five minutes later, the nurse thoroughly rinsed her eye. The amount of peritoneal fluid that touched the eye was small. The nurse was wearing a protective faceshield at the time and the fluid touched her eye through a gap in the faceshield.

- A resident in labor and delivery was performing a cesarean section when blood sprayed from the surgical site into his face. The resident was wearing glasses and a mask but was not sure if blood went into his eyes and mouth. The contact lasted between 15 minutes and one hour and involved a moderate amount of blood. The resident was also wearing double gloves and a surgical gown.

- A resident in a patient room was sewing a perineal tear after delivery. While injecting lidocaine, a small amount of the medication splashed up and hit the doctor's face, eyes, nose, and mouth. A possible blood exposure might have occurred as well. The doctor was exposed for about 5-14 minutes. He was wearing double gloves, a surgical gown, and a faceshield and surgical mask, both of which had slipped down slightly.

- An attending M.D. in a delivery room was performing a vaginal delivery and was reaching to deliver the placenta. A small amount of blood and amniotic fluid splashed over the top of his faceshield and soaked through his gown. The exposure lasted for more than one hour. The doctor was wearing double gloves, a surgical mask, a gown, and a faceshield.

- An attending M.D. in a delivery room reached into the patient's uterus to feel the placenta. Blood soaked through her surgical gown, which was supposed to be waterproof, and also splashed into her eye. The exposure lasted for less than five minutes. In addition to the gown, she was wearing double gloves, a faceshield, and a surgical mask.

- A respiratory therapist in an intensive care unit was emptying ventilator circuit condensate. When the tubing was disconnected from the expiratory limb, a small amount of condensate splashed through the side opening of her goggles. The contact lasted 5-14 minutes. The respiratory therapist was wearing single gloves, goggles, and a surgical mask.

- A resident in a delivery room during a vaginal delivery was suctioning the baby's mouth and nose on the perineum at delivery. A large amount of blood and amniotic fluid squirted over his protective eye gear and face mask and ran down his face. The exposure lasted less than five minutes. The resident was wearing single gloves, a faceshield, surgical mask, and surgical gown.

- A resident in an operating room was removing a piece of the patient's bone. A bone chip and a small amount of blood flew up and hit the resident in the eye. The exposure lasted 5-14 minutes. The resident was wearing a faceshield, surgical mask, and surgical gown.

- An R.N. in an operating room was assisting with an arteriovenous fistula. During the procedure, a bulldog clamp slipped off the artery. Blood sprayed across the nurse's surgical cap (a mask with a shield). A small amount of blood dripped off the cap onto the nurse's forehead, near her eyes. The exposure lasted less than five minutes. The nurse was wearing double gloves, a faceshield, surgical mask, and surgical gown.

- A nurse was assisting in an operating room when a small amount of blood splashed into her right eye through a gap at the top of her goggles. The goggles may have slid down. The exposure lasted less than 5 minutes. The nurse was wearing double gloves, goggles, a surgical mask, and surgical gown.

- An attendant in an operating room was forcefully flushing a used suction tip. A small amount of blood splashed under the faceshield and hit the attendant's face and eyes. The exposure lasted 5-14 minutes. The attendant was wearing a faceshield, mask, gown, and double gloves.

- A surgery attendant was in an operating room when tubing disconnected and a small amount of peritoneal fluid splashed into his eyes. The exposure lasted less than 5 minutes. The surgery attendant was wearing double gloves, a faceshield, mask, and surgical gown.

- A resident was performing angioplasty. While placing the injector syringe on the pressure manifold, contrast dye contaminated with blood squirted and a small drop went under the resident's protective eyewear and into his left eye. The exposure occurred during a cath lab procedure.

- A nurse who had just completed a plasmapheresis procedure was bent over the centrifuge while cutting the apheresis tubing, in order to free the collection bag. While cutting the tubing, she was squirted in the eye with plasma. The nurse was wearing a face mask with eyeshield, but because she was bent over there was a gap between the shield and her eyes.

- A nurse assisting with an endoscopy was injecting cleansing solution through the contaminated scope. Stool blew through the air channel over the plastic shield on her mask, under her glasses, and into her eyes.

- A nurse in a patient's room was attempting to flush a J tube with warm water. The contaminated water splashed up into her face and she felt a drop in her right eye, despite wearing eyeglasses and goggles.

- A nurse in an ICU/CCU had finished suctioning a baby. The respiratory therapist pulled out the tube and a small amount of nasal secretion went over the nurse's goggles and into her right eye. The exposure lasted more than one hour.

- While in a clinical lab, a lab worker used a 5cc syringe to inoculate a bottle with a bone marrow specimen; 5cc syringes used for inoculation of tissue are not aerosol-free. A small amount of specimen backed up and out of the syringe and went into the lab worker's left eye. The exposure lasted less than five minutes. The lab worker was wearing single gloves, a hood with a plastic faceshield, a surgical mask, and a cloth lab coat.

Percutaneous Injuries in Pediatric Healthcare Workers

Patti Miller Tereskerz, M.S., J.D., Ph.D., Melanie Bentley, B.S., Betty Joe Coyner, R.N., M.S.N, and Janine Jagger, M.P.H., Ph.D.

Vol. 2, no. 5, 1996

Drew Stevenson, U.Va. Health System

While the risk of percutaneous injuries in healthcare workers has been well documented,[1,2] relatively little attention has been focused on needlestick injuries occurring in pediatric healthcare workers. Pediatricians have been considered to be at low risk of exposure to bloodborne pathogens when compared to surgeons or anesthesiologists.[3] Yet, difficulties in using gloves, which may discourage their use, and the challenge involved in controlling a child while administering care, may actually place pediatric healthcare workers at considerable risk of exposure to bloodborne pathogens.[3] Furthermore, the spread of HIV among the heterosexual population also puts pediatric healthcare workers at increased risk. During a one-year period, 1,500-2,000 infants infected with HIV were born in the U.S.[4]

We studied sharp-object injuries recorded in the Exposure Prevention Information Network (EPINet) database for the University of Virginia (U.Va.) Medical Center from August 1992 through August 1995. EPINet is a standardized system for recording percutaneous injuries and blood and body fluid exposures.

Of the 3,813 healthcare workers at U.Va. at risk for needlestick injuries, 346 (9%) provide care to children. During the study there were 1,146 sharp-object injuries, with 96 (8%) sustained by persons caring for children.

Injuries to pediatric healthcare workers occurred more often in patient rooms and critical care units, in contrast with injuries to non-pediatric workers, which occurred more often during surgical procedures [see **Table 1,** next page].

The palmar side of the left hand was injured more often in pediatric healthcare workers, while the back of the left hand was injured more often in non-pediatric workers. Injuries to pediatric healthcare workers occurred more often after use of a sharp device or while withdrawing a needle from resistant material, such as a vacuum tube stopper or an intravenous port, as compared to other healthcare workers, who sustained percutaneous injuries more often between steps of a procedure [Table 1]. When injuries occurring in the operating room were excluded from these analyses, the differences remained significant.

After disposable syringes, the device causing the most injuries to pediatric healthcare workers was a winged steel (or butterfly) needle I.V. set (11/96 or 11%). This device was responsible for only 48/1050 (5%) of injuries to all other healthcare workers (p=.01). The winged steel needles causing injury were used most often to draw venous blood from pediatric (6/11 or 55%) and non-pediatric (34/48 or 71%) patients. They were also used frequently to draw arterial blood (4/11 or 36%) in pediatric patients.

The mechanism of injury for winged steel needles was different in pediatric vs. non-pediatric healthcare workers. In 4/11 (36%) of the pediatric healthcare worker cases, the injury occurred while disposing of the needle. This was the mechanism of injury in only 1/48 (2%) of the cases involving all other healthcare workers (p=.003).

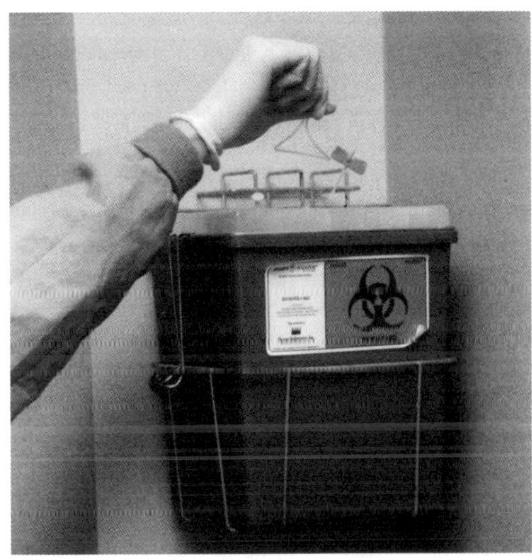

Disposing of butterfly needles in wall-mounted disposal containers is a common exposure risk for pediatric heathcare workers.

Table 1. Sharp-Object Injury Comparisons Between Pediatric Healthcare Workers and Other Healthcare Workers
(Three most frequently occurring categories; University of Virginia, 1992 – 1995)

	Pediatric Workers Number of Injuries = 96			Non-Pediatric Workers Number of Injuries = 1,050		
	Category	#	%	Category	#	%
Job Description	Nurse	51	53%	Nurse	459	44%
	Resident	15	16%	Resident	246	23%
	Attending Physician	7	7%	Attending Physician	79	8%
Place of Occurence	Patient Room*	46	48%	Patient Room*	347	33%
	ICU/CUU*	19	20%	Operating Room*	306	29%
	Outside Patient Room	6	6%	Procedure Room	77	7%
Was the device contaminated?	Yes	89	93%	Yes	947	90%
	Unknown	6	6%	Unknown	88	8%
	No	1	1%	No	14	1%
For what purpose was the device originally used?	IV Injection into I.V. Port	21	22%	Suturing*	191	18%
	Drawing Blood*	20	21%	IV Injection into I.V. Port	173	16%
	Other	10	10%	Other	150	14%
When did the injury occur?	Other After Use*	35	36%	During use	306	29%
	During use	25	26%	Other After Use*	224	21%
	Resistant Material* †	10	10%	Between Steps*	170	16%
How severe was the injury?	Moderate	58	60%	Moderate	590	56%
	Superficial	36	38%	Superficial	401	38%
	Severe	2	2%	Severe	52	5%
What body part was injured?	Left Hand, Palm*	52	54%	Left Hand, Palm*	422	40%
	Right Hand, Palm	26	27%	Right Hand, Palm	262	25%
	Right Hand, Dorsum	10	10%	Left Hand, Dorsum*	155	15%

If there were more than three possible answers to a question, the table shows only the three most frequent categories.
* Indicates significant difference at $p<.05$ either by Chi Square or Fisher Exact Test between pediatric and non-pediatric workers for that item.
Note that the same items were not always in the top three categories for each group; therefore some items, such as suturing, only appear for one group of workers.
† Withdrawing needle from resistant material such as vacuum tube stopper or I.V. port.

Upon further investigation, we discovered that in pediatric units, needle disposal systems were often mounted on the walls and underneath beds or bassinets, while in non-pediatric units disposal containers were placed on the floor.

These results suggest that there is an increased risk of percutaneous injuries from winged steel needles when wall- or bed-mounted disposal systems are used. Such systems are often used in pediatric units to prevent children from gaining access to the needles.

Most other winged steel needle injuries occurred after use but before disposal, in both pediatric (4/11 or 36%) and non-pediatric (20/48 or 42%) healthcare workers.

In other respects, sharps injuries occurring in pediatric healthcare workers were similar to those occurring in all other healthcare workers. Most of the injuries were sustained by nurses (51/96 or 53%) in patient rooms (46/96 or 48%) and were moderate in severity (58/96 or 60%). In most instances the source patient was known (89/96 or 93%), and the injured worker was the original user of the sharp object (71/96 or 74%), which was contaminated in 89/96 (93%) of the cases.

Based on this study, we make the following recommendations:

(1) The design of needle disposal systems used in pediatric units should be modified so that the risk of needlestick injuries to healthcare workers is minimized.
(2) Needleless devices and protective devices on needles should always be used unless clinically contraindicated.
(3) Healthcare workers should refrain from recapping needles and otherwise observe universal precautions.[5]

References

1. Jagger J, Hunt EH, Brand-Elnaggar J, Pearson RD. Rates of needle-stick injury caused by various devices in a university hospital. *N Engl J Med.* 1988;319:284.
2. Ippolito G, De Carli G, Puro V, et al. *JAMA.* 1994;272(8):607.
3. Buss PW, McCabe M, Verrier Jones ER. Attitudes of paediatricians to HIV and hepatitis B virus infection. *Arch Dis Child.* 1991;66(8): 961.
4. Crain EF, Bernstein LJ. Precautions and testing in pediatric HIV infection [review]. *Ped Emerg Care.* 1991;7(3):177.
5. Williams, WW. Centers for Disease Control: guidelines for infection control in hospital personnel. *Infect Control Hosp Epidemiol.* 1983; 4(suppl): 326–349.

Risk of HIV-1 Infection After Human Bites

By Patti M. Tereskerz, M.S., J.D., Ph.D., Melanie Bentley, B.S., and Janine Jagger, M.P.H., Ph.D. *Vol. 2, no. 7, 1996*

The following report, which originally appeared in The Lancet *(1996;348[9040]:1512), was reprinted in* AEP, *and here, with permission from Elsevier.*

The first documented seroconversion of HIV-1 following a human bite[1] raises an important concern regarding occupational transmission of HIV from patient to healthcare worker. The HIV seroconversion described [in *The Lancet* article "Transmission of HIV-1 by human bite"] suggests that for HIV-1 transmission to occur, there must be blood in the mouth of the source patient and a break in the integrity of the skin of the healthcare worker.

Seventy-four hospitals in the U.S. participating in the Exposure Prevention Information Network (EPINet) report their employees' occupational percutaneous injuries and exposures to blood or body fluids to researchers at the University of Virginia. A review of EPINet data from 1993 to 1995 was conducted to determine the rate of bite exposures in healthcare workers and the frequency of associated risk factors that might increase occupational infection risk. There were no occupational HIV-1 seroconversions in participating hospitals, and 50/70 (71%) hospitals reported that overall 1.7% of exposures involved an HIV-1-positive source patient. Fifty (.5%) of 10,125 total incidents involved a healthcare worker who was bitten by a patient, an annual rate of 0.12 reported bites per 100 occupied hospital beds. Based on 518,400 occupied U.S. hospital beds, this yields an estimated annual total of 622 reported bite exposures in U.S. hospitals. The job categories and locations of bites are shown in the table below.

Nineteen of the 50 bites (38%) involved non-intact skin or a percutaneous injury to the healthcare worker. Information concerning presence of blood in the source patient's mouth was available for 36 of the 50 cases. Of these, blood was noted in three cases of exposure to intact skin and in none of the cases in which there was a break in the integrity of the skin.

In contrast to the case reported [in *The Lancet*[1]], of the 28 incidents in which descriptions of the bites were available, none involved an involuntary bite as might occur during a seizure. In the 28 cases, 14 source patients were combative, ten were children, three were psychiatric patients, and one involved the removal of an orthodontic appliance.

These data show that occupational bites are fairly infrequent. Nevertheless, because 86% of bites were to the hands and arms of healthcare workers, the frequency of these exposures can be minimized by consistent glove use and arm protection when healthcare workers are in close contact with pediatric, psychiatric, or combative patients.

Reference

1. Vidmar L, Poljak M, Tomazic J, Seme K, Lavs I. Transmission of HIV-1 by human bite. *Lancet.* 1996;347:1762–1763.

Characteristics of reported bites, 1993–1995

Body Part*	Frequency	Job Category	Frequency
Hand	24	Nurse	25
Arm	23	Attendant	9
Chest	5	Therapist/counsellor/teacher	7
Head	2	Other	7
Leg	1	Physician	2
Total	55	**Total**	50

*Some healthcare workers were bitten on more than one body part.

Implementing the CDC's Recommendations for Postexposure Prophylaxis: A Survey of 31 Hospitals

Melanie Bentley, B.S., and Janine Jagger, M.P.H., Ph.D.

Vol. 3, no. 1, 1997

In the June 7, 1996, issue of *Morbidity and Mortality Weekly Report (MMWR)*[1], the Centers for Disease Control and Prevention (CDC) issued updated recommendations for chemoprophylaxis after occupational exposure to HIV. Under the new guidelines, postexposure follow-up has become more complicated: whether chemoprophylaxis is recommended, offered, or not offered depends upon the type of exposure and the status of the source patient. Furthermore, combination therapy, consisting of a two- or three-drug regimen (ziodovudine [ZDV] + lamivudine [3TC] ± indinavir [IDV]), is recommended for exposures with higher risk for HIV transmission and should be initiated within one-to-two hours of exposure for maximum potential effectiveness.

The latest recommendations offer a new level of hope to today's healthcare workers who are exposed to HIV-infected blood. In order for healthcare workers to benefit, however, institutions have had to integrate these recommendations into their existing policies and procedures for postexposure follow-up and treatment, a task that many have found challenging and labor intensive. To find out what effect the new recommendations for postexposure prophylaxis (PEP) have had, the International Healthcare Worker Safety Center queried 70 hospitals participating in the EPINet data-sharing network.

Thirty-one hospitals responded to our query. The hospitals, representing cities and states with varying degrees of HIV seroprevalance, were located in South Carolina (24), Florida (2), Ohio (1), Pennsylvania (1), Indiana (1), Nebraska (1), and Washington (1). Eleven were teaching and 20 were non-teaching hospitals, with a combined average daily census of 6,847 occupied beds in 1996. Eleven hospitals had an average daily census of less than 100 beds, 12 had an average daily census of between 100 and 300 beds, and eight had an average daily census greater than 300 beds. In 1995, the 31 responding hospitals had an estimated total of 2,753 reported exposures (78% sharp-object injuries and 22% blood and body fluid exposures).

We asked, "In the past year, how many times has your hospital initiated the new PEP protocol? How long did employees stay on it?" Twelve hospitals indicated that the protocol was not initiated in the past year (see **Figure 1**). Of those twelve, four stated that there were no reported exposures meeting the criteria of a high-risk incident. In the remaining 19 hospitals, the protocol was initiated a total of 125 times. Thirty-two healthcare workers received combination chemoprophylaxis for the full four weeks, two stayed on the treatment for two weeks, and 91 stayed on the treatment for less than one week (see **Figure 2**). Not all hospitals surveyed provided information on why PEP treatment was stopped, but in 26 cases it was noted that treatment was discontinued within three days, after source patient test results were obtained. Another reason cited for discontinuing treatment was that side effects from the prophylactic drugs were intolerable. In one case, a healthcare worker was hospitalized because of side-effects; in two cases in which triple drug therapy was given, IDV was stopped but treatment with the other two drugs was continued for the full four weeks.

All hospitals either rewrote or revised their policies on blood and body fluid exposures to incorporate the new CDC recommendations. Hospitals offered or recommended the multi-drug regimen depending on the characteristics of the exposure. With the change in policy, the immediate avail-

Figure 1. Number of Times PEP Protocol Initiated per Hospital

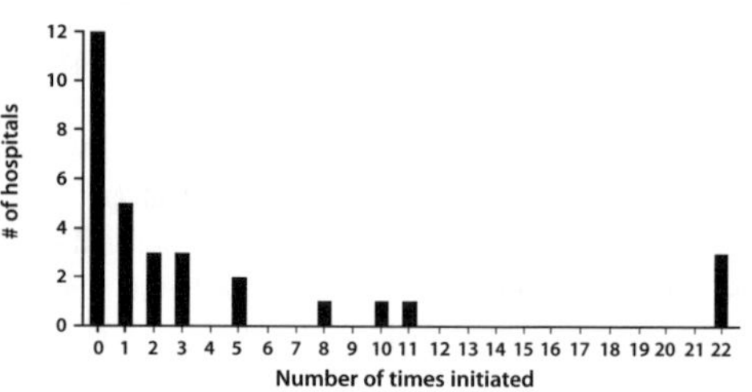

International Healthcare Worker Safety Center, University of Virginia

Figure 2. Length of Time Healthcare Workers Remained on PEP

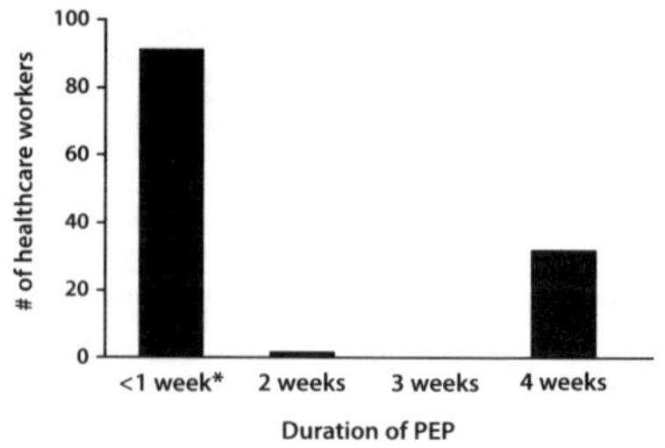

**14 remained on PEP for 1 day, 7 remained on for 2 days, 26 remained on for 3 days, 17 remained on for 2–3 days, 22 remained on for a maximum of 3 days, 1 remained on for 4 days, and 4 remained on for less than 1 week.*

International Healthcare Worker Safety Center, University of Virginia

ability of the CDC-recommended drugs has become an issue. Several hospitals noted that the drugs, provided in starter packets, have been made available in the hospital pharmacy.

Since it is recommended that PEP be initiated within one-to-two hours after exposures, prompt action is called for in the event of an exposure. Hospitals have increased their efforts to encourage employees to report exposures immediately. A key issue mentioned was the need for 24-hour exposure management coverage. Some hospitals utilize the emergency department when the employee health department is closed, while others instruct employees to report to a designated authority after-hours who has been specifically trained to carry out the postexposure protocol.

In order to increase employee awareness about the need for prompt reporting, many hospitals are offering more intensive pre-exposure training on the risks of bloodborne pathogen exposures, the appropriate actions to take when an exposure occurs, and the available options for prophylactic treatment following an exposure. Pre-exposure education has also become important in preparing employees to rapidly provide or withhold informed consent for chemoprophylaxis if an exposure occurs. One hospital mentioned that pre-exposure training has reduced panic among healthcare workers who must make rapid treatment decisions. Even with training, postexposure counseling of the exposed worker remains necessary to support him or her through the decision-making process. The role of counselor is filled in various institutions by employee health or infection control nurses, infectious disease physicians, and emergency department personnel.

When an exposure occurs, the source patient's status (e.g., whether positive for HIV or another bloodborne pathogen and, if HIV-positive, viral load and stage of disease) must be determined quickly in order to make appropriate clinical decisions. Often this assessment requires close work with the lab in order to obtain test results as quickly as possible. Some hospitals have greatly reduced the delay in determining source patient status by the use of a rapid HIV test. The SUDS® (Single-Use Diagnostic System) HIV-1 Test by Murex Diagnostics provides HIV results in 10 minutes. At least nine of the 31 hospitals responding to the survey currently use this test.[2]

Based on the 31 hospitals responding to the survey, with 2,753 total exposures and 125 cases in which PEP was initiated, the total overall estimated exposure rate was 40 exposures per 100 occupied beds per year, and **the rate of PEP initiation was 1.8 times per 100 occupied beds per year**. Twenty-six percent of the cases remained on the treatment for the full four weeks, a rate of .5 times per 100 occupied beds per year. Since only 26% of the cases received the four-week treatment, while 73% received treatment for less than one week, it is possible that the use of the rapid HIV test could greatly reduce the initiation of the CDC-recommended PEP among healthcare workers who have not been exposed to HIV. In cases where a negative HIV test result is obtained from the source patient with the 10-minute test, the cost, side effects, and anxiety that attend the two-to-three day wait for results would be eliminated.

Training employees to take prompt action following an exposure, providing 24-hour exposure management, and assuring the availability of the CDC-recommended drugs are major challenges that healthcare institutions have had to address over the past year. The incorporation of the new guidelines has been, in the words of one respondent, "a very time consuming, difficult protocol to implement for a low-volume, high-risk situation."

References

1. CDC. Update: provisional Public Health Service recommendations for chemoprophylaxis after occupational exposure to HIV. *MMWR.* 1996; 45(22):468–472.
2. For a clinical evaluation of the SUDS HIV-1 test see: Kassler WJ, Haley C, Jones WK, et al. Performance of a rapid, on-site human immunodeficiency virus antibody assay in a public health setting. *J Clin Microbiol.* 1995;33:2899–2902. Technical information on the SUDS HIV-1 test is available from Murex Diagnostics.

Direct Cost of Follow-Up for Percutaneous and Mucocutaneous Exposures to At-Risk Body Fluids: Data From Two Hospitals

Janine Jagger, M.P.H., Ph.D., Melanie Bentley, B.S., and Edwina Juillet

Vol.. 3, no. 3, 1998

The International Healthcare Worker Safety Center has received many requests for information on the cost of needlestick injuries and other occupational blood exposures. Hospitals participating in the EPINet data-sharing network do not routinely forward cost information to us; therefore, cost data recorded in the Center's research databases are limited and until now we have declined to publish it. But because of the continuing demand for this information and a lack of new and better data, we have compiled a brief report in hopes that it will contribute a realistic, if imperfect, picture of direct costs of post-exposure follow-up.

A section on the EPINet report forms is provided for recording post-exposure follow-up charges. The data fields on the forms are broken down into four categories, including: (1) lab charges for blood tests; (2) charges for treatments such as hepatitis B immunoglobulin, hepatitis B vaccine, chemoprophylactic drugs for HIV, and tetanus vaccine; (3) service charges for emergency department or employee health department visits or other services; and (4) other costs such as surgery or any costs not falling into another category. We selected two hospitals, out of approximately 70 hospitals in the EPINet data-sharing network, that provide complete information in the cost fields and forward that information to the Center. Both hospitals are large, exceeding 450 occupied beds. Hospital A is a community hospital in a high-HIV prevalence region, while hospital B is a teaching hospital in a low-HIV prevalence region. In this report we present cost data from these two hospitals from the period June 1, 1995 to May 31, 1997. This time frame was chosen to reflect the most current costs and also to capture differences in treatment costs that might be attributed to the new chemoprophylaxis regimens recommended in June 1996 by the Centers for Disease Control and Prevention (CDC) for HIV-exposed healthcare workers.

These data have several limitations. First, there were no standardized definitions for what constituted a charge or a cost. For instance, hospital B recorded only direct charges to

Table 1. Average and Range of Direct Cost of Percutaneous Injuries and Mucocutaneous Exposures in Two Hospitals

June 1, 1995 – May 31, 1997

	Percutaneous injuries	Mucocutaneous exposures
Hospital A	**cases = 345**	**cases = 114**
Average	$672	$660
Range	$340–$1,025	$265 –$975
Hospital B	**cases = 594**	**cases = 334**
Average	$539	$546
Range	$197–$1,094	$0.0 – $1,232

**Note: Numbers in all tables have been rounded to the nearest dollar.*

Table 2. Average Direct Cost of Percutaneous Injuries in Two Hospitals During Two Time Periods

	June 1, 1995 – May 31, 1996	June 1, 1996 – May 31, 1997
Hospital A	**cases = 185**	**cases = 160**
Lab tests	$163	$161
Treatment	$ 14	$ 19
Service	$245	$242
Other	$250	$249
TOTAL	**$672**	**$671**
Hospital B	**cases = 311**	**cases = 283**
Lab tests	$525	$523
Treatment	$ 4	$ 6
Service	$ 9	$ 11
Other	$ 0	$ 0
TOTAL	**$537**	**$540**

Table 3. Average Direct Cost of Percutaneous Injuries in Two Hospitals for Cases in which the Source Patient was Known vs. Cases in which the Source Patient was Unknown

June 1, 1995 – May 31, 1997

	Source Known	Source Unknown
Hospital A	**cases = 329**	**cases = 16**
Lab tests	$164	$129
Treatment	$ 17	$ 11
Service	$243	$277
Other	$249	$250
TOTAL	**$673**	**$667**
Hospital B	**cases = 501**	**cases = 93**
Lab tests	$515	$574
Treatment	$ 2	$ 21
Service	$ 10	$ 10
Other	$ 0	$ 0
TOTAL	**$527**	**$605**

Table 4. Average Direct Cost of Percutaneous Injuries in Two Hospitals for High-Risk Injuries* vs. Low-Risk Injuries

	High-Risk Injuries	Low–Risk Injuries
Hospital A	cases = 157	cases = 188
Lab tests	$167	$159
Treatment	$ 20	$ 13
Service	$254	$236
Other	$250	$249
TOTAL	**$691**	**$657**
Hospital B	cases = 81	cases = 513
Lab tests	$520	$525
Treatment	$ 0	$ 5
Service	$ 12	$ 10
Other	$ 0	$ 0
TOTAL	**$532**	**$540**

**Note: "High-risk injuries" were defined as injuries caused by needles that had been used to draw blood or to establish intravenous access; all other injuries were classified as low risk.*

Table 5. Average Direct Cost of Mucocutaneous Exposures in Two Hospitals During Two Time Periods

	June 1, 1995 – May 31, 1996	June 1, 1996 – May 31, 1997
Hospital A	cases = 63	cases = 51
Lab tests	$160	$154
Treatment	$ 17	$ 10
Service	$245	$235
Other	$250	$250
TOTAL	**$672**	**$649**
Hospital B	cases = 173	cases = 161
Lab tests	$530	$523
Treatment	$ 2	$ 12
Service	$ 9	$ 17
Other	$ 0	$ 0
TOTAL	**$541**	**$552**

Table 6. Charges for Specific Items in Follow-up Protocol in Hospital A and Hospital B

Hospital A

Laboratory	Employee HBs Ab	$ 15
	Employee HIV antibody panel (ELISA)	25
	Source HBsAg	15
	Source HIV antibody panel (ELISA)	25
	(between June 1, 1995 – May 31, 1997 no HCV tests were performed)	
Treatment	HBIG	$ 100
	HBV vaccine (3 doses + blood tests)	150
	Tetanus	10
	AZT + 3TC ± IDV (4 week supply)	650
Service	Emergency Department visit	$ 85
	Employee Health visit (simple)	50
	Employee Health visit (moderate)	150
	Employee Health visit (extensive)	250
Other	Employee time	$ 250

Hospital B

Laboratory	Employee hepatitis profile*	$ 141
	Employee HIV antibody panel (ELISA)	56
	Employee HCV panel	56
	Source hepatitis profile*	141
	Source HIV antibody panel (ELISA)	56
	Source HCV panel	56
	**(HBsAg, anti-HBs, anti-HBc)*	
Treatment	HBV vaccine (3 doses)	$ 127
	Gamma globulin (5cc)	8
	AZT + 3TC ± IDV (4 week supply)	598
	Tetanus	2
	HBIG (5 doses)	465
Service	Emergency Department visit	$ 40
	Employee Health visit	60
Other	HBV booster	$ 30

Note: The costs cited above were in effect between June 1, 1995 – May 31, 1997

the employee health department, while hospital A added an across-the-board estimate of the cost of lost time for the exposed worker. Also, charges to departments other than the employee health department may not be accounted for in data recorded on the EPINet form. We have not adjusted the data from the two hospitals to redress these limitations or to improve comparability. Second, the data do not provide a breakdown of specific tests performed or treatments provided, so the impact of specific cost components cannot be evaluated. Third, these data do not include any indirect costs or the cost of occupational infections, both of which may be significant. Fourth, the hospitals included in this report may not be representative of other hospitals.

The following tables show cost comparisons of the two hospitals for the time period before and after the implementation of post-exposure chemoprophylaxis guidelines, and for cases of percutaneous injury and mucocutaneous exposure with different characteristics that might impact the cost of follow-up.

Although the total direct cost of follow-up was similar in the two hospitals, there were considerable differences between the hospitals in the cost of laboratory tests and in service charges. Another important difference, as noted previously, was that hospital A added a cost component for the lost work time of exposed employees, which accounted for about one-third of the total recorded cost. A similar amount of time may have been lost by employees in hospital B, but that cost was not accounted for.

The comparison of costs before and after the implementation of post-exposure chemoprophylaxis showed little cost impact of the new policy in these two institutions, despite the fact that hospital A is a high HIV-prevalence region. One circumstance resulting in higher cost, but only in hospital B, was an exposure involving an unknown source patient. Cost comparisons were also carried out to determine if injuries to employees in different job classifications or from different types of devices resulted in different follow-up costs, but these comparisons yielded no remarkable differences.

In summary, the ways in which these two hospitals accounted for direct costs of post-exposure follow-up differed greatly. Further studies of these costs will need to clearly identify the full spectrum of cost parameters and develop standard definitions for each parameter in order to make direct comparisons among hospitals and to develop more accurate extrapolations of the global cost impact of healthcare workers' occupational exposures to bloodborne pathogens.

Injuries from Vascular Access Devices: High Risk and Preventable

Janine Jagger, M.P.H., Ph.D., and Melanie Bentley, B.S.

Vol. 3, no. 4, 1998

Reprinted in AEP, and here, from Journal of Infusion Nursing (1997;20[65]533-539), with permission of Lippincott Williams & Wilkins.

Source: www.mrprotocols.com. Used with permission.

In May 1996 the Centers for Disease Control and Prevention (CDC) published a landmark report showing that occupational exposures to human immunodeficiency virus (HIV) varied in the level of transmission risk they posed to healthcare workers.[1] The report identified a list of characteristics that increase HIV transmission risk, including exposures that involved: (1) a deep percutaneous injury; (2) a needle that was used for vascular access; (3) visible blood on the device causing injury, and/or (4) a source patient with either end-stage AIDS or in the acute phase of retroviral illness (when HIV concentration in the blood is highest). These factors all suggest that HIV transmission risk increases as the amount of virus the healthcare worker is exposed to increases. An additional unexpected finding was that HIV transmission risk among healthcare workers receiving postexposure prophylactic zidovudine (ZDV) was reduced by an estimated 79%. Therefore, the absence of postexposure ZDV (and possibly other anti-HIV drugs) may also be considered a risk factor for HIV transmission.

In this publication, the CDC provided an important confirmation of transmission patterns that were inferred by CDC surveillance data on healthcare workers with documented and possible occupational HIV infection. As of June 1997, 166 cases had been reported in the United States by the CDC.[2] Although there are likely to be many more such cases than have been identified officially by the CDC, the identified cases provide useful information. It is to be expected that nurses are the professional group most often occupationally infected by HIV because there are more nurses in the healthcare workplace (approximately 2 million) than any other professional group, and nurses often perform procedures requiring needles.

Clinical laboratory technicians are the professional group with the second highest number of occupational HIV infections, which requires an explanation. In the United States, phlebotomists usually are classified as laboratory technicians, and in this case, most of the injuries leading to HIV infections were associated with blood-drawing procedures. When a phlebotomist is stuck by a needle, it is most likely to be a blood-filled needle that was used for vascular access (i.e., a high-risk injury). Phlebotomists contracted about two-thirds as many HIV infections as nurses, even though there are only about one-twentieth the number of phlebotomists in the work force as nurses.[3]

The CDC data do not identify HIV-infected nurses who specialize in intravenous (IV) catheter placement as a separate professional group. Like phlebotomists, these nurses' occupational risk of contracting HIV, or other bloodborne pathogens, is likely to be higher than that of other nurses, because when they sustain a needlestick it usually is caused by a blood-filled I.V. catheter stylet, a high-risk injury. Therefore, available epidemiologic data on occupational HIV transmission reinforce the need to identify and implement effective measures to reduce needlestick injury rates from vascular access needles. The following analysis focuses on injuries from I.V. catheter stylets.

EPINet Data

Nationwide data on occupational needlesticks and blood exposures in the U.S. are available for 1993, 1994, and 1995 from a network of 77 U.S. hospitals (the number of hospitals varies slightly from year to year) using a standard exposure tracking system, the Exposure Prevention Information Network (EPINet), in collaboration with researchers at the University of Virginia. EPINet is designed to identify medical products associated with either causation or prevention of occupational blood exposures. The three-year database contains a total of more than 10,000 percutaneous injuries, with information on the kinds of devices causing injuries, procedures the devices were used for, and how the injuries occurred.

Figure 1 (next page) shows the distribution of items most frequently causing percutaneous injuries in 63 participating hospitals in 1995. The shaded areas indicate high-risk injuries involving blood-filled needles. It can be seen that the hierarchy of high-risk injuries does not follow the overall frequency of injuries. For example, the disposable syringe causes more injuries than any other device, but ranks third among high-risk injuries. Ranking first among blood-filled needles causing injury is the I.V. catheter stylet. The remaining injuries from blood-filled needles were associated with blood-drawing procedures, including butterfly-type needles, needles on syringes, vacuum tube phlebotomy needles, and blood gas syringe needles.

The three-year EPINet database includes 77 hospitals

Figure 1. Items Most Frequently Causing Sharp-Object Injuries

EPINet 1995 • 63 healthcare facilities • 3,003 injuries

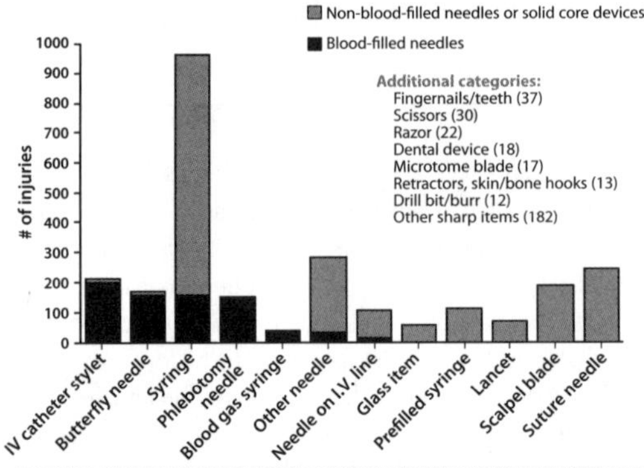

International Healthcare Worker Safety Center, University of Virginia

Figure 2. Job Category of Workers Injured by Conventional I.V. Catheters

EPINet • 77 hospitals • 3 years • 571 injuries

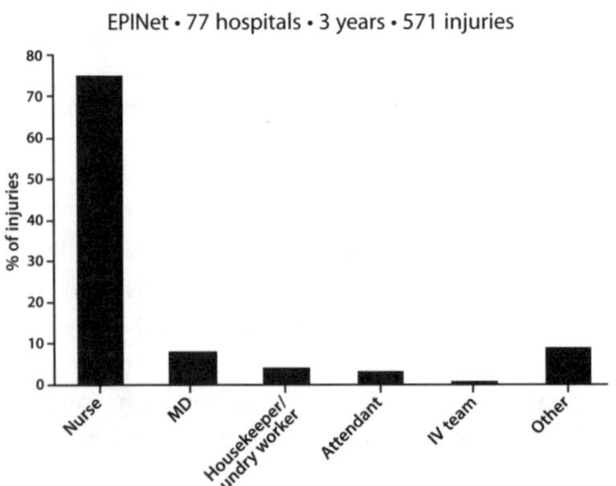

International Healthcare Worker Safety Center, University of Virginia

Figure 3. Location of Injuries from Conventional I.V. Catheters

EPINet • 77 hospitals • 3 years • 571 injuries

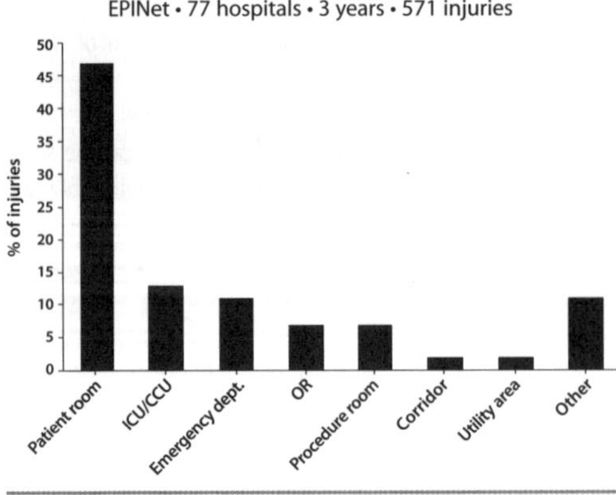

International Healthcare Worker Safety Center, University of Virginia

and contains a total of 571 injuries from conventional I.V. catheter stylets and 73 injuries from safety I.V. catheter stylets. **Figure 2** shows the job categories most frequently sustaining injuries from conventional (non-safety) I.V. catheter stylets. Nurses sustained 75% of injuries, which is notable because across all devices nurses sustained 50% of injuries.[4] Members of I.V. teams sustained 1% of injuries; however, because there are few I.V. teams represented in the 77-hospital network, it cannot be determined, based on these data, whether the 1% figure suggests that members of I.V. teams have a lower injury risk, or simply reflects few individuals in this professional category. Physicians and attendants sustained a total of 11% of injuries. Housekeepers and laundry workers sustained 4% of injuries, all of which were caused by others who mishandled or inappropriately disposed of used stylets.

The locations of injuries from conventional I.V. catheter stylets are shown in **Figure 3.** Forty-seven percent of injuries occurred in patient rooms. Fewer injuries occurred in clinical areas including intensive care units (13%), emergency departments (11%), operating rooms (7%), and procedure rooms (7%). Four percent of injuries occurred in nonclinical areas such as corridors and utility areas. These data are useful for assessing the potential impact of selectively introducing safety I.V. catheters into limited areas of the hospital. In particular, some hospitals have introduced safety I.V. catheters selectively into emergency departments in the belief that there is a higher risk of I.V. catheter stylet injuries in the emergency setting. These data support widespread rather than selective implementation, showing that only 11% of injuries hospital-wide would be addressed by this selective implementation in the emergency department.

Figure 4 shows the locations of injuries from safety I.V. catheter stylets. There were only a small number of injuries from safety I.V. catheters (73) in comparison to conventional I.V. catheters (571); **it should be kept in mind that a safety device can be very effective and still cause a small number of injuries** (as further discussed in relation to Figure 5, next page). The main differences in comparison to locations for conventional devices shown in Figure 3 were a higher percentage of injuries occurring in the emergency department and no injuries occurring in the operating room. **This distribution does not indicate that the risk of injury is higher for safety devices in certain locations, but instead shows where the safety devices are in use in the hospital.** Anesthesiologists are the professional group that is most likely to use I.V. catheters in the operating room, but safety I.V. catheters have not yet been widely accepted by anesthesiologists owing to special design requirements that are unique to anesthesia practice. The absence of injuries from safety I.V. catheter stylets in the operating room reflects that the devices were not used there at the time of data collection.

Figure 4. Location of Injuries from Safety I.V. Catheters
EPINet • 77 hospitals • 3 years • 73 injuries

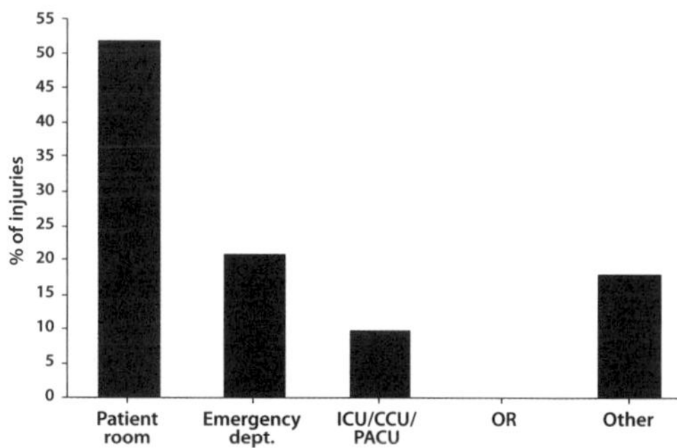

International Healthcare Worker Safety Center, University of Virginia

Figure 5. I.V. Catheter Stylet Injury Rates
3 EPINet hospitals • 1 year, September 1992 – August 1993

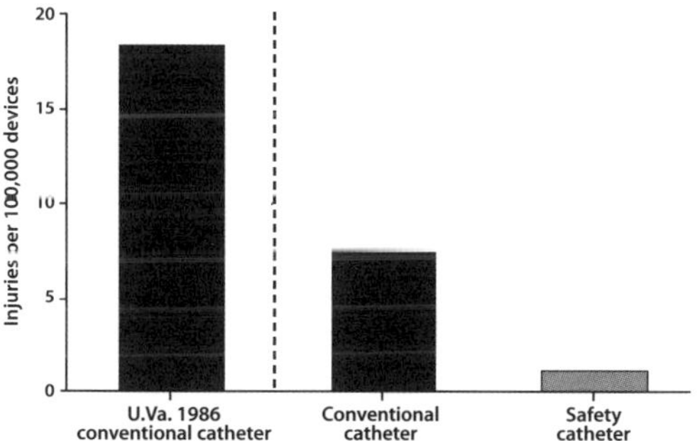

International Healthcare Worker Safety Center, University of Virginia

Figure 6. Safety vs. Conventional I.V. Catheter Injuries
EPINet •77 hospitals • 3 years • 644 injuries

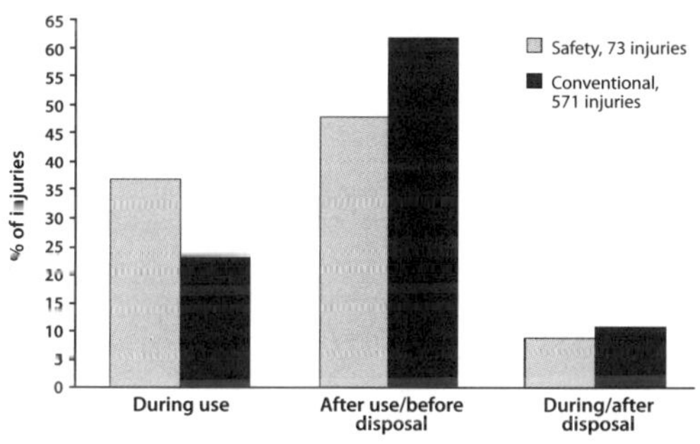

International Healthcare Worker Safety Center, University of Virginia

Conventional vs. Safety I.V. Catheters

An opportunity occurred that permitted the direct comparison of injury rates from conventional versus safety I.V. catheters. Safety I.V. catheters were introduced in three different network hospitals simultaneously in September 1992. The safety device differed from the conventional device in that the stylet of the safety catheter retracted into a cylindrical protective shield as the stylet was removed from the catheter. With the conventional catheter, the sharp stylet remained exposed until it was placed in a disposal container. All three institutions used the same brand of conventional catheter and introduced the same brand of safety I.V. catheter at the same time. In specific areas (pediatrics, anesthesiology) of the three hospitals, the conventional I.V. catheter continued to be used concurrently with the safety catheter. This allowed the direct comparison of needlestick rates for the conventional device versus the safety device across the three hospitals. All hospitals provided records of the number of conventional and safety catheters purchased during the interval under study. These numbers provided an approximation of the number of devices used in calculating device-specific needlestick rates per 100,000 devices used.

Figure 5 shows that in three hospitals in 1992 and 1993, the needlestick rate for the conventional I.V. catheter was 7.5 injuries per 100,000 conventional catheters used. This compared to an injury rate of 1.2 injuries per 100,000 safety catheters used, which was an 84% lower rate than for the conventional catheter. Although historic data were not available from all three hospitals, comparable data were available from one of the three hospitals for 1986 and had been previously published.[5] In 1986, University of Virginia Hospital reported an injury rate from conventional I.V. catheters (same brand as above) of 18.4 per 100,000 devices used. The higher rate, which is more than twice that of the same conventional I.V. catheter in 1992 and 1993, reflects the circumstances that were prevalent before the implementation of the Occupational Safety and Health Administration (OSHA) Bloodborne Pathogens Standard.[6] The changes that took place after 1987 included educational safety programs for hospital staff, placement of disposal containers closer to the point of use, and significant improvements in the design of disposal containers.

These data suggest that safety education and improved disposal systems can result in substantial declines in injury risk (59%) and are important measures. However, after the implementation of good disposal systems and safety education, even greater reductions in injury risk can be achieved by the introduction of an effective safety device (84%). All safety measures together (education, improved disposal systems, safety device) resulted in a 94% reduction in injury risk as compared to 1986, before the implementation of these measures.

The mechanism of injury from safety I.V. catheters differs somewhat from that of conventional catheters. The percentage distribution of injury mechanism for both types of catheters is shown in Figure 6. Because safety I.V. catheters provide a shield for the sharp stylet when it is withdrawn from the patient, it is to be expected that injuries occurring after use of the stylet and during and after disposal of the stylet would be prevented. Injuries occurring during the placement of the catheter might not be affected since the differences between the two devices are minimal during placement. **Figure 6** shows that injuries from safety I.V. catheters are more likely to occur during use (during catheter placement) and less likely to occur after use in comparison to conventional I.V. catheters. Safety I.V. catheters did not eliminate injuries occurring after removal of the stylet, but did decrease the proportion of injuries occurring after use. In most instances in which injuries occurred after removal of the stylet, the user did not completely retract the stylet into locked position in its sheath.

Conclusion

Nationwide EPINet data show that among all devices causing percutaneous injuries, the I.V. catheter stylet is the number one device causing high-risk needlesticks, that is, needlesticks involving a blood-filled, large-bore needle. This occupational risk is concentrated among nurses whose duties include frequent placement of peripheral I.V. catheters. The demonstrated potential for preventing injuries from conventional I.V. catheter stylets is great. Our data show that educational safety programs and improved disposal systems that were introduced when the OSHA Bloodborne Pathogens Standard was implemented were associated with a 59% decrease in device-specific injury rates from I.V. catheter stylets in one EPINet hospital. An additional 84% reduction in I.V. catheter injury rates was achieved in three EPINet hospitals after the introduction of a safety I.V. catheter designed to prevent needlesticks. The prevention of high-risk needlesticks from I.V. catheter stylets will not only prevent the occupational transmission of HIV but of all bloodborne pathogens. Because of the high risk involved in I.V. catheter stylet injuries and the seriousness of occupational infection from bloodborne pathogens, I.V. catheters incorporating needlestick prevention features should be implemented in all health care settings, to the maximum extent clinically feasible, without delay.

References

1. Centers for Disease Control and Prevention. Update: provisional Public Health Service recommendations for chemoprophylaxis after occupational exposure to HIV. *MMWR.* 1996;45:468–472.
2. Centers for Disease Control and Prevention. *HIV/AIDS Surv Rep.* 1997; 9(1):15.
3. Jagger J. Report on blood drawing: risky procedures, risky devices, risky job. *Adv Exposure Prev. 1994;*1(1): 4–9.
4. Ippolito G, Puro V, Petrosillo N., Pugliese G, Wispelwey B, Tereskerz P, Bentley M, Jagger *J. Prevention, Management and Chemoprophylaxis of Occupational Exposure to HIV.* Charlottesville, VA: International healthcare worker Safety Center/University of Virginia;1997,19.
5. Jagger J, Hunt EH, Brand-Elnagger J, Pearson RD. Rates of needle-stick injury caused by various devices in a university hospital. *N Engl J Med.* 1988;319:284–288.
6. Occupational Safety and Health Administration. 29CFR Part 1910.1030: Occupational exposure to bloodborne pathogens. *Fed Regist.* 1991;56(235):64004–65182.

Glass Capillary Tubes: Eliminating an Unnecessary Risk to Healthcare Workers

by Janine Jagger, M.P.H., Ph.D., Melanie Bentley, B.S., and Jane Perry, M.A.

Vol. 3, no. 5, 1998

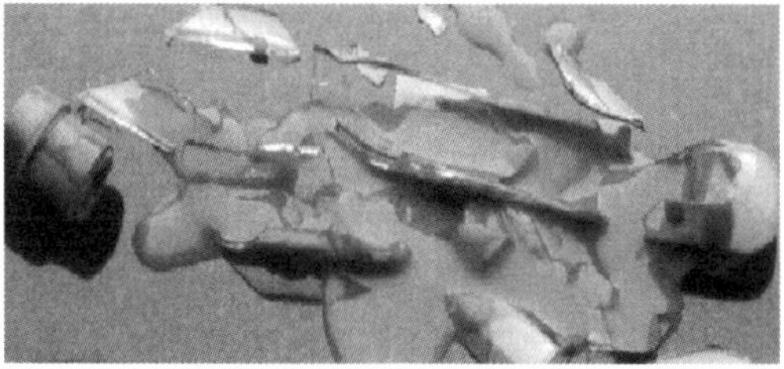

Since the International Healthcare Worker Safety Center began tracking occupational exposures to bloodborne pathogens, microbore glass capillary tubes, most frequently used for hematocrit determination, have persistently shown up as a device causing injuries to healthcare workers.

The exact number of glass capillary tube injuries in the U.S. is not known. In 1992, when the University of Virginia was still using glass capillary tubes, we calculated, based on EPINet data, that approximately 2.6 injuries occurred per 100,000 glass tubes purchased. According to industry estimates, approximately 108 million glass capillary tubes are sold annually in the U.S. (personal communication, William Kendrick, 1998). Based on this figure, the U.Va. rate extrapolates nationally to approximately 2,800 glass capillary tube injuries per year in healthcare settings. The majority of such injuries probably occur outside of hospitals, in blood donation facilities, dialysis centers, private physicians' offices, and blood testing laboratories, where such exposures may not be carefully tracked.

Glass capillary tubes are both fragile and high risk. They can contain more blood than a needle, and can inflict a large laceration with the potential of introducing a large inoculum of blood into the wound. Often the glass fractures near the fingers while the tube is being pushed into sealing clay; this was the case with physician Hacib Aoun, who was infected with HIV in 1983. Dr. Aoun has since died of AIDS. Glass tubes also may shatter during centrifugation, posing a risk to staff when removing glass fragments and cleaning spilled blood. All injuries from glass capillary tubes are preventable by the substitution of safer product alternatives on the market.

EPINet data from 81 hospitals for four years, 1993 through 1996, documented 38 injuries from glass capillary tubes. **Figure 1** shows injuries by place of occurrence: 20 (53%) occurred in clinical laboratories, 6 (16%) in intensive or critical care units, 3 (8%) in outpatient clinics, and 9 (24%) in other areas that included emergency departments, blood banks, dialysis units, procedure rooms, utility areas, and labor and delivery units.[1]

Of the 38 injuries, 21 included descriptions of the exposure event. These descriptions are printed below. One point of interest is that 12 of the 21 injuries (57%) occurred while the healthcare worker was attempting to seal the tube; self-sealing capillary tubes would reduce the probability of tube breakage.

Figure 1. Location of Injuries from Glass Capillary Tubes

EPINet • 81 hospitals • 4 years • 38 injuries

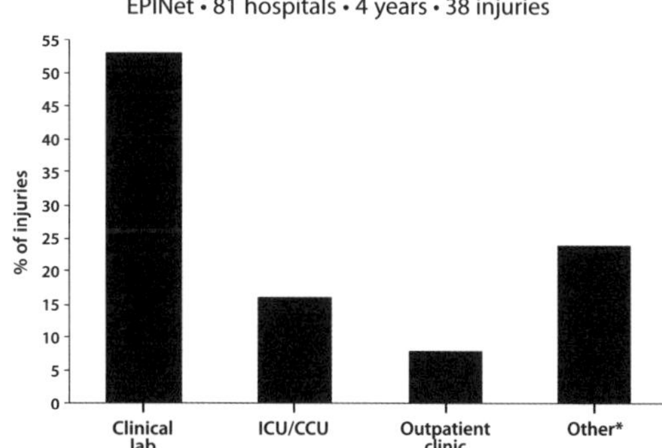

** "Other" includes emergency departments, blood banks, dialysis units, procedure rooms, utility areas, and labor and delivery rooms*

International Healthcare Worker Safety Center, University of Virginia

The Center has previously recommended, and continues to recommend, that the use of microbore glass capillary tubes be discontinued. Safer devices include plastic capillary tubes, mylar-wrapped glass capillary tubes, or alternative methods of measuring hematocrit such as hemoglobin readers, which do not require centrifugation of blood samples. The FDA has also recommended that "health professionals consider using safer alternatives to glass capillary tubes" (FDA Medical Bulletin, June 1993). The Center has requested that the FDA issue a safety alert or public health advisory on the risks of glass capillary tubes. *[Editor's note: This alert was issued in February 1999.]*

To help speed the transition from glass capillary tubes to safer alternatives, the Center is contacting manufacturers of glass capillary tubes and asking that they discontinue making them. To date, the response has been encouraging: most companies that make glass tubes also produce plastic or mylar-wrapped alternatives, and several manufacturers have said that the elimination of glass capillary tubes would not impact negatively upon their business. We are hopeful that with the cooperation of manufacturers and distributors, glass capillary tubes will soon be a thing of the past.

Descriptions of injuries from glass capillary tubes: EPINet, 1993–1996

- A unit clerk in an ICU/CCU was picking up a broken blood-filled capillary tube. A sliver of glass penetrated the clerk's right gloved thumb. The injury was superficial.

- A clinical lab worker in a hematology laboratory was attempting to seal one end of a glass capillary tube with clay so that it could be spun down for daily quality control. The microhematocrit tube broke, injuring the worker's right index finger. The injury was moderate.
- A clinical lab worker in an outpatient dental clinic was sealing a glass hematocrit tube with clay and pushing the tube to make sure the clay was tightly packed to avoid leakage. The tube broke, cutting the worker's right index finger. The injury was superficial.
- A clinical lab worker in a stat lab was cleaning off counters with a cloth when a piece of broken microhematocrit capillary tube punctured the gloves in the palm of the right hand, causing bleeding. The injury was moderate.
- A clinical lab worker in a hematology laboratory was making up capillary tubes for spun HCT. After putting blood in the capillary tubes, the worker was pressing the tubes into the sealant when the tubes broke in half. The glass went through the gloves of the worker's right middle finger. The injury was superficial.
- A nurse in labor and delivery was taking fetal scalp samples. When the nurse went to put the "flea" in the capillary tube, the tube broke. Glass went through the glove and into the nurse's left thumb. The injury was moderate.
- A clinical lab coordinator in a stat lab was reviewing reports and was stuck by a contaminated fetal scalp capillary tube that had been taped to the lab slip. The broken capillary tube stuck the coordinator's right thumb. The injury was moderate.
- A clinical lab worker was sealing a capillary tube when the tube broke and cut his right index finger through a single pair of gloves. The injury was superficial.
- A blood gas technician was preparing to do a blood gas analysis. The capillary tube broke when the technician was attaching the clot catcher to it. The technician cut his right middle finger through his glove. The injury was moderate.
- A resident was in the ICU/CCU when a capillary tube broke during use, causing glass fragments to become embedded in his left thumb. The injury was moderate.
- A non-lab technician in a service/utility area was testing and cleaning a blood warming machine. Some water spilled, and while cleaning up, the technician was cut by a capillary tube on his right middle finger. The injury was superficial.
- A patient hematocrit needed verification by spun hematocrit procedure. A clinical lab worker drew patient blood into a capillary tube from an EdTA tube; when placing one end of the tube in the clay, it snapped in her hand and glass penetrated her left ring finger. The injury was superficial.
- A technologist in clinical labs was attempting to seal a capillary tube filled with blood. The tube broke and punctured the technologist's right thumb, which bled. The injury was moderate.
- A lab worker in a clinical laboratory was preparing to spin a hematocrit. The capillary tube broke as it was being sealed. The broken tube pierced the glove and cut the lab worker's right thumb. The injury was moderate.
- A clinical lab worker was cleaning a countertop in the laboratory with a germicidal wipe. As she was wiping in a corner, her right ring finger was cut by a glass capillary tube that hadn't been disposed of. The injury was moderate.
- A respiratory therapist in the nursery was handed a capillary tube by a nurse. While the RT was in the process of sealing the capillary tube in order to do a blood gas, the tube broke and superficially injured the RT's right thumb.
- A nurse in the emergency department was completing a hematocrit reading while disposing of two capillary tubes. She dropped one; after looking and not being able to find it, she removed her gloves and started to leave the room. She then saw the capillary tube at the entrance of the door and picked it up. It had broken and glass punctured her finger. The injury was moderate.
- A clinical lab worker was in the laboratory running a batch of stats (CBCs) from the ER, NICU, and kidney center. The worker's left index finger was cut on a glass capillary tube. The injury was moderate.
- A clinical lab worker at an off-site lab performed a fingerstick and drew blood into a non-teflon capillary tube. As she pressed the end into clay, the tube broke and lacerated her right thumb. The injury was moderate.
- A technologist in the dialysis facility was preparing to spin a patient's hematocrit. The capillary tube containing the patient's blood shattered while the technologist was sealing one end of the tube. A shard from the broken tube pierced her glove and right index finger. The injury was moderate.
- A clinical lab worker was in an outpatient office doing a donor fingerstick. As the worker was putting the capillary tube into clay, the tube broke and pierced the worker's glove and right thumb. The injury was moderate.

Reference

1. Jagger J, Deitchman S. The hazards of glass capillary tubes to healthcare workers [letter]. *JAMA.* 1998;280:31.

Patterns and Prevention of Blood Exposures in Operating Room Personnel: A Multi-Center Study

by Janine Jagger, M.P.H., Ph.D., Melanie Bentley, B.S., and Patti M. Tereskerz, J.D., Ph.D.

Vol. 3, no. 6, 1998

Reprinted in AEP, and here, with permission from the AORN Journal *(1998; 67[5]: 979-96). ©Copyright 1998, AORN, Inc.*

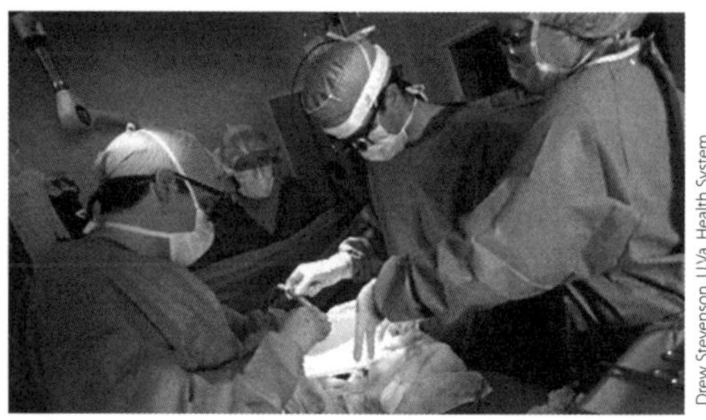

Drew Stevenson, U.Va. Health System

This study was a collaborative research initiative of the International Healthcare Worker Safety Center and the Association of Operating Room Nurses (AORN). Center staff collaborated with AORN members from six institutions and with two members of AORN's professional staff in the implementation of an occupational blood exposure surveillance system. The purpose of the study was to build a descriptive multi-center database to be used as a foundation for identifying the causes of exposures to bloodborne pathogens in the surgical setting, and promoting successful prevention measures.

Introduction

The operating room (OR) is recognized for the high frequency of occupational blood contact and percutaneous injuries that occur there.[1-9] Factors that set the operating room apart from other healthcare settings include prolonged contact of surgical personnel with open surgical sites, frequent manipulation of sharp instruments, and the presence of relatively large quantities of blood. Although national data on occupational infections with human immunodeficiency virus (HIV) in the U.S. are not complete, cases compiled through 1997 show that, at a minimum, four surgical technicians and six surgeons have contracted or possibly contracted HIV from occupational exposures.[10] The 49 nurses enumerated in the report were not identified by work location; therefore, the number of OR nurses among them could not be ascertained. One case of an occupational HIV infection in an anesthesiologist has been reported in the literature.[11] One case of HIV transmission to an Italian surgeon after a scalpel blade injury was documented.[12] Although the hepatitis B virus (HBV) and the hepatitis C virus (HCV) are more readily transmitted than HIV, there is no surveillance system in the U.S. to document how many of these cases may have occurred among OR personnel. The risk of acquiring bloodborne pathogens in the surgical setting is not limited to operating room personnel; there are several reports in the literature documenting healthcare worker-to-patient transmission of HIV, HBV and HCV during invasive procedures.[13-17]

A number of studies have reported the rates and types of blood exposures that place OR personnel at risk of infection from bloodborne pathogens.[1-9] These studies employed observers in the OR or assigned circulating nurses to fill out data collection forms when exposure events were observed during surgical procedures. Occupational risk of blood contact or percutaneous injury in these studies was associated with procedures in which there was increased blood loss, increased operative time, or in which inadequate barrier garments were worn. The types of surgeries with highest exposure risk were trauma, burn, and orthopedic emergency procedures, as well as major vascular, intraabdominal, and gynecologic surgeries. Investigators reported that the devices that most commonly caused injuries were suture needles and scalpel blades, and that a high proportion of blood contact events were associated with inadequate liquid resistance of surgical gowns, a failure to double glove, or a lack of protective eyewear.[1-9]

Prevention recommendations resulting from these studies have focused on increasing the use of protective garments and on the selection of barrier materials with a high degree of liquid resistance.[2,4,8,9] The practice of double-gloving has been actively promoted as a method for reducing blood exposures to hands, especially among surgeons, and the use of puncture-resistant gloves has been proposed as a method for reducing percutaneous injuries to hands.[1,3] The need for more consistent use of protective eyewear also has been stressed.[9]

Risk-reducing techniques have also been promoted, such as requiring mechanical rather than manual retraction of tissue to keep hands at a greater distance from sharp instruments. Most recently, the introduction of safer surgical devices designed to reduce the likelihood of percutaneous injuries, such as blunt suture needles, has been supported by researchers.[18-20]

The conclusions drawn from previous research have been based upon relatively small numbers of exposures in each study. They reveal significant areas of opportunity for improving occupational safety in the OR and, at the same time, point to the need for larger studies that can provide additional detail on exposure mechanisms to efficiently target specific prevention interventions and measure their impact.

This multi-center study was designed to answer the questions, "What types of surgical devices and circumstances are associated with exposures, given a large enough number of cases to reveal product-specific exposure patterns?" and "What unique exposure profiles are associated with various OR personnel?" "Exposure" was defined as a percutaneous injury or mucocutaneous contact with blood or other potentially infectious biological materials. The study also assessed whether the incident report forms provided an appropriate level of detail for describing the OR environment and could be filled out easily and accurately by OR staff members.

Methods

Candidate hospitals indicated their interest in participating as study sites in response to an announcement by AORN in April 1995. Six hospitals were selected for inclusion in the final study: Baptist Hospital of Miami, Florida; Queen's Medical Center of Honolulu, Hawaii; Tucson Veteran's Administration Hospital, Arizona; the University of Pennsylvania Medical Center, Philadelphia; University of Texas Medical Branch, Galveston; and the University of Virginia Health Sciences Center, Charlottesville. Hospitals were selected for participation based on the following criteria:

- availability of an OR nurse study coordinator
- surgical service with volume greater than 10,000 cases per year
- capacity to do on-site data entry
- geographic diversity

Five of the six participating hospitals were teaching hospitals.

The data collection method employed was an adaptation of the Exposure Prevention Information Network (EPINet), a surveillance system developed at the University of Virginia in 1990 and used by approximately 1,500 U.S. hospitals to record and track percutaneous injuries and other at-risk blood and body fluid exposures reported by healthcare workers.[21] Modifications in the EPINet data collection forms were made to provide detail on unique aspects of the OR environment, such as identification of surgical instruments, surgical techniques associated with injuries, surgical equipment related to blood exposures, and the types of barrier equipment in use at the time of blood exposures. Factors related to specific prevention strategies, such as whether hands-free or hand-to-hand passing techniques were in use when injuries occurred during the passing of instruments, were incorporated. The forms were reviewed, revised, and approved by two designated AORN members. The finalized version of EPINet/OR used in the study included a two-sided form for percutaneous injuries, and a two-sided form for mucocutaneous blood contact incidents, such that only one two-sided form was completed for each exposure incident, depending on exposure type.

Before the initiation of data collection, a designated AORN staff member provided on-site training of OR nurses in the surgical departments of study hospitals. All nurses eligible to record exposure events were oriented to the research objectives and study methods. They were provided with written instructions and definitions. The study protocol and data collection forms were also prominently posted in surgical areas.

Circulating nurses were given the responsibility of recording on the appropriate EPINet/OR report forms any percutaneous injuries or mucocutaneous blood contacts occurring in OR personnel during surgical procedures. Exposure incidents included any percutaneous injury to OR personnel, and any mucocutaneous incident involving blood contact with non-intact skin or mucous membranes. Participants also recorded blood exposures to intact skin, but these incidents were included in the study only if a large amount of blood (i.e., more than 50 mL) made contact with intact skin. This restricted definition was intended to exclude a large number of incidents involving a small amount of blood on intact skin, which is believed to present a negligible risk of occupational infection. Incidents involving other fluids that were visibly contaminated with blood were treated as blood exposures. All surgical subspecialties at each hospital were included in the study.

The circulating nurses were instructed to observe all OR personnel during surgical procedures and to ask, "Did you have an injury or a blood exposure?" if it appeared that an individual had incurred an exposure. They also were instructed to capture any additional exposures by asking all OR staff at the end of each procedure to verify whether anyone present had sustained an injury or a blood exposure.

Identities of exposed OR personnel were not recorded on the data collection forms. However, the identity of the circulating nurse who filled out the form was recorded on the form to allow the site coordinator to verify information, if necessary. All OR personnel were instructed to continue reporting exposure incidents to their employee health service in compliance with standard hospital protocol.

Data collection began in July 1995 and continued for 15 months in five hospitals, and for 12 months in one hospital. Data were entered at each hospital using software provided by investigators. Each site copied data to a diskette quarterly and sent it to researchers at the University of Virginia where the data were reviewed, corrected if necessary, and merged into the multi-center database. Data entry and analysis were carried out using Epi Info, a statistical package developed by the Centers for Disease Control and Prevention (CDC).[22]

Results

A total of 481 exposure events were reported, including 386 percutaneous injuries and 95 mucocutaneous blood exposures. The number of exposures reported by individual hospitals ranged from 25 exposures by the hospital reporting the fewest to 175 by the hospital reporting the highest number, as seen in **Figure 1.** The frequency of exposures by service is shown in **Figure 2**. Cardiovascular surgery had the highest number of exposures, accounting for over 15.8% of cases. The top three services, including cardiovascular, general surgery, and orthopedic surgery, together reported 41.4% of all exposure incidents. Fewer incidents, no more than 5.0% each, were reported by the remaining services, including: plastic surgery; peripheral vascular surgery; neurosurgery; ear, nose and throat; obstetrics and gynecology; thoracic surgery; ophthalmology; transplant surgery; anesthesia; oral surgery; trauma surgery; pediatric surgery; and dermatology.

Among occupational groups, attending and resident surgeons had the highest frequency of both percutaneous injuries and mucocutaneous exposures, together accounting for 55.1% of all exposures **(Figure 3).** Obstetrics and gynecology residents and attending physicians together accounted for 4.0% of exposures. Anesthesia personnel, including attending and resident anesthesiologists and nurse anesthetists, accounted for 6.2% of reported exposures. Of interest was the finding that for all physician categories, residents' injuries outnumbered those of attending physicians, accounting for 36.4% and 28.3% of total injuries, respectively. Scrub nurses ranked after surgeons, with 19.1% of exposures, and circulating nurses and other OR nurses added 6.0% more to the figure representing those in OR nursing roles. Of those classified in the roles of scrub and circulating nurses, 31 out of 107 were identified as OR technicians. Eight additional OR technicians did not specify their role during the procedure and were classified with technicians in the "other" category. Of additional interest was the fact that while scrub nurses had a significantly higher frequency of percutaneous injuries than circulating nurses (81 versus 6), the frequency of mucocutaneous blood exposures was similar for the two groups (11 versus 9). Medical students accounted for 3.1% of injuries, and OR attendants for .8%.

The characteristics of percutaneous injury events were evaluated separately from mucocutaneous exposures. The locations where injuries occurred followed a gradient which was highest near the patient and lower at a distance from the patient. **Figure 4** (next page) shows that the highest proportion of injuries occurred on the surgical field (33.4%). Injuries in the surgical site ranked second (25.1%); such injuries have an added significance because they also have the potential of exposing patients to healthcare workers' blood. Injuries occurring at the mayo stand and at the back table accounted for 7.3% and 7.0% of injuries, respectively. Injuries occurring at the anesthesia cart and the anesthesia machine together accounted for 1.5% of injuries. More remote areas from the patient, including the OR floor and areas adjoining the OR, such as the pre-operative area and the OR utility room, accounted for the remainder of injuries.

The distribution of injuries by body location in **Figure 5** shows that 93.3% of injuries were to the hands. Of all

Figure 1. Frequency of Exposures in Surgical Settings by Hospital

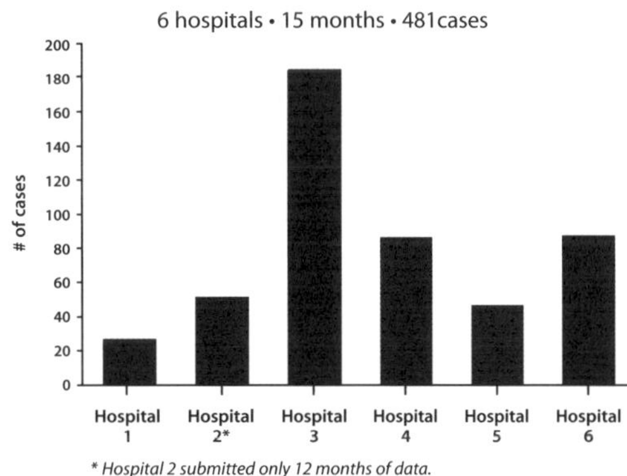

Figure 2. Frequency of Exposures in Surgical Settings by Surgical Service

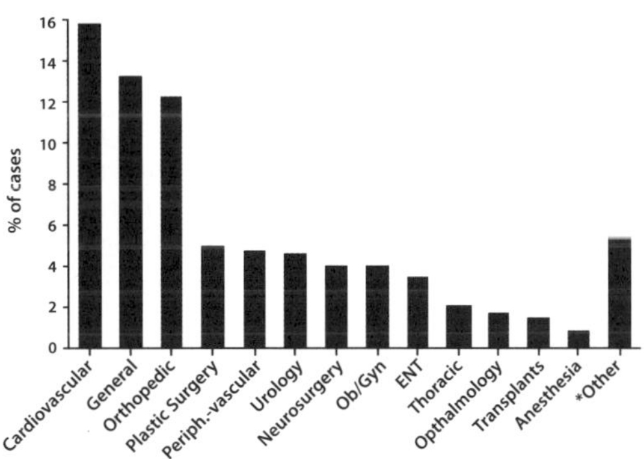

* "Other" includes 4 oral, 3 trauma, 3 pediatric, 2 dermatology, 1 sterile processing and 14 unspecified.

Figure 3. Frequency of Percutaneous Injuries and Blood Exposures in Surgical Settings by Job Category

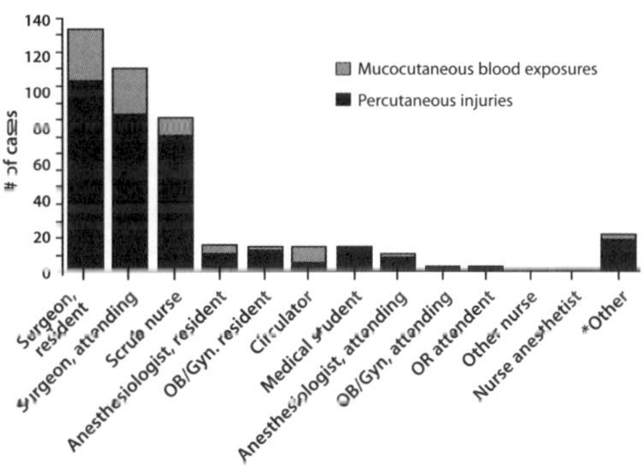

* "Other" includes 15 technologists, unclassified as to type; 2 audiologists; 2 radiologists; 1 clinical lab worker; 1 housekeeper; 1 nurse anesthetist student; and 1 physician, unclassified as to type.

Figure 4. Location of Sharp-Object Injuries in Surgical Settings

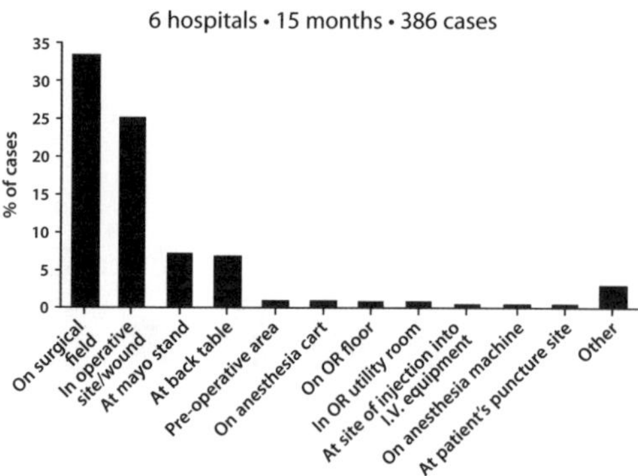

injuries to the hands (360 cases), the left hand was more frequently injured than the right hand (59.7% versus 40.3%). These findings are relevant to the optimal use of puncture-resistant gloves and finger guards. In cases in which handedness was known (289), the non-dominant hand was injured in 55.7% of cases and the dominant hand was injured in 44.3% of cases. The discrepancy in injuries to hands was also looked at in 198 cases in which the user of the device injured himself or herself, as opposed to being injured by another person holding a device. In this circumstance, 60.6% of injuries were to the non-dominant hand and 39.4% of injuries were to the dominant hand.

The devices causing injuries are shown in **Table 1.** More than three-fourths of injuries were caused by only five types of devices. Suture needles were at the top of the list, accounting for more than half of injuries, followed by scalpel blades, syringes, retractors/skin or bone hooks, and electrocautery devices. The OR, however, is also characterized by the use of a wide variety of sharp equipment, which is reflected in the remaining 21.2% of injuries that were distributed across 32 device categories. Hollow-bore needles caused 13.2% of injuries, but only 1.8% of injuries were caused by hollow-bore needles specifically used for vascular access, which are classified as high-risk injuries for HIV transmission by CDC criteria.[23]

Characteristics of injuries by the two top devices, suture needles and scalpel blades, were evaluated separately in order to assess the potential benefits of different prevention strategies. The mechanism of 197 suture needle injuries are shown in **Figure 6.** In contrast to other needle devices, the majority of injuries from suture needles occurred during use, that is, during suturing (67.5%). A prevention strategy would have to provide protection during the suturing process to be effective. Blunt suture needles are designed to penetrate muscle and fascia, but do not penetrate denser cutaneous tissue easily, in either the patient or the healthcare worker. We found that in cases where the specific purpose for which suture needles were used could be determined, 59% of injuries were caused by suture needles used to suture muscle or fascia. This indicates the maximum proportion of suture needle injuries that are potentially preventable by using blunt suture needles.

The mechanism of scalpel blade injuries differed from

Figure 5. Distribution of Sharp-Object Injuries by Body Location

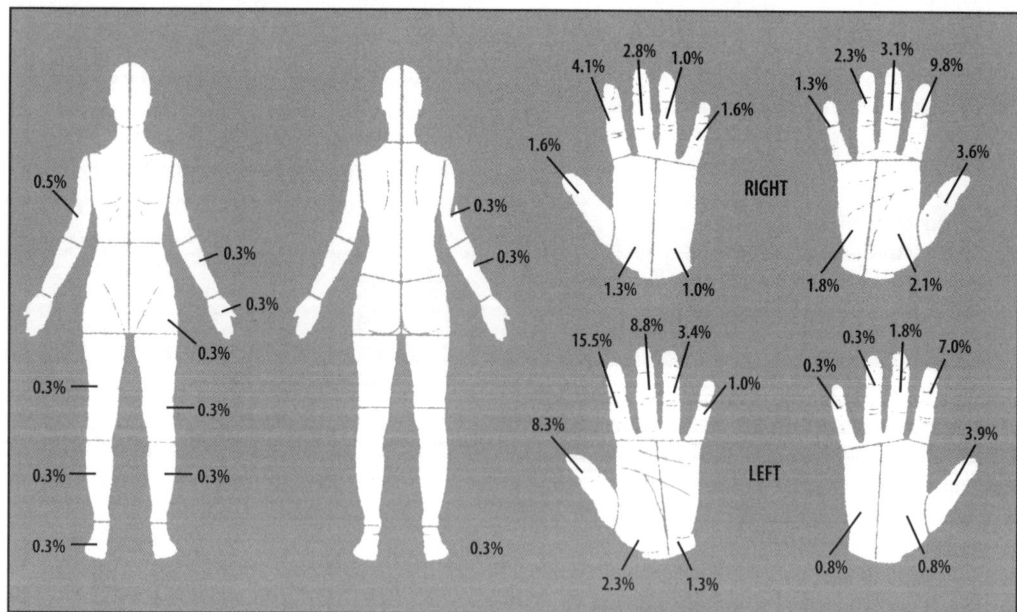

Table 1. Devices and Other Sharp Items Causing Percutaneous Injuries

6 hospitals • 15 months

Device	Injuries	% of total
Needles (hollow-bore)		
Syringe	34	8.8%
IV catheter stylet	4	1.0%
Unattached hypodermic needle	4	1.0%
Butterfly-type needle	2	0.5%
Spinal/epidural needle	2	0.5%
Other unspecified hollow-bore	2	0.5%
Needle on I.V. line	1	0.3%
Arterial catheter needle	1	0.3%
Nonvascular catheter needle	1	0.3%
Surgical Instruments/Equipment		
Suture needle	197	51.0%
Scalpel blade	45	11.7%
Retractor, skin/bone hook	17	4.4%
Electrosurgical needle/blade	11	2.8%
Wire	8	2.1%
Scissors	7	1.8%
Pickups/forceps	4	1.0%
Orthopedic pin	4	1.0%
Trocar	4	1.0%
Saw blade	3	0.8%
Staples/steel sutures	3	0.8%
Sutures	2	0.5%
Awl/reamer	2	0.5%
Bone fragment	1	0.3%
Curette	1	0.3%
Cutting jig	1	0.3%
Drill bit	1	0.3%
Drill guard	1	0.3%
Edge of OR table	1	0.3%
Emulsifier tip	1	0.3%
Glass ampule	1	0.3%
Nerve stimulator	1	0.3%
Probe	1	0.3%
Razor	1	0.3%
Rod extractor	1	0.3%
Stapler	1	0.3%
Subdural electrode	1	0.3%
Towel clip	1	0.3%
Missing/unclassified	14	3.6%
Total	**386**	**100.7%** *

**Difference from 100.0% is due to rounding error.*

Figure 6. Mechanism of Suture Needle Injuries

6 hospitals • 15 months • 197cases

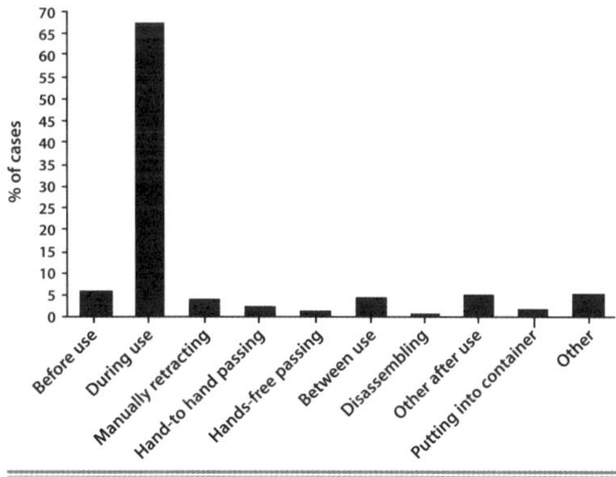

Figure 7. Mechanism of Scalpel Blade Injuries

6 hospitals • 15 months • 45 cases

% of cases: 0, 5, 10, 15, 20, 25, 30

Before use, During use, Manually retracting, Hand-to-hand passing, Hands-free passing, Between uses, Disassembling, Other after use, Pretruded from container, Inappropriate trash, Other

that of suture needles (Figure 7). Reversing the pattern found in suture needle injuries, nearly two-thirds (64.4%) of injuries from scalpel blades occurred after use or during passing, disassembling or disposal. Only 28.9% of injuries occurred during use, that is, during cutting. The potential impact of different approaches to injury reduction can be assessed by determining what stage of the use-reuse-disposal cycle they are relevant to. Prevention strategies that provide blade-shielding features address the phases of handling between uses, after use, and during disposal. Products that eliminate the need to remove blades manually from scalpel handles, such as disposable scalpels and blade removal accessories, address a smaller fraction of injuries, those that occur during disassembly.

Figure 8 compares injury severity for suture needles and scalpel blades. Superficial injuries were defined as those resulting in little or no bleeding, moderate injuries as those in which the skin was broken and some bleeding occurred, and severe injuries as those involving a deep cut or puncture or resulting in profuse bleeding. Although suture needle injuries were most frequent, scalpel blade injuries were more likely to cause moderate or severe injuries: 64.5% of scalpel blade injuries were moderate or severe, compared to 47.2% for suture needles. Scalpel blades were twice as likely to cause severe injuries (8.9% of incidents) as suture needles (4.1% of incidents). Because the transmission risk increases with the severity of the injury, this finding is relevant both to the risk of bloodborne pathogen transmission for the injured healthcare worker, and to the patient's risk of being inoculated with the blood of the healthcare worker.

Injuries caused by vascular access needles fall into the

Figure 8. Severity of Injury: Suture Needles vs. Scalpel Blades
6 hospitals • 15 months

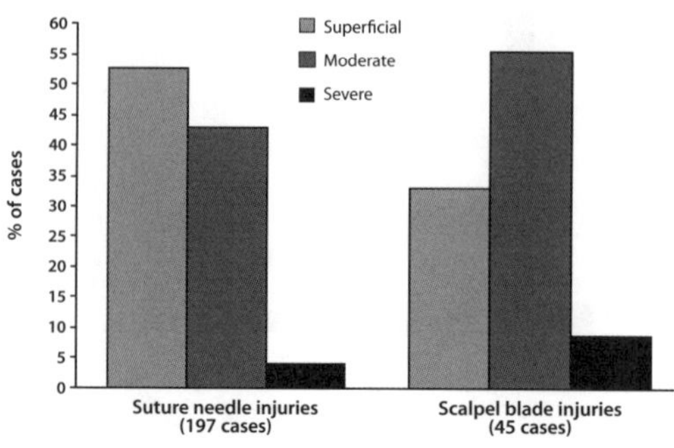

Figure 9. Percentage of Injuries in Surgical Settings from Blood-filled, Hollow-bore Needles by Job Category
6 hospitals • 15 months • 386 cases

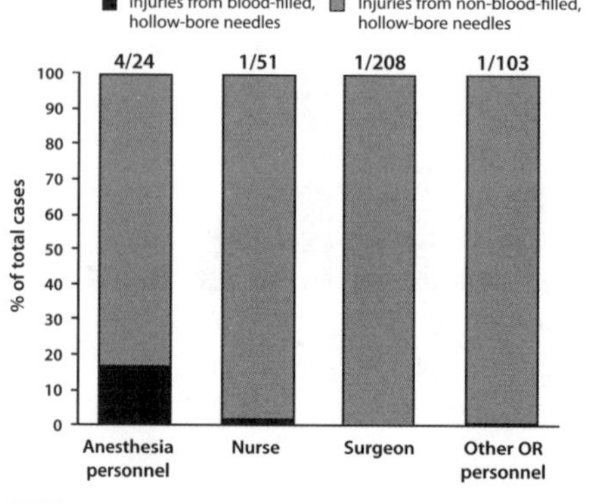

Figure 10. Mechanism of Blood Exposures in Surgical Settings
6 hospitals • 15 months • 95 cases

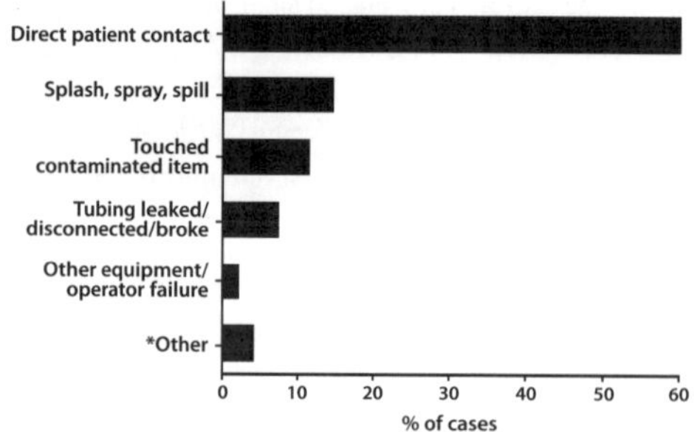

"Other" includes cases with insufficient information to categorize.

CDC's high-risk category for bloodborne pathogen transmission.[23] The higher risk is linked to a larger blood inoculum associated with injuries from blood-filled, hollow-bore needles. **Figure 9** shows which categories of OR personnel are most likely to sustain injuries from blood-filled, hollow-bore needles, including I.V. and arterial catheter introducers. Anesthesia personnel sustain most of these potentially high-risk injuries, which account for 16.7% of their injuries overall. Few such high-risk injuries are sustained by nurses, surgeons, and other OR personnel, accounting for only 2.0%, 0.5%, and 1.0% of their injuries, respectively. By contrast, in a 63-hospital study in which all clinical areas were included, it was found that 25% of all reported percutaneous injuries were from blood-filled, hollow-bore needles, a greater proportion of high-risk injuries than in the OR.[24]

Characteristics of mucocutaneous blood exposures were also evaluated. The low number of cases relative to percutaneous injuries (95 versus 386, respectively) is due to the narrow definition employed that excluded incidents in which intact skin was exposed to less than 50 mL of blood (unless the incident also involved blood exposure to non-intact skin or to a mucous membrane). A blood exposure incident could involve more than one type of skin or mucous membrane and could affect more than one body area.

Figure 10 shows the mechanism of blood contact. In 60% of cases, direct patient contact was involved. However, in the remaining 40% of cases blood was splashed or sprayed or a product served as a vehicle of exposure, indicating that it is not unusual in the OR environment for personnel to be exposed to blood even when located at a distance from the patient. Products commonly used in the OR that are potential sources of blood exposures include devices that pump blood under pressure, blood bags, suction canisters, irrigation devices, and blood-contaminated sponges, drapes, and instruments.

The body location of blood exposures is shown in **Figure 11.** The face was the most affected body area; this may be the result of the exclusion criteria of the study, which were likely to have eliminated many small blood exposures to other areas. Twice as many blood contacts were recorded to the face as to the hands, which ranked second in frequency, followed in descending order by the torso, chest, legs, and arms.

The type of skin or mucous membrane exposed to blood is shown in **Figure 12.** Consistent with the body locations of blood exposures was the finding that blood contact with the mucosa of the eyes was the most common type of exposure, accounting for 45.3% of incidents. Blood exposure of non-intact skin occurred in 29.5% of cases, followed by 15.8% of cases which involved intact skin exposures to greater than 50 mL of blood. Exposures of less than 50 mL of blood to intact skin, shown in Figure 12, did not in themselves meet study criteria but were secondary to exposures that did.

It is notable that relatively few blood exposures occurred

Figure 11. Number of Blood Exposures by Body Location*

6 hospitals • 15 months • 95 cases

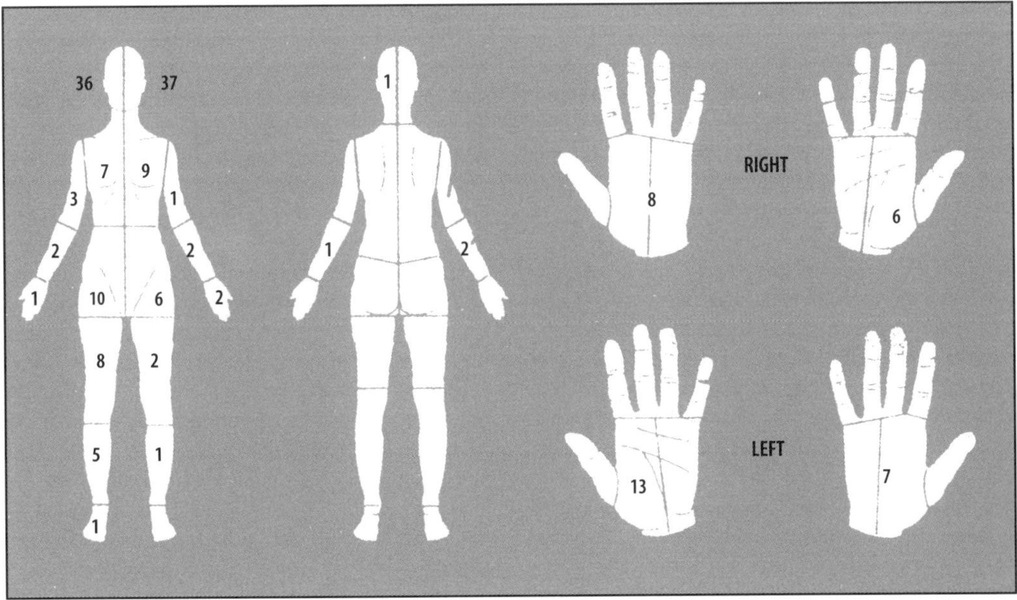

A single exposure event can involve more than one body location.

Figure 12. Type of Mucocutaneous Blood Exposure

6 hospitals • 15 months • 95 cases

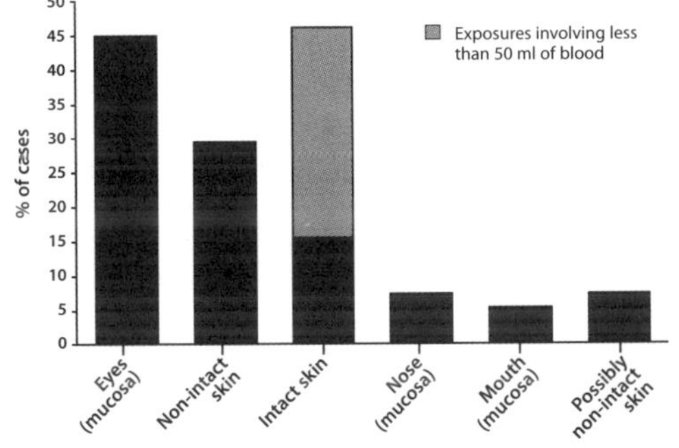

Figure 13. Personal Protective Equipment Worn During Eye Exposures in Surgical Settings

6 hospitals • 15 months • 43 cases

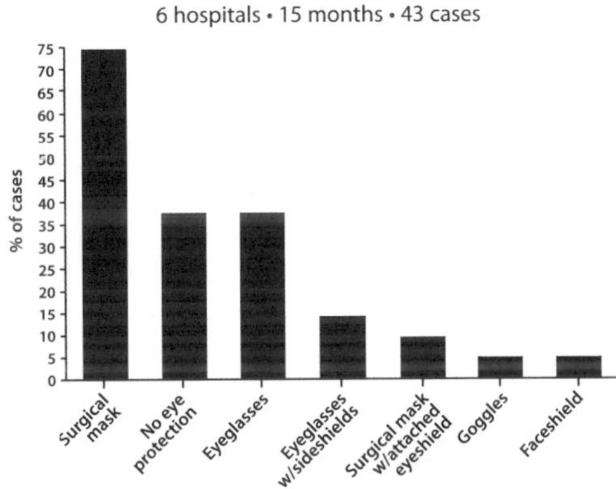

to the mucosa of the nose or mouth (7.4% and 5.3% of cases, respectively) in comparison to the eyes (45.3%). Since blood that splashes or sprays into the faces of OR personnel is not likely to effect just one area of the face (i.e., the eyes), these data suggest that barrier protection for the nose and mouth is more consistently worn by OR personnel than eye protection. **Figure 13,** which shows the types of protective items worn by those who sustained blood exposures to the eyes, supports this interpretation. Among those who sustained eye exposures, 83.7% wore surgical masks, including surgical masks with attached eyeshields, and conversely, 74.4% were not wearing eye protection (this includes those wearing eyeglasses). In the remainder of cases, some form of eye protection was worn, including goggles, faceshields, or eyeshields attached to masks, but the protective eyewear was, in these instances, inadequate to prevent eye exposures. Hand exposures were documented in 23 cases. Among them, 13.0% were wearing no gloves, 82.6% were wearing single gloves, and only 4.3% were wearing double gloves (one case).

Discussion

This study confirmed that EPINet/OR is a useful tool for understanding the causes of blood exposures in the surgical setting and for identifying prevention priorities. Participants in the six study hospitals successfully used the data collection instruments and data entry software and provided quality data to researchers. The data yielded a spectrum of information indicating the types of prevention strategies with the potential for achieving substantial reductions in percutaneous injuries and mucocutaneous blood exposures in the surgical setting.

In relation to percutaneous injuries, the prevention approach which holds the greatest promise is to reduce the use of sharp instruments to the maximum possible extent. The device which caused far more injuries than any other surgical instrument is the suture needle (51.0% of all injuries). In this study, 59% of suture needle injuries were caused by needles used to suture muscle or fascia, for which blunt suture needles could be substituted. Furthermore, research has demonstrated the feasibility of blunt suture needles and their effectiveness in reducing suture needle injuries.[18] The predominance of suture needles as a cause of injury in the OR is such that if a 59% drop in suture needle injuries were realized, this alone would result in an overall 30% drop in percutaneous injuries in the OR. Most of the remaining injuries caused by suture needles during the closure of cutaneous tissue are also potentially addressable by substituting products that present little or no risk of injury. Stapling devices, adhesive strips and tissue adhesives provide alternatives for skin closure; if used in combination with blunt suture needles, they could reduce the use of standard suture needles to a small fraction of current use. Other substitutions for unnecessary sharp devices include: needleless or shielded-needle I.V. infusion systems, blunt towel clips, blunt electrocautery devices, and, when feasible, blunt retractors, blunt scissors, blunt clamps, and blunt precut wire. Sometimes the sharp feature of a device is essential, but the use of sharp equipment should be strictly reserved for the specific applications for which there is no safer alternative.

Scalpel blade injuries should not be considered low-risk if there is blood visible on the blade at the time of injury. Scalpel blade injuries were the second most common type of injury in this study, and were more severe than injuries from suture needles. Many of these injuries are potentially preventable by scalpel designs that incorporate blade-shielding features. The optimal design would provide a shield that not only covers the blade after use, but also shields the blade when passing and disassembling, and during and after disposal. Such a device could potentially reduce scalpel blade injuries by as much as 64.4%, which would result in an overall 8% drop in percutaneous injuries in the OR. Further reductions in scalpel blade injuries can be achieved by using round-tipped blades whenever possible and restricting to a minimum the use of sharp-tipped blades. Safer methods for removing scalpel blades from reusable handles, or the use of disposable scalpels, address the relatively small fraction of injuries that occur during the disassembly process (2.2% of scalpel blade injuries).

The potential benefits of using puncture- or laceration-resistant gloves or fingerguards to prevent percutaneous injuries to the hands should be assessed. The acceptance of this approach in the surgical setting remains low because of the loss of tactile sensation and the less-than-total puncture resistance of most available barrier materials. Nevertheless, given the predictable distribution of percutaneous injuries to the hands, the selective use of puncture-resistant gloves or finger guards should be evaluated under circumstances where the risk of injury to the hands is greatest, and in a manner that least interferes with tactile sensation (that is, worn selectively on the non-dominant hand).

Specific techniques have been recommended by researchers for reducing percutaneous injury risk in the OR. One widely accepted recommendation is to use retraction devices rather than the hands for tissue retraction[3,5] In this study 3.4% of all injuries occurred when retracting tissue manually, while one injury (0.3%) occurred when retracting tissue mechanically. There have also been recommendations to use a hands-free rather than a hand-to-hand technique for passing instruments, in order to avoid collisions between instruments and hands. In this study, 6.0% of injuries occurred during hand-to-hand instrument passing, while 1.6% occurred during hands-free passing. These data do not answer the question of whether the hands-free technique is safer than the hand-to-hand technique, but they do establish that instrument passing accounts for a relatively small fraction of injuries in the OR overall. Future research should answer the question of which technique is safer, recognizing that with increased use of blunted or shielded surgical devices, instrument passing should become less hazardous, regardless of technique.

Our findings also confirm those of previous studies that increased use of barrier precautions and improved liquid resistance of barrier materials are important factors in reducing mucocutaneous blood contact in the OR.[1,2,4,6,9] Our results specifically highlight the vulnerability of the eyes to blood contact and emphasize the importance of improving eye protection in the surgical setting. It is now well documented that conjunctiva serve as a transmission route for HIV and HCV.[25-26] Conjunctival exposure is a particular risk in the OR because, while it is not uncommon for blood to spray or splash significant distances, protective eyewear is often worn only by those working in closest proximity to the operative site. In this study it was significant that circulating nurses had nearly the same number of eye exposures as scrub nurses, despite the generally held belief that circulating nurses are at negligible risk of eye exposure. Because of the possibility of blood splashing and spraying significant distances during surgery, all OR staff, regardless of proximity to the operative site, should wear protective eyewear as routinely as surgical masks.

The cases reported here in which eye exposures occurred despite the use of goggles and faceshields also point to the importance of appropriate design of eyewear. Protective eyewear worn in the OR should have a seal above the eyes. Eye exposures in which blood squirted on the scalp or forehead of healthcare workers have been described elsewhere. Although they were wearing faceshields or goggles the blood ran freely into their eyes because there was no seal above the eyes.[27]

Our findings also highlight the role of OR equipment as potential vehicles of blood exposures. Blood pumping and infusion equipment, blood bags, irrigation devices, suction canisters, and other equipment that contains blood or blood-contaminated fluids are potential sources of occupational blood exposures.[28] All such equipment in the OR should be evaluated for safety features to confirm that junctions in tubing segments are connected with positive locking mechanisms (not friction fit), that canister lids are designed to remain secure under pressure, and that blood pumping equipment has automatic shut-off valves and/or alarms if the pressure of blood being pumped through the unit increases beyond a safe level. The incorporation of these safety criteria into the product evaluation and selection process will prevent many unsafe products from being introduced into the OR.

This study shows that there are substantial opportunities for reducing percutaneous and mucocutaneous exposures to all categories of OR personnel. Reducing occupational blood exposures in the surgical setting will take time, sustained effort, and a multitude of changes, because of the complexity of the environment and the wide variety of products and instruments used in it. Exposure surveillance will play an important role in future prevention initiatives as a means for identifying needs, tracking progress, and as a foundation for clinical trials that will document the efficacy of safer products and procedures. It is an investment that will ultimately benefit not only OR personnel, but also the patients they care for.

References

1. Gerberding JL, Littell C, Tarkington A, Brown A, Schecter WP. Risk of exposure of surgical personnel to patients' blood during surgery at San Francisco General Hospital. *N Engl J Med.* 1990; 322:1788–93.
2. Panlilio AL, Foy DR, Edwards JR, Bell DM, Welch BA, Parrish CM, Culver DH, Lowry PW, Jarvis WR, Perlino CA. Blood contacts during surgical procedures. *JAMA.* 1991;265:1533–1537.
3. Wright JG, McGeer AJ, Chayatte D, Ransohoff DF. Mechanisms of glove tears and sharp injuries among surgical personnel. *JAMA.* 1991;266:1668–1671.
4. Popejoy SL, Fry DE. Blood contact and exposure in the operating room. *Surg Gynecol Obstet.* 1991;172:480–483.
5. Tokars JI, Bell DM, Culver DH, Marcus R, Mendelson MH, Sloan EP, Farber BF, Fligner D, Chamberland ME, McKibben PS, Martone WJ. Percutaneous injuries during surgical procedures. *JAMA.* 1992;267:2899–2904.
6. White MC, Lynch P. Blood contact and exposures among operating room personnel: a multicenter study. *Am J Infect Control.* 1993;21:243–8.
7. Jagger J, Detmer DE, Blackwell B, Litos M, Pearson P. Comparative injury risk among operating room, emergency department, and clinical laboratory personnel. *Infect Control Hosp Epidemiol.* 1994;15:345.
8. Johanet H, Antona D, Bouvet E. Risks of accidental exposure to blood in the operating room. Results of a multicenter prospective study. Groupe d'Etude sur les risques d'Exposition au Sang. *Ann Chir.* 1995; 49(5):403–410.
9. Tokars JI, Culver DH, Mendelson MH, Sloan EP, Farber BF, Flinger DJ, Chamberland ME, Marcus R, McKibben PS, Bell DM. Skin and mucous membrane contacts with blood during surgical procedures: risk and prevention. *Infect Control Hosp Epidemiol.* 1995;16(12):703–711.
10. Centers for Disease Control and Prevention. *HIV/AIDS Surv Rep.* 1998;9(2):21.
11. Busby J. Through the valley of many shadows: HIV infected physicians. *Tex Med.* 1991;87:36–46.
12. Ippolito G. and the Studio Italiano Rischio Occupazionale da HIV (SIROH). Scalpel injury and HIV infection in a surgeon. *Lancet* [letter].1996; 347:1042.
13. French Ministry of Health and Social Security. HIV transmission from an orthopedic surgeon to a patient. Press release. Paris; January 15, 1997.
14. Update: transmission of HIV infection during invasive dental procedures Florida. *MMWR.* 1991;40(23):377–381.
15. Harpaz R, Von Seidlein L, Averhoff FM, Tormey MP, Sinha SO, Kotsopoulou K, Lanbert SB, Robertson BH, Cherry JD, Shapiro CN. Transmission of hepatitis B virus to multiple patients from a surgeon without evidence of inadequate infection control. *N Engl J Med.* 1996;334:549–554.
16. Heptonstall J. Transmission of hepatitis B to patients from four infected surgeons without hepatitis B e antigen. *N Engl J Med.* 1997;336:178–184.
17. Estaban JI, Gomez J, Martell M, Cabot B, Quer J, Camps J, Gonzolez A, Otero T, Moya A, Estaban R, Guardia J. Transmission of hepatitis C virus by a cardiac surgeon. *N Engl J Med.* 1996;224:555–60.
18. Centers for Disease Control and Prevention. Evaluation of blunt suture needles in preventing percutaneous injuries among healthcare workers during gynecologic surgical procedures–New York City, March 1993–June 1994. *MMWR.* 1997;46:25–29.

19. Montz FJ, Fowler JM, Farias-Eisner R, Nash TJ. Blunt needles in fascial closure. *Surg Gynecol Obstet.* 1991;174:147–148.
20. Stafford M, Uthayakumar S, Falder S, Thomas P, Jolly M, Smith JR. Techniques for reducing needlestick injury in surgical practice. *Infect Control Hosp Epidemiol.* 1994;15:350.
21. Jagger J, Cohen MC, Blackwell B. EPINet: a tool for surveillance and prevention of blood exposures in health care settings. In: Charney W, ed. *Essentials of Modern Hospital Safety* (vol. 3). Boca Raton, FL: Lewis Publishers/CRC Press, Inc.;1994:223–239.
22. Dean AG, Dean JA, Coulombier D, Brendel KA, Smith DC, Burton AH, Dicker RC, Sullivan K, Fagan RF, Arner TG. Epi Info, Version 6: a word processing, and statistics program for public health on IBM-compatible microcomputers. Centers for Disease Control and Prevention, Atlanta, Georgia, U.S.A., 1995.
23. Centers for Disease Control and Prevention. Update: provisional Public Health Service recommendations for chemoprophylaxis after occupational exposure to HIV. *MMWR.* 1996;45:468–472.
24. Ippolito G, Puro V, Petrosillo N, Pugliese G, Wispelwey, B, Tereskerz PM, Bentley B, Jagger J. *Prevention, Management and Chemoprophylaxis of Occupational Exposure to HIV.* Charlottesville, VA: International healthcare worker Safety Center, University of Virginia;1997:23.
25. Gioannini P, Sinicco A, Cariti G, Lucchini A, Paggi G, Giachino O. HIV infection acquired by a nurse. *Eur J Epidemiol.* 1988;4:119–120.
26. Sartori M, La Terra G, Aglietta M, Manzin A, Navino C, Verzetti G. Transmission of hepatitis C via blood splash into conjunctiva. *Scand J Infect Dis.* 1993;25:270-271.
27. Bentley M. Blood and body fluid exposures to healthcare workers' eyes while wearing faceshields or goggles. *Adv Exposure Prev.* 1996;2(4):9.
28. Jagger J. Arnold WP. Blood salvage machines cause blood exposures to operating room personnel. *Adv Exposure Prev.* 1995;1(2):3.

Acknowledgements

We would like to recognize the invaluable contributions of all the OR nurses in each hospital providing data to this study. Every completed data collection form entered into the database was a product of their efforts. We especially recognize the dedication and hard work of the study coordinators in each institution. They are: Baptist Hospital of Miami, Catherine Moses, R.N.; The Queen's Medical Center of Honolulu, Susan Slavish, R.N., M.P.H.; Tucson Veteran's Administration Hospital, Suzanne Pear, R.N., M.S.; University of Pennsylvania Medical Center, Philadelphia, Dietra Evans, R.N.; University of Texas Medical Branch, Galveston, Julie Callender, R.N., M.S.N.; University of Virginia Health Sciences Center, Charlottesville, Kathryn Stacy, R.N., M.S.N.

We also acknowledge the contributions of Julie Callender, R.N., M.S.N., and Carol Applegeet, R.N., M.S.N., who assisted in the review and revision of data collection forms, and who provided on-site training to participating hospitals; and of David Edmiston who provided administrative support from AORN headquarters throughout the study.

We would like to thank Becton Dickinson and Company of Franklin Lakes, New Jersey, for their financial support of this study, and Craig Newman, B-D project officer, for his commitment to its success.

Injuries from Huber Needles

by Melanie Bentley, B.S.

Vol. 3, no. 6, 1998

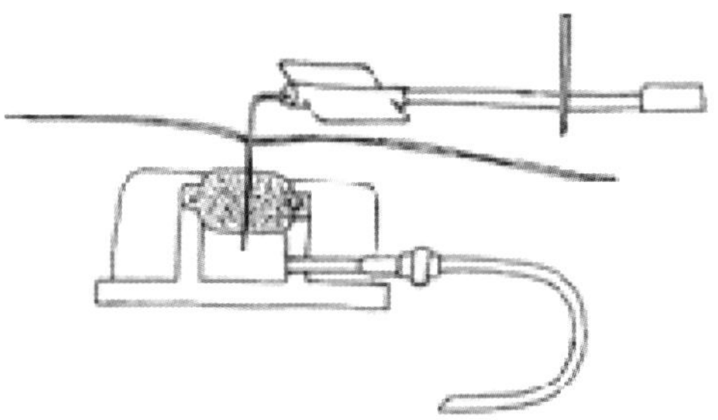

Huber needles, which have a deflective, non-coring tip, are used to administer antibiotics or chemotherapy to implanted I.V. ports. Since they are used for venous access, injuries from huber needles are high risk for bloodborne pathogen transmission. After several AEP readers noted frequent injuries with this device, we searched the EPINet database for more information.

From 1993 through 1997, a total of 84 hospitals participated in the EPINet data-sharing network, providing data to the International Healthcare Worker Safety Center. During that time, 55 injuries from huber needles were reported. With 6 injuries in 1993, 10 in 1994, 8 in 1995, 16 in 1996, and 15 in 1997, the number of huber needle injuries has increased.

All but two of the huber needle injuries were to nurses. The remaining injuries were to housekeepers/laundry workers.

Figure 1 shows the location of huber needle injuries. The majority (69%) of huber needle injuries occurred in patients' rooms. Thirteen percent of huber needle injuries occurred in the home care setting, where they are frequently used. Additionally, 4% of injuries occurred in the ICU/CCU, 4% in the emergency department, and 4% in treatment rooms. Two percent of injuries occurred in the operating room, 2% in outpatient clinics, 2% in radiology, and 2% outside patient rooms.

As seen in **Figure 2,** nearly half (47%) of the huber needle injuries occurred while withdrawing the needle from the I.V. port. Rebound injuries are common with huber needles because one needs to pull with a great deal of force to overcome the resistance of the rubber septum. Since the non-dominant hand is used to secure the implanted port during huber needle withdrawal, it is often stuck on the rebound.

Figure 1. Location of Huber Needle Injuries
EPINet • 84 hospitals • 5 years • 55 cases

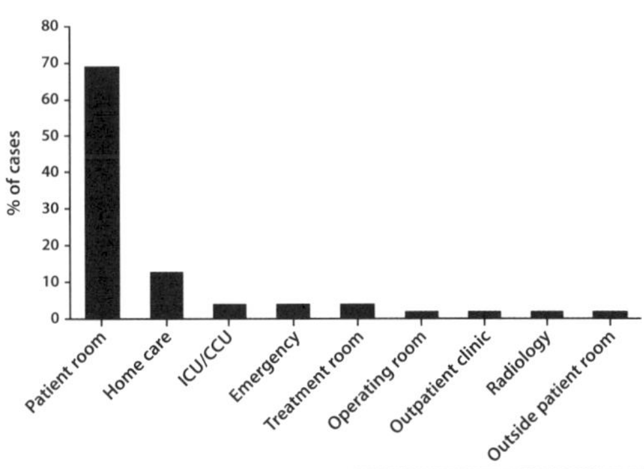

International Healthcare Worker Safety Center, University of Virginia

Figure 2. Mechanism of Huber Needle Injuries
EPINet • 84 hospitals • 5 years • 55 cases

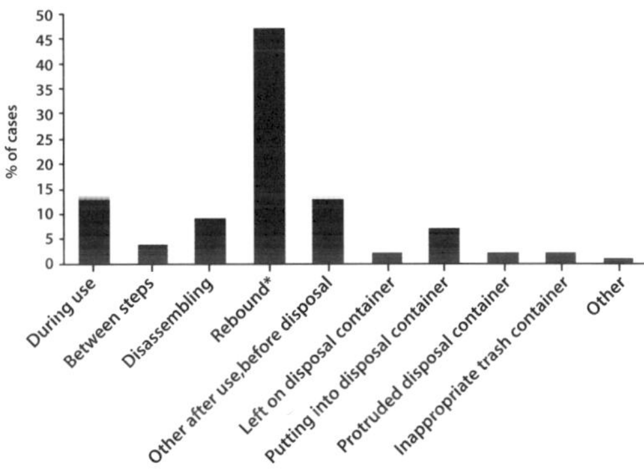

**Resulting from withdrawal of needle from resistant material.*

International Healthcare Worker Safety Center, University of Virginia

Improvements in huber needle design are needed in order to address this cause of high risk injuries that has been overlooked to date.

Safe Disposal of Safety Devices

by Janine Jagger, M.P.H., Ph.D., and Melanie Bentley, B.S.

Vol. 4, no. 2, 1999

The transition from conventional needles to safer needle devices is well on its way in the United States. Devices that either eliminate unnecessary needles or provide needle shielding, retracting or blunting features to prevent needlesticks are in widespread use across the country. Current industry estimates indicate that needleless or protective needle designs account for about 63% of I.V. infusion sets, 28% of butterfly-type needles, 25% of I.V. catheters and less than 10% of vacuum tube phlebotomy needles and disposable syringes.

Much work still needs to be done to quantify the reduction in injuries from this new technology; however, some data documenting reductions in injuries with needle safety technology exist. Three blood-drawing devices with safety features were evaluated in a clinical trial conducted by the Centers for Disease Control and Prevention.[1] All three devices reduced injury rates when compared to conventional devices—a safety butterfly by 25%, a self-blunting phlebotomy needle by 76%, and a hinged-cap phlebotomy needle by 66%. In another study, a shielded-stylet I.V. catheter reduced injury rates by 83% when compared to a conventional I.V. catheter.[2]

Needlestick reductions have also been linked to the implementation of needle-less or recessed-needle I.V. infusion sets. Data from the EPINet data-sharing network, including 84 hospitals over a five-year period, show that in 1993, 30% of needlesticks were caused by unnecessary needles used to access or connect I.V. lines. After the FDA issued a safety alert advising hospitals not to use hypodermic needles in I.V. lines, there was a progressive decline in the number of needlesticks caused by this type of needle. EPINet data shows that by 1997 only 3% of needlesticks were caused by unnecessary needles in I.V. lines, a 90% decline in injuries. These different sources of data show that impressive declines in needlestick injuries can be achieved by introducing protective devices.

As the implementation of safety devices progresses, spurred on by legislative initiatives in several states, further declines in needlesticks are likely. The anticipation of these reductions has raised many questions, among them whether safety devices will eliminate or reduce the need for sharps disposal containers, and whether safety devices can be disposed of in non-sharps waste containers such as red bags.

Safety device evaluations and surveillance data show that even with high rates of implementation of protective needle devices, needlestick injuries will not be completely eliminated. EPINet data from 84 hospitals from 1993-1997 show that of all injuries from hollow-bore needles, conventional needles accounted for 93.7% of injuries, while safety devices (with needle shielding, retracting or blunting features) accounted for 6.3% of injuries **(Figure 1).** Safety devices in this small fraction of injuries included butterfly-type needles, phlebotomy needles, I.V. catheters, disposable syringes, and blood gas syringes. It should be noted that these data do not reflect the effectiveness of specific safety devices, since the number of devices used in each category is not known. With even the best safety devices, the needle must be exposed briefly during use and a small fraction of injuries would be expected to occur. Additionally, if the protective feature must be activated by the user, then

Figure 1. Injuries from Hollow-bore Needles: Conventional vs. Safety Devices

EPINet • 1993–1997 • 84 hospitals

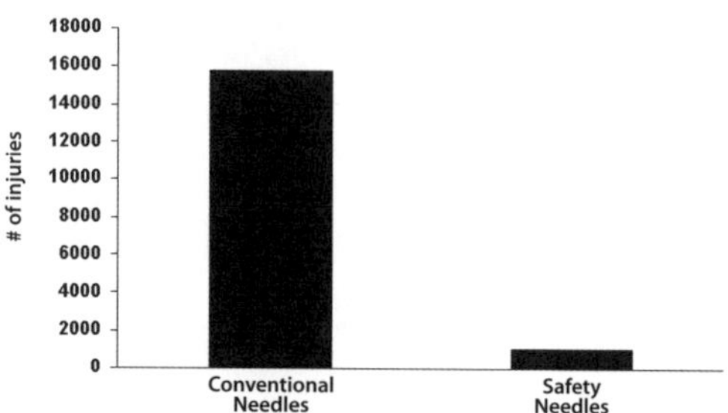

International Healthcare Worker Safety Center, University of Virginia

Figure 2. Injuries from Hollow-bore Needles: Mechanism of Safety Device Injuries

EPINet • 1993–1997 • 84 hospitals

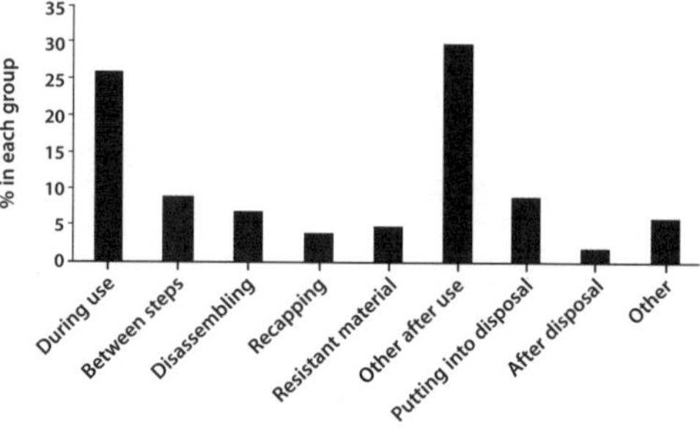

International Healthcare Worker Safety Center, University of Virginia

sometimes it will not be activated, and some needles will remain exposed during and after disposal.

Figure 2 shows how injuries occurred with safety devices. Twenty-nine percent of injuries occurred after use/before disposal and another 11% occurred during and after disposal, indicating that some needles remained exposed and continued to pose a hazard after use. Thus, the safe containment of these needles after use remains a need.

An additional consideration when weighing the pros and cons of altering disposal protocols is that in the hospital setting a variety of sharp devices are in use. Even with the increase in use of protective needle devices, sharps disposal containers will still be needed to safely contain the full array of sharp devices that must be disposed of. EPINet data from 1997 (the most recent year) shows that injuries occurring in patient rooms were classified into 26 different device categories; 24% of the injuries were not in categories clearly addressed by recent needlestick legislation. Therefore, well-designed sharps disposal containers in acute-care settings will continue to be needed even with the increased implementation of safety devices.

In conclusion, these data demonstrate that although injuries from protective needle devices are infrequent, the use of such devices does not completely eliminate health-care workers' exposure to contaminated needles. Furthermore, the wide variety of sharp devices used in clinical settings include many items for which protective alternatives either do not yet exist or are not widely available. Until needles and sharp devices are entirely eliminated from the clinical environment, which is not likely in the near future, the need for well-designed, puncture-resistant sharps disposal containers will remain.

References

1. Centers for Disease Control and Prevention. Evaluation of safety devices for preventing percutaneous injuries among health-care workers during phlebotomy procedures–Minneapolis-St.Paul, New York City, and San Francisco, 1993-1995. *MMWR.* 1997;46:21–25.
2. Jagger J, Bentley M. Injuries from vascular access devices: high risk and preventable. *J Intraven Nurs.* 1997;20(6S):S33–S39.

Exposures from Segment Sampling in Blood Banks

Melanie Bentley, B.S.

Vol. 4, no. 2, 1999

"Segment sampling" can pose a risk of blood exposure for blood bank workers. After donor blood is drawn, routine tests are performed for ABO and Rh typing and detection of specific pathogens. Compatibility testing may also be performed on the blood sample to insure there are no matching problems. The current process for obtaining a sample for testing puts the healthcare worker at potential risk for exposure to bloodborne pathogens.

After the blood is drawn and the needle is removed from the donor, the blood-filled tubing attached to the blood bag is heat-sealed at intervals along the tube to create a series of separate segments. When needed, a segment is cut from the end of the tubing. A blood sample is then obtained by cutting through the segment with scissors and squeezing blood from the open segment into a glass or plastic tube.

The process of cutting segments with scissors presents several possible hazards to the healthcare worker. First, it can cause blood to splatter on work surfaces and on healthcare workers. There is also the potential for percutaneous injuries from sharp pointed scissors.

According to the American Association of Blood Banks, approximately 2%-10% of donor blood is rejected due to positive tests for HTLV-I, HIV, HBV, HCV, or syphilis. Although there are false positives among the rejected units, such statistics indicate that an exposure to donor blood is potentially hazardous.

Several products currently on the market are designed to provide a safer alternative for accessing blood in segments **(Figure 1).** These products utilize a recessed cannula or shielded blade to pierce the segment, thus eliminating the need to cut through the segment with scissors.

Below are some descriptions of how blood segment exposures occurred, drawn from databases of the International Healthcare Worker Safety Center.

- A medical technician in a blood bank was cutting a blood sample segment to put blood in a tube. When she cut the segment it leaked on her ungloved right hand. The blood was in contact with her skin for less than five minutes.
- A medical technician in a blood bank was cutting a blood segment and blood squirted on the index fingers and thumbs of both hands. The blood remained in contact with her intact skin for less than five minutes.
- A clinical laboratory worker in a blood bank was using scissors to cut labels and segments when she sustained a superficial injury to her left index finger from the scissors.

Figure 1. Examples of Segment Sampling Devices

Safety Segment Slitter by Innovative Laboratory Acrylics

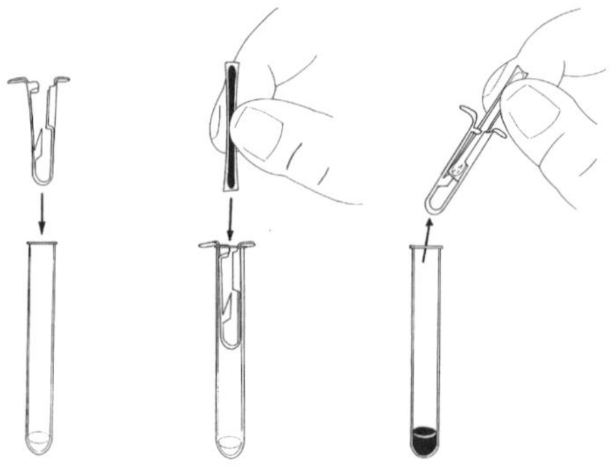

Hematype Segment Device by Baxter

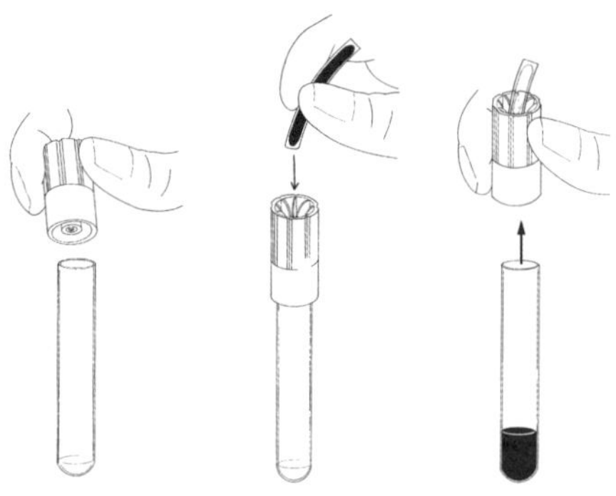

Percutaneous Injuries in Outpatient Settings and Physicians' Offices

by Janine Jagger, M.P.H., Ph.D., and Melanie Bentley, B.S.

Vol. 4, no. 6, 1999

Drew Stevenson, U.Va. Health System

Little has been published on percutaneous injury risks to healthcare workers employed in physicians' offices and outpatient settings. And when it comes to implementing regulations that protect the health and safety of employees in these settings, the maxim "no data, no problem" applies in full force. Healthcare workers in office settings may be more vulnerable than others if there are no data to support their need for protective measures.

Although as many as half of all healthcare workers are employed in non-hospital settings, most published reports on percutaneous injuries to healthcare workers describe exposure risks in hospitals. The main reason for this is the convenient access to hospital data. Hospitals employ large numbers of healthcare workers who are required to report at-risk injuries and blood exposures to a central location, usually the employee health department, where a surveillance database is compiled. Surveys have shown that percutaneous injury rates for different healthcare worker groups vary from a low of 0.15 injuries per year to a high of more than one injury per year for surgeons, the group at highest risk of injury. In a hypothetical hospital that employs 1,000 healthcare workers who each have an average annual risk of 0.2 percutaneous injuries, 200 injuries would occur each year. The accumulated injury data would be sufficient to reveal the risk patterns of the different clinical settings and job categories in that hospital, and create an incentive for the hospital to address these risks and reduce the number of injuries.

However, the situation is different in physicians' offices, where there are relatively few employees. For example, in a practice employing 10 full-time healthcare workers, each with an annual risk of 0.2 percutaneous injuries, only one injury would occur every five years, on average. Workers in this kind of setting may erroneously believe that they are at low occupational risk because they rarely observe percutaneous injuries. Such a viewpoint has been reflected in arguments of small employers seeking exemption from safety regulations or laws because there were few or no recent injuries in their facilities. However, this perspective overlooks the fact that the true risk of exposure in a small office or outpatient setting is not determined by the number of injuries per practice, but rather by the injury rate per device or per at-risk procedure performed. Research on this topic has been hampered by the difficulty in collecting small amounts of data from a large number of clinical sites.

Assessment of the occupational risk of healthcare workers in outpatient settings should take into consideration (1) whether they perform procedures associated with the risk of bloodborne pathogen transmission, and (2) whether their rate of percutaneous injuries is different from that of hospital workers performing the same procedures.

The Exposure Prevention Information Network (EPINet) database provides some information on the occupational risks encountered in physicians' offices and outpatient clinics. From 1993 through 1998, 84 hospitals provided EPINet data on percutaneous injuries to the International Healthcare Worker Safety Center. Most of the data were from in-patient settings, but in some instances physicians' offices and outpatient clinics were affiliated with EPINet hospitals, and percutaneous injuries reported from those sites were entered into the EPINet database. Figures 1, 2 and 3 (next page) compare the characteristics of percutaneous injuries occurring in in-patient settings with those occurring in physicians' offices and outpatient clinics.

There were 17,825 percutaneous injuries reported by hospital workers during the six-year interval. Workers in physicians' offices and outpatient clinics reported 925 injuries during the same period. **Figure 1** shows that the distribution of job categories for workers reporting injuries was similar for office and in-patient settings. There were, proportionately, slightly fewer nurses, and proportionately more physicians, phlebotomists, and attendants reporting injuries from office settings.

More pertinent to the question of risk is the type of devices that cause injury and the procedures being performed. Blood-filled, hollow-bore devices, such as those used for blood drawing and vascular access, carry a higher infection transmission rate than needles used for injections. **Figure 2** compares the patterns of device risks in the two settings. In office settings, there were proportionately more injuries from syringes, phlebotomy needles, butterflies and lancets. In hospitals, there were proportionately more injuries from scalpel blades, suture needles, I.V. catheter stylets, and needles accessing I.V. ports.

Figure 3 compares the procedures associated with the devices causing injuries. This comparison shows that, in office settings, proportionately more injuries were associated with injections and blood drawing, whereas in hospital settings, proportionately more injuries were associated with suturing, cutting, and intravenous access. Overall, 29.0% of injuries in office settings were "high risk," that is, associated with blood drawing or intravenous access. The percentage of high-risk injuries in hospitals was lower, with 24.4% of injuries associated with blood drawing or intravenous access.

Annual percutaneous injury rates for the different job categories could not be calculated or compared using EPINet data. However, a previous survey of phlebotomists conducted in 1994 and reported in AEP (vol. 1, no. 1, pp. 6–7) provides data on percutaneous injury risks to phlebotomists working in hospitals compared with those working in non-hospital settings. When adjusted to a standard 40-hour work week, 76 phlebotomists working in hospitals had an annual needlestick rate of 0.33, while 64 phlebotomists working in non-hospital settings had a higher adjusted needlestick rate of 0.40 injuries per year.

These figures show that when sufficient data are compiled on percutaneous injuries sustained in physicians' offices and outpatient clinics and compared with injuries occurring in hospital settings, a similar spectrum of risks can be seen in both settings. And when the causes of the injuries are compared, workers in physicians' offices and outpatient settings have a proportionately higher frequency of injuries from blood-filled needles—those most likely to transmit bloodborne pathogens. Currently, there are no data that suggest that the risks of occupational blood exposure are less for healthcare workers in non-hospital settings than for those employed by hospitals.

In conclusion, wherever healthcare workers must handle sharp medical devices, including injection equipment, blood-drawing devices, and vascular access needles, they are in need of the protection afforded by the safest technology, regardless of clinical setting.

Figure 1. Job Category of Injured Workers: Physicians' Offices vs. Hospitals

EPINet • 87 hospitals • 1993–1998
Physicians' Offices/Outpatient Clinics: 925 injuries
Hospitals: 17,825 injuries

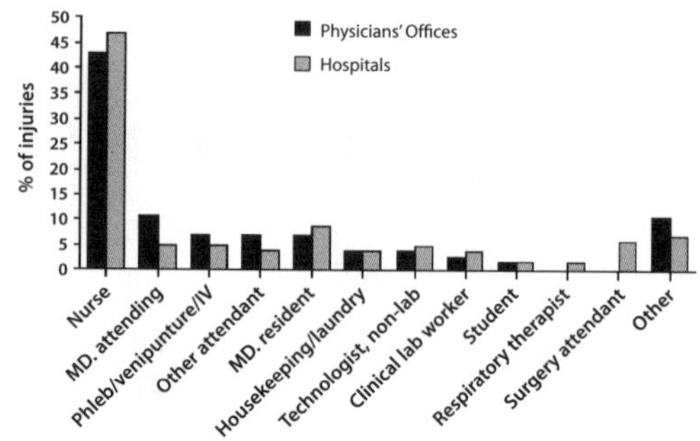

International Healthcare Worker Safety Center, University of Virginia

Figure 2. Device Causing Injury: Physicians' Offices vs. Hospitals

EPINet • 87 hospitals • 1993–1998
Physicians' Offices/Outpatient Clinics: 925 injuries
Hospitals: 17,825 injuries

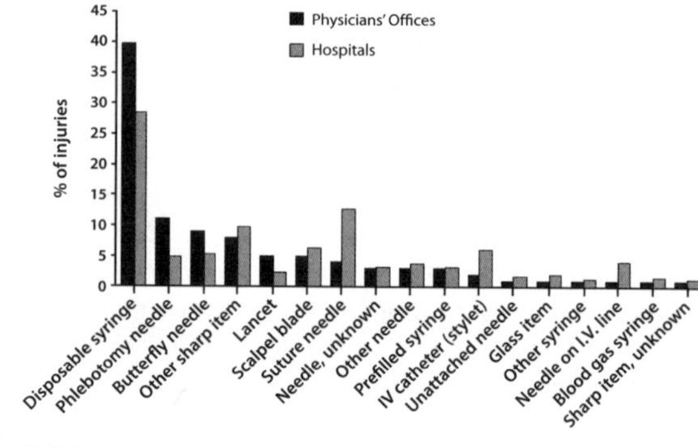

International Healthcare Worker Safety Center, University of Virginia

Figure 3. Original Purpose of Device Causing Injury: Physicians' Offices vs. Hospitals

EPINet • 87 hospitals • 1993–1998
Physicians' Offices/Outpatient Clinics: 925 injuries
Hospitals: 17,825 injuries

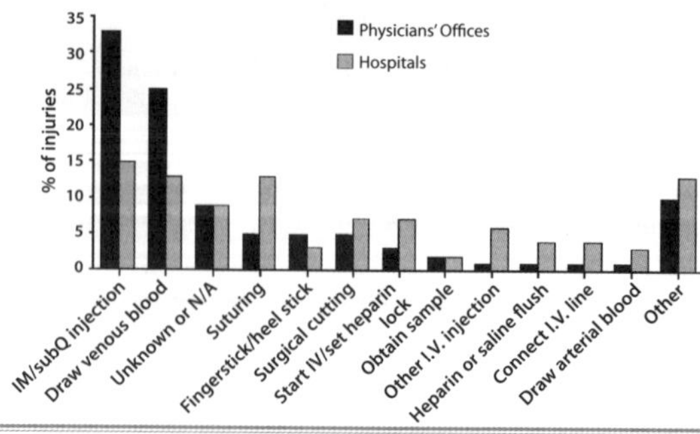

International Healthcare Worker Safety Center, University of Virginia

Drawing Venous Blood With Syringes: A Risky Use of Injection Equipment

Janine Jagger, M.P.H., Ph.D., and Ginger Parker, M.B.A.

Vol. 5, no. 3, 2000

Needles used for blood drawing have long been recognized as presenting a high risk of bloodborne pathogen transmission following needlestick injuries. Among 46 healthcare workers with documented, occupationally acquired HIV, 20 cases (43%) were associated with injuries from blood drawing needles. Conversely, needles used for intramuscular and subcutaneous injections were associated with only one case (2%) of occupational HIV infection[1], despite the fact that injections are administered much more frequently than blood is drawn. While the average transmission rate following percutaneous exposure to HIV has been estimated at .3%[2], an Italian study of 1,610 HIV-exposed healthcare workers showed a .55% (2/365) transmission rate for exposures involving blood-filled needles, and no transmissions (0/840) from exposures involving non-blood filled needles.[3]

Disposable syringes are unique among sharp medical devices in that they are multi-purpose. Their most common use is for subcutaneous or intramuscular injection of medication. They are also used as tools for manipulating body fluid specimens in clinical laboratories and for mixing drugs in the pharmacy. Among their most hazardous uses, however, is venous blood drawing. In 1998, the national EPINet database, including 52 hospitals, showed disposable syringes to be the device causing the most reported percutaneous injuries, accounting for 30% (950/3,180) of all reported injuries. It also showed that 19% (179/950) of syringe injuries involved syringes that had been used for venous blood drawing. Syringes are unique, therefore, in that they can be associated with injuries having either the lowest or the highest risk of bloodborne pathogen transmission, depending on the purpose for which they are used.

Venous blood drawing is a procedure for which a variety of devices is used. **Figure 1** shows that there were 548 injuries related to venous blood drawing in the national EPINet database in 1998. Of those, 38% were associated with winged steel needles, 31% with vacuum tube phlebotomy devices, and 25% with disposable syringes. These data show that drawing venous blood with disposable syringes remains a common practice that results in a significant number of needlesticks.

Since there are better alternatives, why do healthcare workers use syringes for venous blood drawing? The reasons vary. In some cases, healthcare workers prefer to control the vacuum during blood drawing if patients have difficult veins. In many cases, it is just a question of habit; it is what the healthcare worker has always used or it was the device most readily available when blood drawing needed to be performed. With that choice comes a particular set of risks.

Figure 2 compares the way needlesticks occur with syringes used for venous blood drawing versus syringes used for intramuscular or intravenous injections. These data highlight one of the specific hazards caused by drawing blood into syringes: the requirement of transferring the blood from the syringe into a specimen container. The figure shows that injuries resulting from pulling a needle out of a resistant substance such as rubber are uniquely associated with drawing blood into syringes. Once blood is drawn into a syringe it is often injected into a vacuum tube

Figure 1. Devices Causing injuries During Venous Blood Drawing

EPINet • 52 hospitals • 1998 • 548 cases

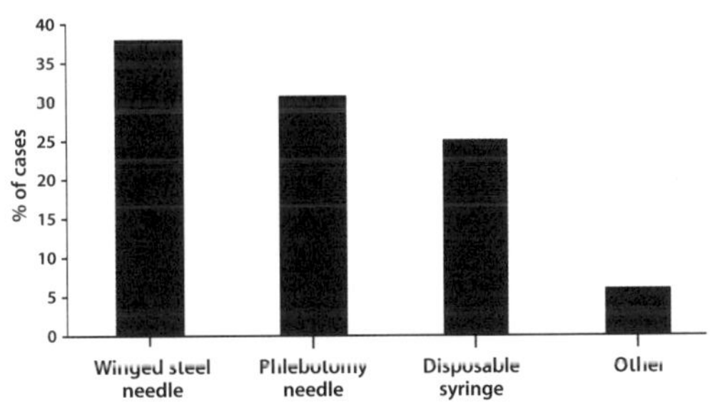

International Healthcare Worker Safety Center, University of Virginia

Figure 2. How Injuries Occurred with Syringes Used for Injections vs. Venous Blood Drawing

EPINet • 52 hospitals • 1998

451 cases (injections), 138 cases (blood drawing)

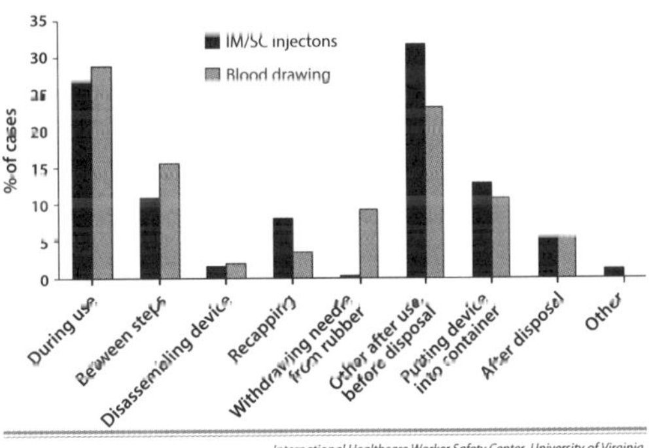

International Healthcare Worker Safety Center, University of Virginia

Figure 3. Once blood is drawn into a syringe it is often injected into a vacuum tube through its rubber stopper.

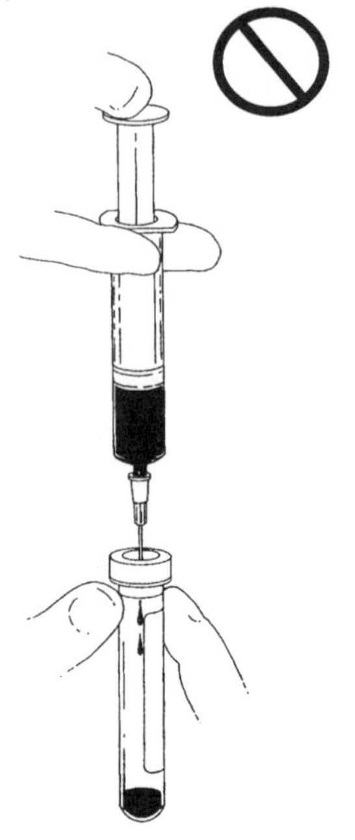

through its rubber stopper. (See Figure 3). This involves an additional manipulation of the needle with an extra set of risks. First, the needle must not miss the rubber stopper, which is a narrow target. If it misses, it is likely to stick the hand holding the tube. But even if the needle is inserted into the tube without incident there remains another risky hurdle: it must be removed. Pulling a needle from a resistant substance can result in a rebound needlestick, when the needle suddenly disengages and the hand holding the needle lurches forward in a reflex motion and sticks the opposing hand with the needle. Another adverse event can occur if the injection of blood into the tube overcomes the vacuum. The stopper can pop off and splash the worker with blood.

Syringes with needles are sometimes inappropriately used to draw blood from rubber ports on intravenous or arterial lines. Again the problem arises of pulling the needle out of the port against resistance and risking a rebound needlestick. Also, there is the additional exposure risk of transferring blood to a specimen container.

Another risk that is unique to drawing blood into syringes is that of accidental blood injections. These rare incidents inoculate healthcare workers with much larger quantities of blood than needlesticks. Some documented cases of occupational transmission of HIV were the result of accidental blood injections. In one case, a syringe full of blood was left on a table. A healthcare worker inadvertently backed into the table, pushing against the syringe. The syringe was pushed back against a rigid surface which caused the plunger to depress, injecting the healthcare worker with blood. This type of incident, which can only happen with syringes, carries a much higher risk of pathogen transmission.

A recent twist on the inappropriate use of syringes is the use of safety-engineered syringes (with shielding or retracting features) for venous blood drawing. **Figure 4** shows that both conventional syringes and safety syringes were associated with injuries during venous blood drawing procedures. Unfortunately, "safety syringes" are unlikely to reduce the hazards of venous blood drawing. Their use for this purpose defeats the benefit of the safety design. Since the protective feature can only be put in place after the blood has been injected from the syringe into a specimen container, the user has already been exposed to the additional risks described with conventional syringes before the safety feature can be activated.

Figure 4 also shows that both conventional and safety syringes caused injuries when they were inappropriately used for injections into intravenous ports and for intravenous flushes, both procedures for which needles are not necessary. In fact, 28.7% of injuries from conventional syringes and 38.7% of injuries from safety-engineered syringes were associated with inappropriate uses of a syringe. These findings emphasize the need for education in limiting the use of injection equipment, whether conventional or safety-engineered, to appropriate applications.

In conclusion, the practice of drawing venous blood into syringes should be reduced to a minimum. Devices that draw blood directly into vacuum tubes or other specimen containers should be preferentially employed. The "needle end" of the blood-drawing device should have an integrated safety mechanism such as a needle-shielding or blunting feature. Such devices have been shown to reduce injury rates from phlebotomy devices by 25% to 76% in a CDC study.[4] If, in specific clinical settings, there is no other alter-

Figure 4. Injuries from Conventional vs. Safety Syringes: Original Purpose of Devices

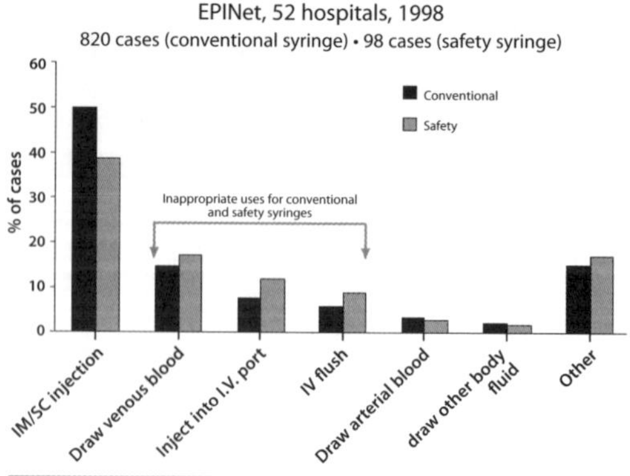

International Healthcare Worker Safety Center, University of Virginia

native to syringes for venous blood drawing, then large-volume syringes with a sliding shield should be employed. The safety shield should have a larger diameter than the vacuum tube into which the blood will be injected. The safety shield should be locked in place over the needle and the tube inserted into the shield for the injection of blood.

References

1. *Adv Exposure Prev.* 1998; 3(3): 33 (table).
2. Henderson DK. HIV-1 in the health care setting. In: Mandell GL, Bennett JE, Dolin R, eds. *Principles and Practice of Infectious Diseases.* New York, NY: Churchill Livingstone;1995: 2632–2656.
3. Ippolito G, Puro V, Petrosillo N, Pugliese G, Wispelwey B, Tereskerz PM, Bentley M, Jagger J. *Prevention, Management and Chemoprophylaxis of Occupational Exposure to HIV.* Charlottesville, VA: International Healthcare Worker Safety Center, 1997, p. 13 [table 6].
4. Centers for Disease Control and Prevention. Evaluation of safety devices for preventing percutaneous injuries among health-care workers during phlebotomy procedures—Minneapolis-St. Paul, New York City, and San Francisco, 1993-1995. *MMWR.* 1997;46;21–2.

Percutaneous Injuries in Home Healthcare Settings

by Jane Perry, M.A., and Ginger Parker, M.B.A.

Vol. 5, no. 3, 2000

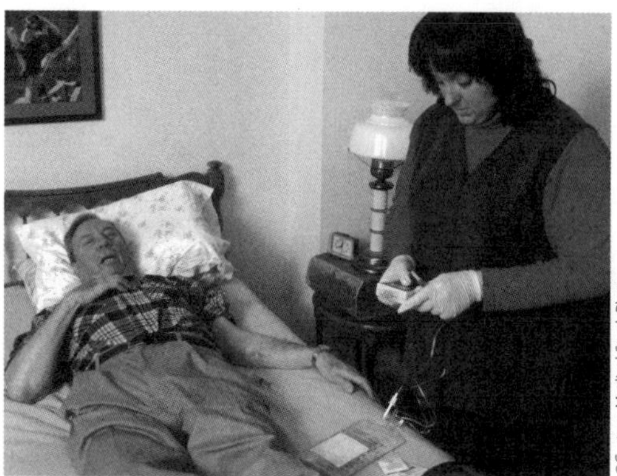

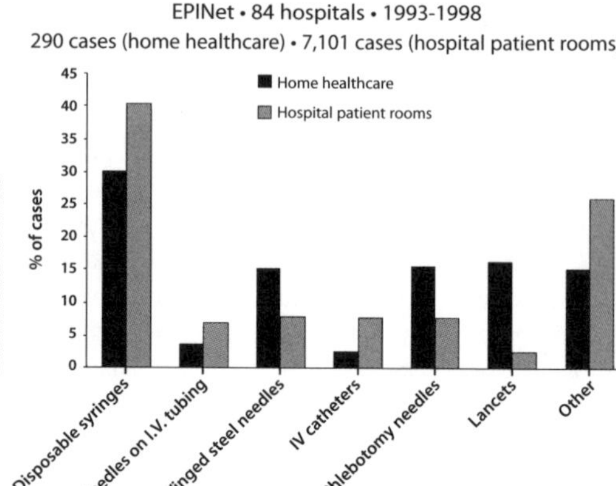

Figure 1. **Devices Causing Injury in Home Healthcare vs. Hospital Patient Rooms**

International Healthcare Worker Safety Center, University of Virginia

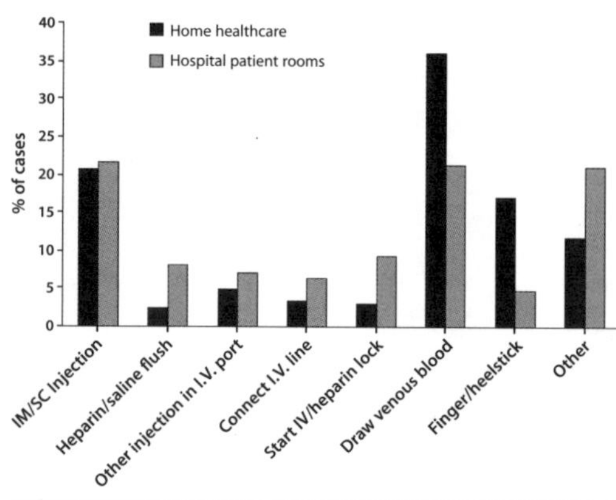

Figure 2. **Original Procedure of Devices Causing Injuries in Home Healthcare vs. Hospital Patient Rooms**

International Healthcare Worker Safety Center, University of Virginia

Home healthcare has been one of the fastest-growing sectors in the healthcare industry, but the prevention of occupational exposures in this setting has not received the attention it deserves. Of the more than 8 million U.S. healthcare workers who work in hospitals and other healthcare settings[1], between 650,000 and 850,000 are employed in home healthcare.[2] There are few data and no national estimates on the number of needlestick injuries that occur each year in home healthcare settings. Perhaps because of this lack of documentation, some recent state bills on needlestick prevention, such as the one passed in August 2000 in Massachusetts, overlook home healthcare and other non-hospital settings. But does the absence of data mean absence of risk?

From 1993 through 1998, 84 hospitals provided EPINet data on percutaneous injuries to the International Healthcare Worker Safety Center. Most of the data were from in-patient settings, but in some instances, home healthcare agencies were associated with EPINet hospitals, and percutaneous injuries reported from those sites were entered into the EPINet database. Figures 1, 2 and 3 compare the characteristics of percutaneous injuries occurring in hospital patient rooms with those occurring in home healthcare settings.

There were 7,101 percutaneous injuries reported by hospital workers in patient rooms during the six-year interval. Workers in home healthcare settings reported 290 injuries during the same time period. Nurses sustained the overwhelming majority of injuries—87%—in the home care setting, compared to 65% for nurses in patient rooms.

Figure 1 compares the pattern of device risks in the two settings. In home healthcare settings, there are proportionately more injuries from winged steel needles, phlebotomy needles and lancets. In patient rooms, there are proportionately more injuries from syringes, I.V. catheters and needles on I.V. tubing.

Figure 2 compares the procedures associated with the devices causing injuries. This comparison shows that in home healthcare settings, proportionately more injuries were associated with blood drawing, including finger and heelsticks, whereas in patient rooms, proportionately more injuries were associated with intravenous access—starting I.V.s and connecting I.V. lines and doing heparin/saline

Figure 3. Mechanism of Injuries Occurring in Home Healthcare vs. Hospital Patient Rooms

EPINet • 84 hospitals • 1993-1998

290 cases (home healthcare) • 7,101 cases (hospital patient rooms)

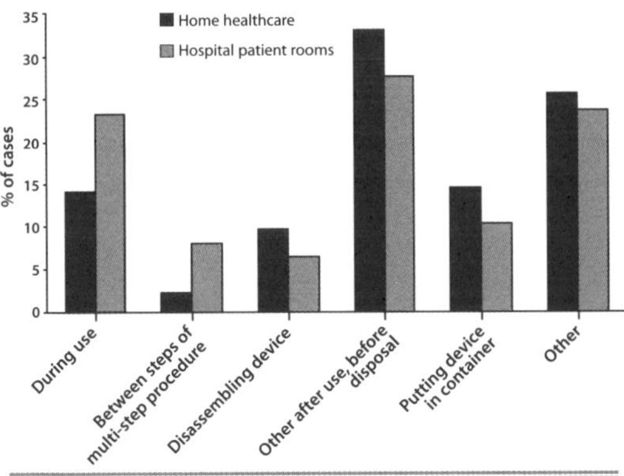

International Healthcare Worker Safety Center, University of Virginia

flushes and other injections involving I.V. ports. Overall, 40% of injuries in home healthcare settings were "high risk," that is, associated with blood drawing or intravenous access, compared with 34% for injuries in patient rooms.

Figure 3 compares the mechanism of injuries for home healthcare settings and patient rooms. Where 38% of injuries occurred either after use or during disposal in patient rooms, 48% of injuries occurred during these steps in home settings. This may reflect the more unpredictable environment of home care settings compared to patient rooms, and problems with inadequate, overfilled or nonexistent disposal containers.

These data show that home healthcare workers have a proportionately higher frequency of injuries from blood-filled needles—those most likely to transmit bloodborne pathogens—compared with hospital healthcare workers in patient rooms. While home healthcare workers have as great or greater need of protection from sharps injuries as hospital workers, the home health sector is less regulated by OSHA than the hospital sector. When OSHA's bloodborne pathogens standard was promulgated in 1991, both the dental and the home healthcare industries sought exemptions from the standard. In a court ruling, American Dental Assn. v Martin, the dental industry was denied an exemption, but one was allowed for home healthcare. The 1999 revised compliance directive for the bloodborne pathogens standard states that, "In implementing this decision, OSHA determined that the employer will not be held responsible for the following site-specific violations: housekeeping requirements, such as the maintenance of a clean and sanitary worksite and the handling and disposal of regulated waste; ensuring the use of personal protective equipment; and *ensuring that work practices are followed... and ensuring the use of engineering controls*" [emphasis added]. The reasoning behind the court's decision, of course, was that private homes cannot be regulated by OSHA. However, OSHA *does* state that the employer is responsible for all non-site-specific requirements of the standard, such as maintaining an exposure control plan and providing personal protective equipment and engineering controls to employees, as well as postexposure evaluation and follow-up.

Because of the unpredictable and sometimes chaotic environment in which home healthcare is provided, home health employers should ensure that workers have sharp devices with engineered sharps injury protection, especially safety blood-drawing and I.V. access equipment. By providing sharps that are automatically covered after use, both workers and home healthcare patients and their families will be protected, since these patients are often responsible for disposing of their own medical waste, and since proper, puncture-resistant disposal containers are sometimes in short supply.

References

1. National Institute for Occupational Safety and Health, U.S. Department of Health and Human Services. NIOSH Alert: Preventing Needlestick Injuries in Health Care Settings. (DHHS [NIOSH] Publication No. 2000 108.) November 1999, p. 2.
2. Sources: U.S. Department of Labor, Bureau of Labor Statistics, 1998; Service Employees International Union, personal communication.

Percutaneous Injuries in the Dialysis Setting

Jane Perry, M.A., Ginger Parker, M.B.A., and Janine Jagger, M.P.H., Ph.D.

Vol. 5, no. 5, 2001

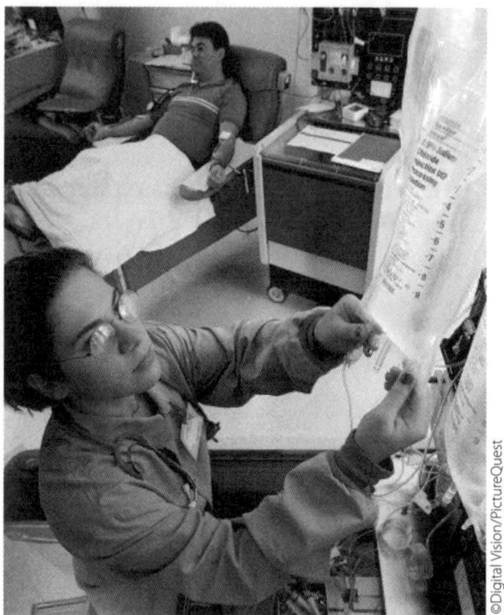
©Digital Vision/PictureQuest

In 1999, during the comment period for California OSHA's revised bloodborne pathogens standard mandating the use of safety-engineered devices in healthcare facilities, a state dialysis association sought an exemption from the safety device requirement, arguing that "there is a very low rate of exposure incidents in the dialysis industry."

Data on percutaneous injuries from the multi-hospital EPINet network, coordinated by the International Healthcare Worker Safety Center at the University of Virginia, show that the types of exposures sustained by dialysis workers are more likely to involve large-bore, blood-filled needles, and are therefore more apt to transmit bloodborne pathogens, than exposures sustained by non-dialysis workers. Working in a blood-intensive environment, dialysis personnel also treat patients who are at higher risk of being infected with bloodborne pathogens, particularly hepatitis C, than the general population.

National statistics are not available on how many needlesticks occur in dialysis settings each year, nor for the total number of dialysis workers infected with a bloodborne pathogen, but the Centers for Disease Control and Prevention (CDC) has identified four dialysis workers among healthcare workers occupationally infected with human immunodeficiency virus (HIV). While there are safety devices on the market to address the types of high-risk exposures that dialysis workers experience, dialysis facilities have been significantly behind acute-care facilities in adopting this technology.

To understand more about percutaneous injuries in dialysis settings, we looked at six years of data (1993-1998) from the EPINet network, with 84 hospitals contributing data. During this period, there were 119 sharp-object injuries sustained by healthcare workers in dialysis settings.

Figure 1, showing the job categories of injured dialysis workers, indicates that nurses sustained the majority of injuries (58%). Not only do nurses in dialysis facilities sustain more injuries than any other workers in that setting, they also have a higher rate of injury compared to nurses overall. At the American Nephrology Nurses Association's annual symposium, the International Healthcare Worker Safety Center presented data showing that the percutaneous injury (PI) rate for dialysis nurses was more than twice that for all other types of nurses: .39 PI/year for dialysis nurses compared to .17 PI/year for all other nurses.[1]

Technicians sustained 23% of PIs; of this group, about one-third identified themselves specifically as dialysis technicians (some of the others may also have been dialysis techs, but did not label themselves as such). Jobs in the "other" category (9%) included renal therapist, respiratory therapist, lab specialist, certified nurse assistant, medical student, and attendant.

Figure 2 shows the devices causing injuries in dialysis facilities. Syringes were responsible for the largest number of injuries (46%); of these, only about one-fourth (or 13% of injuries overall) involved syringes used for blood-drawing, which would be considered "high-risk" for bloodborne pathogen transmission. (High-risk injuries are those involving blood-filled needles.) Dialysis fistula needles caused the most high-risk injuries (20%); of these, none involved a safety-engineered device. Fistula needles are large bore (typically 14- and 15-gauge), so percutaneous injuries from these needles involve a larger inoculum of blood—and a greater risk of bloodborne pathogen transmission—than smaller-gauge needles. Other blood-filled, hollow-bore devices causing dialysis injuries were needles on I.V. tubing (6%), I.V. catheters (5%), and vacuum tube blood collection needles (3%). The "other" category (15%) included injuries from medication ampules and unattached hypodermic needles.

Data from the EPINet network indicate that across all healthcare settings, 23% of percutaneous injuries involve blood-filled, hollow-bore needles and thus are in the high-risk category. This fraction is much higher for the dialysis setting. Overall, 47% of percutaneous injuries to dialysis workers were from blood-filled, hollow-bore needles. This means that when a dialysis worker sustains a needlestick injury, it is twice as likely to involve a blood-filled needle as injuries sustained by other healthcare workers.

Figure 3 shows how injuries occurred with dialysis fistula needles. Injuries were equally divided between three categories: during use (25%), which includes cannulation with the fistula needle; disassembling device (25%), which includes removal of the fistula needle; and other, after use (25%). Disposing of fistula needles accounted for 17% of injuries. Injuries in the "disassembling device" category were 21% higher for fistula needles compared to other sharps in the dialysis setting. Removing fistula needles and disassembling dialysis sets can be particularly hazardous because the healthcare worker is often trying to apply pressure to the access site to stem bleeding while simultaneously handling the needle. Another unique characteristic of the dialysis setting is that two fistula needles are used when establishing access for dialysis—one for arterial and one for venous access. The two-needle procedure means increased risk of needlestick.

Although this report does not include data on blood splashes, spraying and other mucocutaneous exposures to blood and body fluids in the dialysis setting, this is clearly an additional serious risk that dialysis workers face, as documented in other studies.[2] Mucocutaneous blood exposures typically occur while applying pressure to puncture sites to stem bleeding, which can result in blood sprays, as well as during emergency interventions or from inadvertent breaks in blood circuits. Dialysis workers must be provided with, and trained to use consistently, barrier garments that prevent blood contact (not cotton lab coats), and eye protection, such as faceshields and goggles, that prevents blood dripping into the eyes.

In conclusion, EPINet data show that there is a serious and urgent need to implement safety-engineered sharp devices in the dialysis setting, particularly for high-risk devices such as fistula and blood-drawing needles, in order to reduce dialysis workers' exposure risk. Fistula needles that provide a protective shield to cover the sharp after use and during disassembly and disposal have the potential for preventing as many as 75% of injuries from this device. Blood drawing from dialysis lines should be carried out with needleless equipment.

Many dialysis facilities operate independently, outside the sphere of major healthcare institutions, and have been slower to adopt safety devices. Clearly, however, with the passage of the federal Needlestick Safety and Prevention Act in November 2000, the time has come to accelerate the transition to safety devices and intensify efforts to provide the best protective equipment for healthcare workers in the dialysis setting.

Figure 1. Job Category of Injured Dialysis Workers
EPINet • 84 hospitals • 1993–1998 • 119 cases

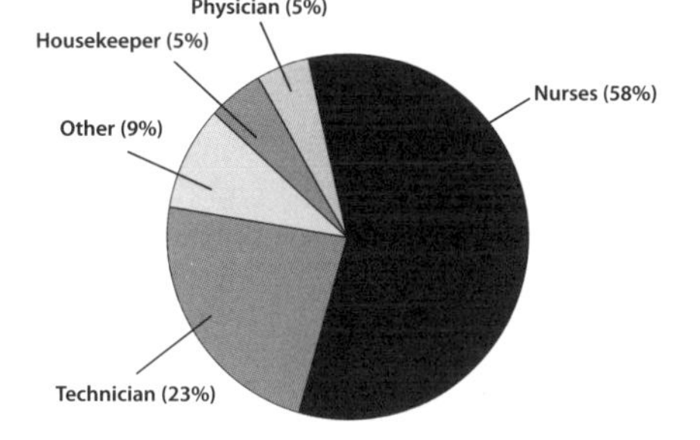

International Healthcare Worker Safety Center, University of Virginia

Figure 2. Devices Causing Injuries in Dialysis Facilities*
EPINet • 84 hospitals • 1993-1998 • 119 cases

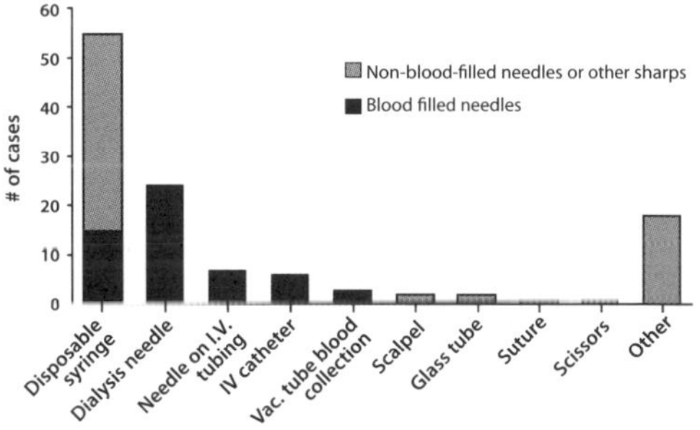

*47% of injuries were from blood-filled, hollow-bore needles

International Healthcare Worker Safety Center, University of Virginia

Figure 3. How Injuries Occurred with Dialysis Fistula Needles
EPINet • 84 hospitals • 1993-1998 • 24 cases

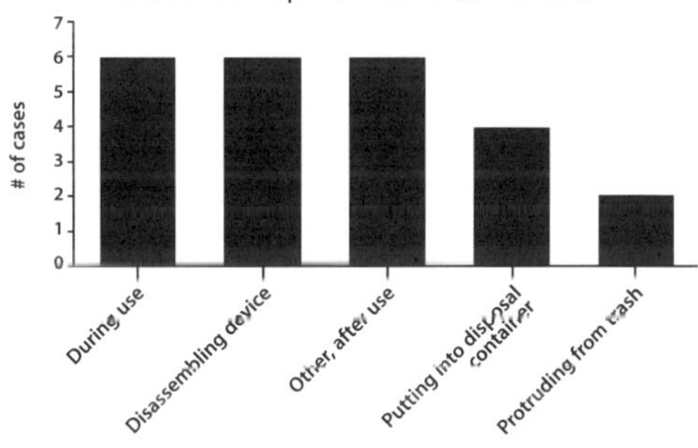

International Healthcare Worker Safety Center, University of Virginia

References

1. Jagger J, Bentley M. Occupational blood exposures among renal dialysis workers. Poster presentation at 30th National Symposium of the American Nephrology Nurses' Association. Baltimore, MD: April 1999.
2. Petrosillo N, Puro V, Jagger J, Ippolito G, and the Italian Multicenter Study on Nosocomial and Occupational Risk of Infections in Dialysis. The risks of occupational exposure and infection by human immunodeficiency virus, hepatitis B virus, and hepatitis C virus in the dialysis setting. *Am J Infect Control.* 1995;23:278–`85.

EPINet Report: 1999 Percutaneous Injury Rates

By Ginger Parker, M.B.A., Jane Perry, M.A., and Janine Jagger, M.P.H., Ph.D.

Vol. 6, no. 1, 2002

In 1999, the International Healthcare Worker Safety Center at the University of Virginia collected percutaneous injury data from 21 U.S. healthcare facilities (20 hospitals and one outpatient facility). Four of the hospitals and the outpatient facility are part of the Palmetto Hospital Trust network in South Carolina; eight hospitals are members of the Sisters of Providence network in the Pacific Northwest. The other eight hospitals are located in states in the eastern half of the U.S. Twelve (including the outpatient facility) are teaching institutions, and nine are non-teaching. Seven hospitals had an average daily census (ADC) of less than 100 occupied beds (with "occupied beds" defined as the ADC for the same year in which the data were collected); seven had an ADC of 100 to 300; and six had an ADC of more than 300. The total number of percutaneous injuries reported for 1999 was 2,025.

The 1999 data yielded these findings:

- *The average percutaneous injury (API) rate for non-teaching hospitals was 34 injuries per 100 occupied beds.*
- *The API for teaching hospitals and facilities was 40 injuries per 100 occupied beds.*
- *There was no correlation between hospital size and institutional injury rates for non-teaching hospitals.*

By comparison, in 1998 the API rates for non-teaching and teaching hospitals were 22 and 38, respectively; 52 hospitals reported data in that year, and the total number of injuries was 3,180. (Most hospitals that dropped out of the network between 1998 and 1999 were small non-teaching hospitals.) The 1999 API rate for non-teaching hospitals (34, versus 22 in 1998) does not represent an increase in injury rates among the healthcare facilities remaining in the network, but rather reflects the loss of one large non-teaching hospital that had a low injury rate.

Institutional injury rates can be affected by a variety of factors, such as changes in reporting patterns (for example, a higher level of injury reporting among surgeons after a sharps prevention education program), which may increase *reported* injuries even though *actual* injuries did not increase. Another factor that can contribute to higher rates, when average daily census is used as the denominator, is the declining length of hospital stays; shorter stays mean that more procedures (which can result in injuries) may be performed in less time. While this is a shortcoming in using ADC as a denominator, in the U.S. the information needed to calculate ADC rates for multi-institutional comparisons is readily available. The fact that institutional information systems are not standardized in the U.S. limits the types of denominator data that can be easily obtained from different institutions.

EPINet data from 1999, as in previous years, revealed great variation among individual facilities in annual percutaneous injury rates: from less than 15 injuries to more than 125 injuries per 100 occupied beds. While we do not fully understand the reasons for such variation, it may be due to a number of factors, including mix of patients, injury underreporting rates, safety systems implemented, and whether a facility is teaching or non-teaching.

Because of these variables, we cannot assume that a healthcare facility with a low rate of reported injuries per year necessarily has a better safety record than a hospital with a higher rate. For example, hospital A, with a higher injury rate, may do a better job of educating its employees about the need to report needlestick injuries than hospital B with a lower rate. For that reason, benchmarking against other hospitals may not be very meaningful. It is more reliable to track injury trends within a single hospital over several years, and make historical comparisons as prevention measures are implemented, than to compare one institution to another.

1999 U.S. EPINet Percutaneous Injury Rates

21 healthcare facilities (20 hospitals, 1 outpatient facility)

Average daily census = 5,118 • Total injuries = 2,025

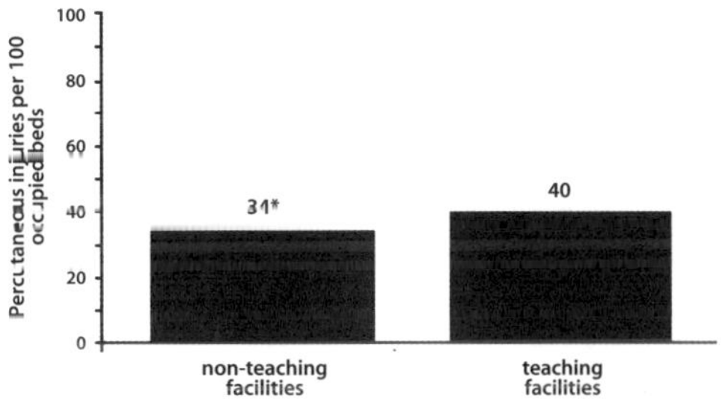

**Reflects loss from EPINet network, in 1999, of one large, non-teaching hospital with low injury rates.*

International Healthcare Worker Safety Center, University of Virginia

Percutaneous Injuries and Blood Exposures in Emergency Department Settings

By Jane Perry, M.A., and Janine Jagger, M.P.H., Ph.D.

Vol. 6, no. 2, 2002

Emergency department (ED) staff are especially vulnerable to bloodborne pathogen exposures. Like operating room (OR) personnel, ED workers are more likely to be exposed to large quantities of blood than healthcare workers in other settings; unlike the OR, however, such exposures are more apt to occur under unpredictable circumstances. Compounding the risk are combative or uncooperative patients, all-too-familiar to ED staff. Moreover, a higher proportion of emergency patients are infected with bloodborne pathogens than the general population. One study tested patients in an inner-city ED for human immunodeficiency virus (HIV), hepatitis B, and hepatitis C, and found that 24% of patients were infected with at least one of these pathogens.[1] Studies in Italy and the U.S. found that ED staff failed to identify the majority of HIV-infected patients (59% not identified in the Italian study, 69% in the U.S. study).[2,3] Another study reported that ED personnel do not consistently use personal protective equipment during trauma cases.[4] And a survey of 95 ED workers at a university teaching hospital found that while they had an average of 56.5 blood and body fluid (BBF) contacts per year, only 4% reported their most recent contact.[5]

This already risky environment for healthcare workers has been exacerbated by overcrowding in EDs. A recent survey conducted for the American Hospital Association found that "a majority of EDs were full and often operated at or over capacity," and concluded that "overcrowding problems in the nation's emergency departments are high and getting worse."[6] AHA officials say that overcrowding is related to problems with nurse shortages. With not enough nursing staff and too many patients, the potential for sharps injuries and blood exposures only increases.

To better understand the risks of percutaneous injuries and blood exposures in the ED, we analyzed five years of EPINet data (1996-2000), with 1,060 percutaneous injuries (PI) and 338 BBF contacts. The data showed that ED workers are more likely to sustain percutaneous injuries involving blood-filled needles—those with the highest risk of bloodborne pathogen transmission—than other hospital personnel. In all other hospital settings, 24% of injuries are from blood-filled needles; in EDs, the fraction is 42%.

Figure 1 shows the job categories of ED workers sustaining injuries. Nurses reported the highest proportion of injuries (44%), followed by physicians (19%), phlebotomy personnel (6%) and housekeepers (5%). A higher proportion of physicians in EDs reported sharps injuries compared to physicians in other areas of the hospital (19% compared to 15%).

Figure 1. Job Categories of Emergency Department Workers Reporting Percutaneous Injuries

EPINet • 72 healthcare facilities • 1996-2000 • 1,060 cases

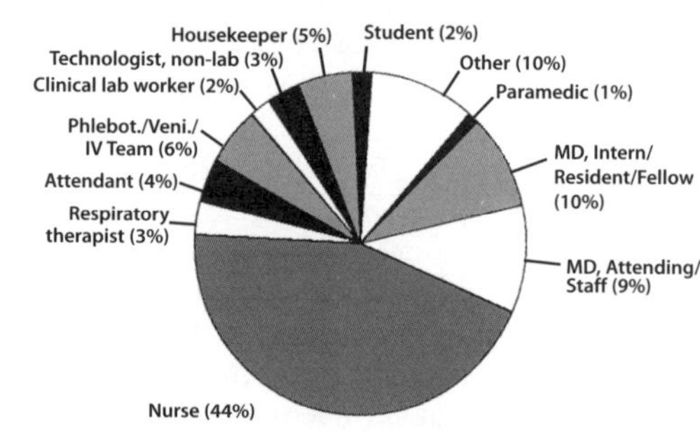

International Healthcare Worker Safety Center, University of Virginia

Figure 2. Devices Causing Percutaneous Injuries in Emergency Departments vs. Patient Rooms

EPINet • 72 healthcare facilities • 1996-2000
ED: 1,060 cases • Patient rooms: 4,295 cases

Emergency Depts.
Patient Rooms

% of injuries

Syringe, Suture needle, Butterfly needle, I.V. catheter, Other needle, Phleb. needle, Other sharp, Prefilled syringe, Scalpel, Glass item, Blood gas, Needle on I.V. line, Lancet

International Healthcare Worker Safety Center, University of Virginia

Figure 3. Original Purpose of Devices Causing Percutaneous Injuries in Emergency Departments vs. Patient Rooms

EPINet • 72 healthcare facilities • 1996-2000
ED: 1,060 cases • Patient rooms: 4,295 cases

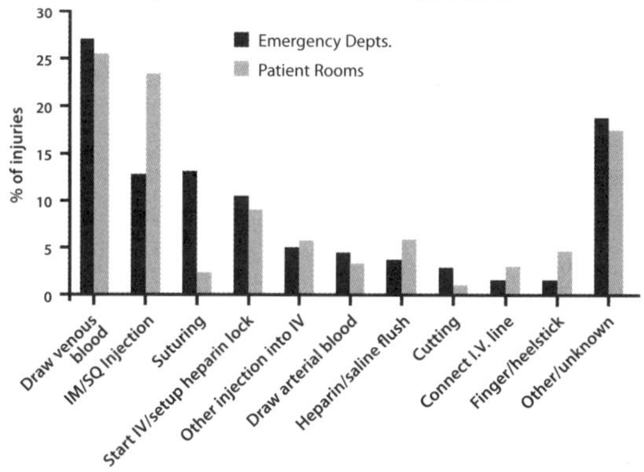

International Healthcare Worker Safety Center, University of Virginia

Figure 4. Percutaneous Injuries in Emergency Departments vs. Patient Rooms: When Injuries Occurred

EPINet • 72 healthcare facilities • 1996-2000
ED: 1,060 cases • Patient rooms: 4,295 cases

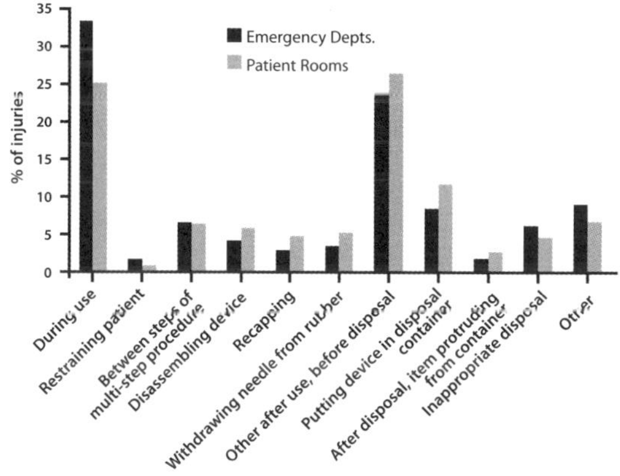

International Healthcare Worker Safety Center, University of Virginia

Figure 5. Mechanism of Blood and Body Fluid Exposures in Emergency Departments vs. Patient Rooms

EPINet • 65 healthcare facilities • 1996-2000
ED: 338 cases • Patient rooms: 1,365 cases

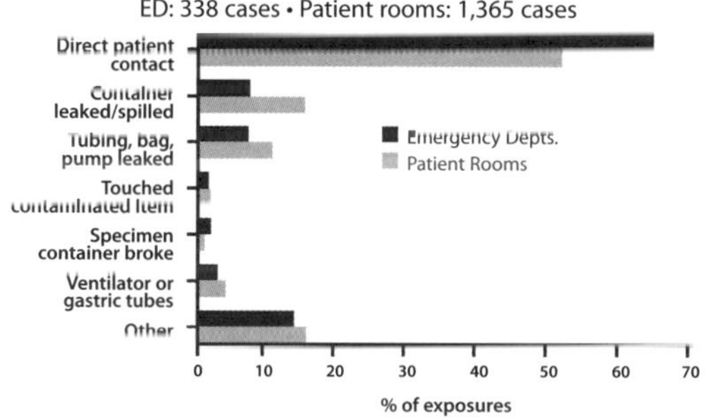

International Healthcare Worker Safety Center, University of Virginia

Figure 2 shows the devices causing percutaneous injuries in EDs and patient rooms. Needles on syringes were the most frequent cause of injury in EDs (29%), followed by suture needles (13%), butterfly needles (12%), I.V. catheter needles (10%), and phlebotomy needles (8%). Other devices causing injury in EDs included prefilled syringes (5%), syringes used for arterial blood draws (3%), and scalpel blades (3%). There were many more injuries caused by glass in EDs (3%) compared to patient rooms (0.3%) and all other hospital settings (1.5%). Similarly, the proportion of injuries from suture needles in EDs (13%) was five times higher than in patient rooms (2.4%), and more than three times higher than in all other settings (4%), excluding operating rooms.

Figure 3 shows the original purpose of devices causing percutaneous injuries in EDs as compared with patient rooms. Procedures most frequently involving injuries in EDs were venous blood drawing (27%), intramuscular or subcutaneous injections (13%), suturing (13%), starting an I.V. (10%), injecting into an I.V. injection site or port (5%), and drawing arterial blood (4%). Suturing and cutting were both associated with many more injuries in EDs (13% and 3%, respectively) than in patient rooms (2% and 1%, respectively).

Figure 4 compares the mechanism of injuries in EDs compared to patient rooms. A higher proportion of injuries occurred during use of the device in EDs compared to patient rooms (33% vs. 25%); this may reflect the greater number of combative or uncooperative patients found in EDs compared to other healthcare settings. Overall, 60% of sharp-object injuries in EDs occurred after use; that translates into 60% of ED injuries that could potentially be prevented by needleless devices or effective safety-engineered sharp devices that cover the sharp after use. Of that "after use" fraction, 6% were caused by devices that were not disposed of properly (i.e., left on a floor or bed; left on or near—but not in—a sharps disposal container; or put in a trash bag or other inappropriate container). In EDs, 1.6% of injuries occurred while restraining patients, compared to 0.7% for patient rooms.

Figure 5 shows the mechanism of BBF exposures in EDs compared to patient rooms. A higher proportion of BBF exposures involved direct patient contact in EDs compared to patient rooms (66% vs. 53%). This may reflect an environment in which, due to trauma or other factors, there are more sprays and splashes from blood and body fluids compared to patient rooms.

In these data, which were collected before the full implementation of the Needlestick Safety and Prevention Act and the revised bloodborne pathogens standard, 86% of ED injuries were from conventional (non-safety) devices, indicating that they were in common use. Implementing safety devices is especially urgent in the ED because of a higher risk of encountering uncooperative patients. Highest priority should be given to safety butterfly needles and I.V. catheter stylets, because they are the most frequent source of high-risk injuries in EDs. Syringes and sharp-tip suture needles, the two devices in the ED that cause the largest proportion of injuries overall, should be replaced with safety-engineered syringes and blunt-tip suture needles wherever clinically feasible; tissue adhesives and other alternative methods of skin closure should also be employed wherever possible.

Particular attention should be paid to training ED staff adequately in the use of new safety-engineered devices. In high-stress emergency situations, ED staff need to feel as comfortable with the new devices as they did with the conventional ones. They also need to test the compatability of the new devices with other equipment used in the ED.

Arnold Berry, M.D., professor of anesthesiology at Emory University School of Medicine, has pointed out a unique challenge in implementing safety devices in EDs. Drugs used for codes and cardiac arrests, such as epinephrine and sodium bicarbonate, are packaged in prefilled syringes to save time during an emergency by precluding the need to draw up the drug from a vial. However, these prefilled syringes often come with needles attached. This creates the potential for a mismatch between needleless I.V. systems and prefilled syringes. Berry states: "In choosing drugs for emergency use, the prefilled syringes must be compatible with the injection ports on the I.V. tubing used in the hospital. If the committee choosing the I.V. equipment is not the same as the one that chooses emergency drugs, there may be incompatibility that could lead to disastrous consequences during an emergency. Planning teams need to make sure that emergency carts are stocked with prefilled syringes that can be easily administered with the I.V. system in use." Equipment compatibility may also be an issue between EDs and the emergency medical services that deliver patients to them.

ED personnel should be instructed not to use syringes for blood-drawing, which requires transferring blood from the syringe to a vacuum tube and increases needlestick risk; safety-engineered phlebotomy devices should be used instead. Another important prevention measure is substituting plastic for glass capillary and vacuum tubes and other glass items. Finally, with the frequency of blood and body fluid sprays and splashes in EDs, barrier garments and protective eyewear with a seal above the eyes should be consistently worn when carrying out exposure-prone procedures.

These measures will help significantly reduce the risk that ED personnel face in an unpredictable and fast-paced environment.

References

1. Kelen GD, Green GB, Purcell RH, Chan DW, Qaqish BF, Sivertson KT et al. Hepatitis B and hepatitis C in emergency department patients. *N Engl J Med.* 1992; 326(21):1399–1404.
2. De Carli G, Puro V, Binkin NJ, Ippolito G. Risk of human immunodeficiency virus infection for emergency department workers. Italian Study Group on Occupational Risk of HIV Infection. *J Emerg Med.* 1994;12(6):737–44.
3. Marcus R, Culver DH, Bell DM, Srivastava PU, Mendelson MH, Zalenski RJ et al. Risk of human immunodeficiency virus infection among emergency department workers. *Am J Med.* 1993;94(4):363–70. *(See also*: Clark SJ, Kelen GD, Henrard DR, Daar ES, Craig S, Shaw GM, Quinn TC. Unsuspected primary human immunodeficiency virus type 1 infection in seronegative emergency department patients. *J Infect Diseases.* 1994;170(1):194–7.)
4. Kim LE, Evanoff BA, Parks RL, Jeffe DB, Mutha S, Haase C, Fraser VJ. Compliance with Universal Precautions among emergency department personnel: implications for prevention programs. *Am J Infect Control.* 1999;27(5):453–455.
5. Jagger J. Powers RD, Day JS, Detmer DE, Blackwell B, Pearson RD. Epidemiology and prevention of blood and body fluid exposures among emergency department personnel. *J Emerg Med.* 1994;12:75–765.
6. Anonymous. Overcrowding in ER expected to worsen. *Mater Manag Health Care.* 2002;11(5):6. Survey findings available on-line at: www.aha.org/info/EmergencyDeptOverload.asp.

Comparison of EPINet Data for 1993 and 2001 Shows Marked Decline in Needlestick Injury Rates

By Janine Jagger, M.P.H., Ph.D., and Jane Perry, M.A.

Vol. 6, no. 3, 2003

For more than a decade the United States has been the leader in the development, testing and implementation of safety-engineered sharp medical devices. The new devices became widely available in the U.S. in the early 1990s, and their acceptance and implementation in the workplace has been gradual but steady. The Needlestick Safety and Prevention Act of 2000, which became fully enforceable in July 2001, turned a trend into a requirement and made the use of safety devices mandatory. The benefits of the new technology have been documented in numerous ways, including clinical trials and demonstration projects comparing conventional needles to their safety counterparts, and in reports from specific institutions showing downward trends in percutaneous injury rates following the adoption of a variety of safety-engineered devices. These focused reports have been encouraging, but there has been a lack of documentation showing an across-the-board impact of both the new technology and the Needlestick Safety Act in a multihospital sharps injury surveillance network. In this report we present data from the EPINet Multihospital Sharps Injury database, coordinated by the International Healthcare Worker Safety Center at the University of Virginia, which documents the impact on needlestick injury rates associated with the widespread adoption of safety devices.

The interpretation of needlestick data from healthcare facilities is complicated by a variety of factors. Many different healthcare worker occupational groups are represented in the data, and some groups report their injuries more reliably than others. The reporting patterns for specific worker groups may change over time, and an increase in reporting rates for certain groups may mask a decline in the overall injury rates in an institution or network. Further, the devices causing sharps injuries are numerous. While there has been a widespread conversion to safety in some device categories, such as phlebotomy needles and intravenous catheters, in others, such as laboratory equipment and surgical instruments, relatively small numbers of safety devices are in use. Reductions in overall injury rates will be diluted by device categories in which safety-engineered alternatives have not been widely adopted. One example is sharp-tip suture needles, where the safety alternative, blunt-tipped suture needles, has yet to be accepted by most surgeons.

Figure 1. Comparison of 1993 and 2001 Percutaneous Injury Rates for Nurses, By Device (Conventional Only)

EPINet • 1993 • 18 teaching hospitals
cumulative average daily census = 7,000 • 1,366 total injuries

EPINet • 2001 • 11 teaching hospitals
cumulative average daily census = 4,174 • 401 total injuries

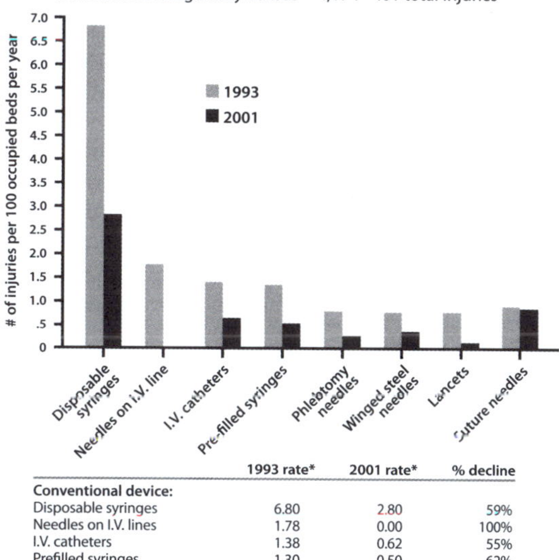

	1993 rate*	2001 rate*	% decline
Conventional device:			
Disposable syringes	6.80	2.80	59%
Needles on I.V. lines	1.78	0.00	100%
I.V. catheters	1.38	0.62	55%
Prefilled syringes	1.30	0.50	62%
Phlebotomy needles	0.77	0.23	70%
Winged steel needles	0.73	0.33	55%
Lancets	0.70	0.09	87%
Suture needles	0.84	0.80	5%

*per 100 occupied beds per year

International Healthcare Worker Safety Center, University of Virginia

Figure 2. Comparison of Percutaneous Injury Rates for Nurses, 1993 and 2001

EPINet • 1993 • 18 teaching hospitals
cumulative average daily census = 7,000 • 1,366 total injuries

EPINet • 2001 • 11 teaching hospitals
cumulative average daily census = 4,174 • 401 total injuries

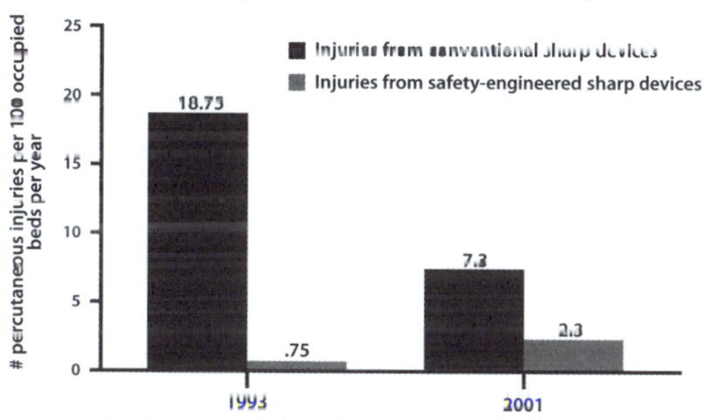

Percutaneous Injury (PI) rates for nurses, based on data for conventional and safety devices combined, declined from 19.50 per 100 occupied beds in 1993 to 9.6 per 100 occupied beds in 2001 (50.76% decrease)

International Healthcare Worker Safety Center, University of Virginia

Another factor affecting before-and-after comparisons of needlestick data within a long-term multihospital surveillance network is the number of teaching hospitals relative to the number of community hospitals in the two years being compared. Since teaching hospitals have higher rates of reported percutaneous injuries than community hospitals, a change in the balance of teaching/nonteaching facilities in the two time periods introduces a confounding factor that could mask true changes in rates.

All of these complicating factors make it difficult to attribute observed changes in percutaneous injury (PI) rates within a multihospital network to a specific cause, and also may preclude real changes from being observed.

In order to document the impact of safety-engineered sharp devices on PI rates for hospital workers, we compared rates from 1993 and 2001 in a network of participating EPINet hospitals. We selected 1993 because it was the first year for which EPINet data were available, and 2001 because it is the most recent complete year of EPINet data. In order to avoid the confounding factors described above, we specified several parameters for the data used in this report.

First, we used only data from teaching hospitals, to avoid the problem of changing ratios of teaching and nonteaching hospitals in the two time periods. Second, we selected data for nurses only: historically, they are the professional group with the largest proportion of PIs (44% of all cases in the 2001 EPINet data), and their reporting patterns have remained relatively stable over time. Third, we compared PI rates for conventional and safety devices separately, so that the relative effects of conventional versus safety device usage could be distinguished. Also, the different device categories were compared separately so that reductions in categories where there has been a high penetration of safety devices (such as intravenous catheters) would not be obscured by other categories in which there has been a low penetration of safety devices (such as suture needles).

We expressed PI rates as the number of injuries per 100 occupied hospital beds (# PIs/100 occupied beds). To obtain a denominator, we added together the annual average daily census (i.e., the average number of occupied beds per day) for all the hospitals in each of the comparison groups. For 1993, the cumulative average daily census (ADC) for the 18 teaching hospitals reporting data that year was 7,000; in 2001, the ADC for the 11 teaching hospitals reporting data was 4,174. There are other relevant denominators that could be used, such as the number of full-time equivalent nurses employed in each group; these may be considered in subsequent reports.

Percutaneous Injury Rates for Conventional Devices

Disposable syringes caused the largest proportion of PIs to nurses in both the 1993 and 2001 EPINet data for conventional devices (38% and 40%, respectively). However, in 1993 the PI rate for conventional disposable syringes was 6.8/100 occupied beds; in 2001, the rate declined by 59% to 2.8/100 occupied beds **(Figure 1).**

Percutaneous injuries from needles on I.V. lines were eliminated, with 1.78/100 occupied beds in 1993 and no injuries in 2001. This dramatic decline may in part reflect the impact of the 1992 U.S. Food and Drug Administration safety alert advising against use of needles on I.V. lines, and the widespread implementation of needleless and recessed needle I.V. systems that followed the alert.

Substantial declines in PI rates for nearly all the conventional needles that commonly injure nurses are seen in Figure 1. PIs from intravenous (IV) catheters decreased by 55%; PIs from phlebotomy needles decreased by 70%; PIs from prefilled syringes decreased by 62%; and PIs from winged steel needles decreased by 55%. PIs from lancets dropped by 87%. Only one device category, suture needles, did not show a similar marked decline; it decreased by only 5%.

Location-related rates

In both the 1993 and the 2001 data, PIs to nurses from conventional devices most frequently occurred in patient rooms, followed by operating rooms (OR), intensive/critical care units (ICU/CCU) and emergency departments (ED). Again, a significant decrease in PI rates for each of these settings was seen: 65% for patient rooms; 50% for ORs; 79% for ICU/CCU; and 54% for EDs. However, there was a 58% increase in the rate of PIs in labor and delivery settings.

Procedure-related rates

Percutaneous injuries to nurses from conventional devices most frequently occur while giving injections (intramuscular or subcutaneous); injection-related PIs decreased by 39%. Other procedure-related PI rates dropped as well: for drawing venous blood, a 50% decrease; for starting IVs, a 59% decrease; for heparin or saline flushes, a 96% decrease. The PI rate for suturing remained the same (1/100 occupied beds).

Rates for when injuries occurred

In both the 1993 and 2001 data, PIs to nurses from conventional devices most frequently occurred during use or between steps of a multi-step procedure (32% in 1993 for both categories combined, and 31% in 2001). Rates in these combined categories decreased by 61% from 1993 to 2001. Rates for injuries occurring after use but before disposal decreased by 72%; for injuries occurring while withdrawing a needle from a resistant substance such as a rubber stopper, by 60%; from recapping, by 47%; and while disassembling a device, by 79%. The rate for injuries occurring while putting a device in a disposal container decreased by 47%, and for injuries from sharps protruding from disposal containers, by 17%. A category that was added after 1993—injury from a "device left on floor, table, bed or other inappropriate place"—had a 2001 rate of .55/100 occupied beds. Such data indicate a need for continued attention to disposal safety.

Injuries from Safety Devices

In 1993, when nurses were asked whether the device causing injury was a safety design, 4% answered yes and 96% answered no. In 2001, 25% answered yes, 72% said no, and 3% responded "unknown." These data reflect the sharp increase in adoption of safety devices over the last 10 years. When we consider PI data for safety-engineered sharp devices, it is important to keep in mind that safety devices that have a needle attached (in contrast to needleless devices) will still cause a residual fraction of sharps injuries. These are PIs that may be more difficult to eliminate, such as those that occur during use, when the sharp is, by necessity, exposed. Also, PIs from safety devices can occur *after* use if the protective feature is not activated.

As the number of safety-engineered sharp devices in the healthcare workplace increases, we can expect a rise in the number of PIs associated with them. The EPINet data for nurses from 1993 and 2001 reflect this reality: in 1993, the PI rate for all safety devices was .75/100 occupied beds; in 2001, it was 2.3/100 occupied beds. By contrast, the PI rate for all conventional devices declined from 18.75/100 occupied beds in 1993 to 7.3/100 occupied beds in 2001.

Conclusion

The most important finding from this comparison of 1993 and 2001 percutaneous injury rates for nurses is that **the overall rate (including injuries from both conventional and safety devices) declined from 19.5 PIs per 100 occupied beds in 1993 to 9.6 PIs per 100 occupied beds in 2001—a decrease of 51%**. The decline in PI rates from conventional devices—reflecting the decline in conventional device usage overall—far outweighs the increase in PI rates for safety devices that has accompanied their widespread implementation in healthcare facilities. Such data support the benefit of the new technology in reducing percutaneous injury risk to nurses, and most likely to other healthcare workers—benefits that will continue to increase as compliance with the Needlestick Safety and Prevention Act comes closer to 100%.

EPINet Report: 2001 Percutaneous Injury Rates

By Jane Perry, M.A., Ginger Parker, M.B.A., and Janine Jagger, M.P.H., Ph.D.. *Vol. 6, no. 3, 2002*

In 2001, the International Healthcare Worker Safety Center at the University of Virginia collected data on percutaneous injuries and blood and body fluid exposures from 58 healthcare facilities in the United States that use the EPINet surveillance program to track exposure incidents. These facilities voluntarily participate in the collaborative EPINet network coordinated by the Center, and their exposure data are combined into an aggregate database. The 2001 percutaneous injury report and blood and body fluid exposure report accompany this article.

Forty-three of the facilities that contributed data in 2001 are part of a state-wide network in South Carolina coordinated by Palmetto Hospital Trust Services; nine facilities are members of the Sisters of Providence network in the Pacific Northwest; the other six facilities are located in the eastern half of the U.S., except for St. Joseph Hospital in Omaha, Nebraska. Thirteen of the facilities are teaching hospitals, and 45 are nonteaching hospitals. In 2001, 33 facilities had an average daily census (ADC) of less than 100 occupied beds*; 12 facilities had an ADC of 100 to 300; and 11 facilities had an ADC of greater than 300. Most of the facilities are acute-care or tertiary-care hospitals or medical centers, some of which have physicians' offices, home health agencies and other outpatient settings affiliated with them. Among the participating facilities is an alcohol and drug abuse agency with a detoxification unit, a home hospice agency, a long-term acute-care facility, a skilled nursing facility, and a rehabilitation hospital.

In 2001, a total of 1,929 percutaneous injuries (PIs) were reported by network facilities. The 2001 data yielded these findings:

- The average percutaneous injury (PI) rate for teaching hospitals was 26 injuries per 100 occupied beds.
- The average PI rate for nonteaching facilities was 18 injuries per 100 occupied beds.
- The average PI rate for hospitals with an ADC of less than 100 was 22.79 per 100 occupied beds; for hospitals with an ADC of 100-300, 25.82 per 100 occupied beds; and for hospitals with an ADC of greater than 300, 19.85 per 100 occupied beds.

By comparison, in 1999 the average PI rate for teaching hospitals was 40 per 100 occupied beds, and for nonteaching facilities, 34 per 100 occupied beds. Twenty-one facilities reported data in 1999; the total number of PIs was 2,025. (Most of the hospitals that were added into the network between 1999 and 2001 were small nonteaching hospitals.)

"Occupied beds" is defined as the ADC for the same year in which the data were collected.

The decline in PI rates for 2001 compared to 1999 may be attributed to several factors. Foremost is the increased use of safety devices, especially since OSHA issued a revised compliance directive for the bloodborne pathogens standard in November 1999 which explicitly stated, for the first time, that use of safety devices was required. Other factors that may have contributed to the decline in PI rates include better education of healthcare workers about the risks associated with sharps injuries (which in turn helps to lessen resistance to use of safety devices) and increased training in the proper use of safety devices. Also, injuries from needles used to access intravenous lines appear to have been nearly eliminated: in 1999, 1% of injuries involved needles used for this purpose; in 2001, that figure had declined to 0.3%.

EPINet data from 2001, as in previous years, revealed great variation among individual facilities in PI rates: six facilities had a zero PI rate, while three facilities had rates between 50 and 60 per 100 occupied beds. We do not fully understand the reasons for such variation, but they may include the mix of patients, injury underreporting rates, the extent to which a facility has converted to safety devices, and whether it is a teaching or nonteaching institution.

Because of these variables, we cannot assume that a healthcare facility with a low PI rate necessarily has a better safety record than a hospital with a higher rate. For example, a hospital with a high PI rate may educate its employees about the need to report needlestick injuries, or may have more patients requiring invasive procedures than another

U.S. EPINet 1999 and 2001 Percutaneous Injury Rates

1999 • 21 healthcare facilities
(12 teaching hospitals, 9 nonteaching)
cumulative average daily census = 5,118 • total injuries = 2,025

2001 • 58 healthcare facilities
(13 teaching hospitals, 45 nonteaching)
cumulative average daily census = 8,703 • total injuries = 1,929

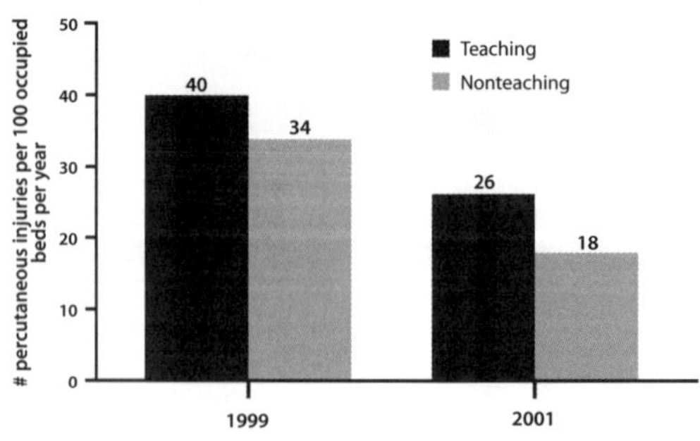

International Healthcare Worker Safety Center, University of Virginia

Uniform Needlestick and Sharp-Object Injury Report
U.S. EPINet Network • 2001 • 58 healthcare facilities*

Total cases = 1,929 (excluding injuries before use); total avg. daily census = 8,703 (*13 teaching/45 non-teaching hospitals)

JOB CATEGORY:		
M.D. (attending/staff)	152	8.1%
M.D. (intern/resident/fellow)	124	6.6%
Medical student	13	0.7%
Nurse RN/LPN	818	43.6%
Nursing student	12	0.6%
Respiratory therapist	29	1.5%
Surgery attendant	166	8.8%
Other attendant	27	1.4%
Phlebotomist/ venipuncture/I.V. team	111	5.9%
Clinical laboratory worker	36	1.9%
Technologist (non-lab)	90	4.8%
Dentist	2	0.1%
Dental hygienist	0	0%
Housekeeper	64	3.4%
Paramedic	7	0.4%
CNA/HHA	48	2.6%
Laundry worker	2	0.1%
Security	2	0.1%
Other student	26	1.4%
Other	148	7.9%

WHERE INJURY OCCURRED:		
Patient room	592	31.5%
Outside patient room	45	2.4%
Emergency department	176	9.4%
Intensive/critical care unit	91	4.8%
Operating room	541	28.8%
Outpatient clinic/office	91	4.8%
Blood bank	1	0.1%
Venipuncture	16	0.9%
Dialysis facility	13	0.7%
Procedure room	76	4.0%
Clinical laboratories	34	1.8%
Autopsy/pathology	14	0.7%
Service/utility area	28	1.5%
Labor and delivery	38	2.0%
Home-care	26	1.4%
Other	97	5.2%

ORIGINAL PURPOSE OF SHARP ITEM:		
Unknown, N/A	148	7.9%
Injection, IM/subcutaneous	390	20.9%
Heparin or saline flush	32	1.7%
Other injection/aspiration I.V.	83	4.4%
Connect I.V. line	16	0.9%
Start I.V. or heparin lock	91	4.9%
Draw venous blood sample	295	15.8%
Draw arterial blood sample	41	2.2%
Obtain body fluid/tissue sample	29	1.6%
Fingerstick/heelstick	43	2.3%
Suturing	324	17.3%
Cutting (surgery)	145	7.8%
Electrocautery	9	0.5%
Contain specimen/ pharmaceutical	10	0.5%
Place arterial line	24	1.3%
Drilling	12	0.6%
Other	177	9.5%

NOTE: The needlestick and sharp-object injury report and blood and body fluid exposure report that appear on this page and the next are based on 2001 data from the EPINet data-sharing network coordinated by the International Healthcare Worker Safety Center at the University of Virginia.

WHEN INJURY OCCURRED:		
During use of item	564	30.1%
Between steps of multistep procedure	257	13.7%
Disassembling device	79	4.2%
Preparing instrument for reuse	36	1.9%
Recapping device	68	3.6%
Withdrawing from resistant material	67	3.6%
Other after use, before disposal	314	16.8%
Item left on disposal container	12	0.6%
Putting item into disposal container	122	6.5%
After disposal, item protruding from disposal container	53	2.8%
Pierced side of disposal container	5	0.3%
Item pierced side of inappropriate disposal container	39	2.1%
Restraining patient	10	0.5%
Device left on floor, table or other inappropriate place	116	6.2%
Other	130	6.9%

TYPE OF DEVICE CAUSING INJURY:		
Disposable syringe	650	36.1%
Prefilled cartridge syringe	49	2.7%
Blood gas syringe	20	1.1%
Syringe, other type	14	0.8%
Needle on I.V. tubing	9	0.5%
Winged steel needle	120	6.7%
I.V. catheter (stylet)	65	3.6%
Vacuum tube blood collection needle	72	4.0%
Spinal or epidural needle	5	0.3%
Unattached hypodermic needle	17	0.9%
Arterial catheter introducer needle	10	0.6%
Central line catheter introducer needle	6	0.3%
Other vascular catheter needle	4	0.2%
Other non-vascular catheter needle	1	0.1%
Needle, unknown type	42	2.3%
Needle, describe	65	3.6%
Lancet	40	2.2%
Suture needle	307	17.0%
Scalpel, reusable	77	4.3%
Razor	6	0.3%
Scissors	14	0.8%
Bovie electrocautery device	9	0.5%
Bone cutter	6	0.3%
Towel clip	4	0.2%
Microtome blade	6	0.3%
Trocar	2	0.1%
Vacuum tube, plastic	1	0.1%
Fingernails/teeth	4	0.2%
Scalpel, disposable	48	2.7%
Retractors, skin/bone hooks	14	0.8%
Staples/steel sutures	3	0.2%
Wire	15	0.8%
Pin	11	0.6%
Drill bit	3	0.2%
Pickups/forceps/hemostats	8	0.4%
Sharp item, not sure what kind	8	0.4%
Other sharp item (describe)	44	2.4%
Medication ampule	1	0.1%
Vacuum tube, glass	4	0.2%
Specimen/test tube, glass	5	0.3%
Capillary tube	1	0.1%
Glass slide	2	0.1%
Glass item, unknown type	2	0.1%
Other glass item	9	0.5%

SOURCE PATIENT IDENTIFIABLE?		
Yes	1704	90.7%
No	122	6.5%
Unknown	46	2.4%
Not available	7	0.4%

INJURED WORKER ORIGINAL USER OF SHARP ITEM?		
Yes	1066	57.3%
No	731	39.3%
Unknown	35	1.9%
N/A	29	1.6%

SHARP ITEM CONTAMINATED?		
Yes	1689	90.3%
No	28	1.5%
Unknown	154	8.2%

IF INJURY WAS CAUSED BY A NEEDLE, WAS IT A SAFETY DESIGN?		
Yes	393	22.4%
No	1260	71.8%
Unknown	103	5.9%

IF YES, WAS SAFETY MECHANISM ACTIVATED?		
Yes, fully	44	12.1%
Yes, partially	61	16.8%
No	258	71.1%

IF YES, DID INJURY HAPPEN:		
Before activation	182	56.9%
During activation	84	26.3%
After activation	54	16.9%

DEPTH OF INJURY:		
Superficial (little/no bleeding)	891	49.0%
Moderate (skin punctured, some bleeding)	854	46.9%
Severe (deep stick/cut, profuse bleeding)	75	4.1%

BODY PART INJURED:		
Arm	46	2.5%
Back	2	0.1%
Face/head	5	0.3%
Foot	7	0.4%
Front	5	0.3%
Hand, left	1011	55.8%
Hand, right	715	39.4%
Leg	22	1.2%

GLOVES–*Did sharp item penetrate:*		
Single pair of gloves	1210	68.9%
Double pair of gloves	210	12.0%
No gloves	335	19.1%

TYPE OF FACILITY:		
Non-teaching hospital	703	37.2%
Teaching hospital	1185	62.8%

Uniform Blood and Body Fluid Exposure Report
U.S. EPINet Network, 2001, 49 health care facilities*

Total cases=463; total avg. daily census = 7,716 (*37 teaching/12 nonteaching hospitals)

JOB CATEGORY:		
M.D. (attending/staff)	25	5.4%
M.D. (intern/resident/fellow)	25	5.4%
Medical student	1	0.2%
Nurse RN/LPN	250	54.2%
Nursing student	3	0.7%
Respiratory therapist	16	3.5%
Surgery attendant	23	5.0%
Other attendant	5	1.1%
Phlebotomist/venipuncture/ I.V. team	5	1.1%
Clinical laboratory worker	12	2.6%
Technologist (non-lab)	31	6.7%
Housekeeper	4	0.9%
Paramedic	7	1.5%
Other student	5	1.1%
CNA/HHA	17	3.7%
Other, describe	32	6.9%

WHERE EXPOSURE OCCURRED:		
Patient room	175	37.8%
Outside patient room	9	1.9%
Emergency department	61	13.2%
Intensive/critical care unit	34	7.3%
Operating room	81	17.5%
Outpatient clinic/office	8	1.7%
Blood bank	3	0.6%
Venipuncture	1	0.2%
Dialysis facility	1	0.2%
Procedure room	31	6.7%
Clinical laboratories	7	1.5%
Service/utility area	2	0.4%
Labor and delivery	8	1.7%
Home-care	6	1.3%
Other, describe	36	7.8%

BBF† INVOLVED IN EXPOSURE: (more than one item can be checked)‡		
Blood or blood products	332	71.7%
Vomit	10	2.2%
Sputum	36	7.8%
Saliva	29	6.3%
Cerebro-spinal fluid	2	0.4%
Peritoneal fluid	3	0.6%
Pleural fluid	6	1.3%
Amniotic fluid	12	2.6%
Urine	23	5.0%
Other body fluid	50	10.8%

WAS THE BODY fLUID, OTHER THAN BLOOD, VISIBLY CONTAMINATED WITH BLOOD?		
Yes	220	67.1%
No	81	24.7%
Unknown	27	8.2%

EXPOSED PART(s): (more than one item can be checked)‡		
Intact skin	187	40.4%
Non-intact skin	82	17.7%
Eyes (conjunctiva)	268	57.9%
Nose (mucosa)	7	1.5%
Mouth (mucosa)	37	8.0%
Other exposed parts	12	2.6%

DID THE BLOOD OR BODY fLUID: (more than one item can be checked)‡		
Touch unprotected skin	355	82.9%
Touch skin through gap between protective garments	48	11.2%
Soak through protective garment	13	3.0%
Soak through clothing	12	2.8%

BARRIER ITEMS WORN AT TIME OF EXPOSURE: (more than one item can be checked)‡		
Single pair latex/vinyl gloves	283	61.1%
Double pair gloves	28	6.0%
Goggles	18	3.9%
Eyeglasses (not protective)	50	10.8%
Eyeglasses with sideshields	2	0.4%
Faceshield	13	2.8%
Surgical mask	42	9.1%
Surgical gown	39	8.4%
Plastic apron	2	0.4%
Lab coat, cloth (not protective)	25	5.4%
Lab coat, other	4	0.9%
Other item	50	10.8%

CAUSE OF EXPOSURE:		
Direct patient contact	218	47.7%
Specimen container leaked/ spilled	21	4.6%
Specimen container broke	10	2.2%
IV tubing/bag/pump leaked	38	8.3%
Other body fluid container spilled/leaked	23	5.0%
Touched contaminated equipment/surface	5	1.1%
Touched contaminated drapes/ sheets, gown	3	0.7%
Feeding/ventilator/other tube separated/leaked/spilled	45	9.8%
Other, describe	89	19.5%
Unknown	5	1.1%

SOURCE PATIENT IDENTIFIABLE?		
Yes	427	96.6%
No	12	2.7%
Unknown	3	0.7%

LENGTH OF TIME BBF IN CONTACT WITH SKIN OR MUCOUS MEMBRANES:		
Less than 5 minutes	357	80.8%
5-14 minutes	49	11.1%
15 minutes-1 hour	30	6.8%
More than 1 hour	6	1.4%

AMOUNT OF BBF THAT CAME IN CONTACT WITH SKIN OR MUCOUS MEMBRANES:		
Small amount (up to 5 cc)	387	88.4%
Moderate amount (up to 50 cc)	37	8.4%
Large amount (more than 50 cc)	14	3.2%

EXPOSURE LOCATION		
Largest exposure:		
Arm	23	5.3%
Face/head	305	70.9%
Front	9	2.1%
Hand, left	26	6.0%
Hand, right	61	4.2%
Leg	6	1.4%

MEDIUM-SIZED EXPOSURE:		
Arm	11	5.5%
Face/head	138	69.3%
Foot	1	0.5%
Front	20	10.1%
Hand, left	10	5.0%
Hand, right	17	8.5%
Leg	2	1.0%

SMALLEST EXPOSURE:		
Arm	11	25.0%
Face/head	10	22.7%
Front	7	15.9%
Hand, left	11	25.0%
Hand, right	3	6.8%
Leg	2	4.5%

TYPE OF FACILITY:		
Non-teaching hospital	193	41.7%
Teaching hospital	270	58.3%

†BBF = blood or body fluids
‡Because more than one item can be checked in this category, percentages total more than 100%.

facility with a lower rate. For that reason, comparing rates between hospitals may not be very meaningful. It is more reliable to track injury trends within a single institution over several years, and make historical comparisons as prevention measures are implemented.

2001 Percutaneous Injury Data for Safety Devices

The third EPINet report provides 2001 data on PIs involving safety devices. It is important to note that, while PIs can still occur with safety devices—either during use or, if the safety mechanism is not activated, after use—the widespread implementation of safety devices over the last decade (and concommitant decrease in the use of conven-

Needlestick and Sharp-Object Injury Report, EPINet 2001, 58 Healthcare Facilities*, Injuries from Safety Devices Only

Total cases = 393 (excluding injuries before use); total avg. daily census = 8,703
(*13 teaching/45 nonteaching hospitals)

JOB CATEGORY:		
M.D. (attending/staff)	4	1.0%
M.D. (intern/resident/fellow)	5	1.3%
Nurse RN/LPN	240	61.1%
Nursing student	4	1.0%
Respiratory therapist	5	1.3%
Surgery attendant	7	1.8%
Other attendant	7	1.8%
Phlebotomist/venipuncture/ I.V. team	53	13.5%
Clinical laboratory worker	9	2.3%
Technologist (non-lab)	9	2.3%
Housekeeper	6	1.5%
CNA/HHA	18	4.6%
Security	1	0.3%
Other student	3	0.8%
Other	22	5.6%

WHERE INJURY OCCURRED:		
Patient room	198	50.5%
Outside patient room	10	2.6%
Emergency department	38	9.7%
Intensive/critical care unit	30	7.7%
Operating room	29	7.4%
Outpatient clinic/office	19	4.8%
Venipuncture	8	2.0%
Dialysis facility	4	1.0%
Procedure room	10	2.6%
Clinical laboratories	6	1.5%
Autopsy/pathology	2	0.5%
Service/utility area	1	0.3%
Labor and delivery	5	1.3%
Home-care	11	2.8%
Other	21	5.4%

ORIGINAL PURPOSE OF SHARP ITEM:		
Unknown, n/a	11	2.8%
Injection, IM/subcutaneous	99	25.2%
Heparin or saline flush	11	2.8%
Other injection/aspiration I.V.	18	4.6%
Connect I.V. line	1	0.3%
Start I.V. or heparin lock	44	11.2%
Draw venous blood sample	153	38.9%
Draw arterial blood sample	12	3.1%
Obtain body fluid/tissue sample	4	1.0%
Fingerstick/heel stick	16	4.1%
Suturing	3	0.8%
Cutting (surgery)	5	1.3%
Contain specimen/ pharmaceutical	1	0.3%
Other, describe	15	3.8%

WHEN INJURY OCCURRED:		
During use of item	126	32.2%
Between steps of multistep procedure	24	6.1%
Disassembling device	12	3.1%
In preparation for reuse of reusable instruments	1	0.3%
Recapping used needle	15	3.8%
Withdrawing from resistant material	16	4.1%
Other after use, before disposal	100	25.6%
Item left on or near disposal container	3	0.8%
Putting item into disposal container	27	6.9%
After disposal, stuck by item protruding from disposal container	11	2.8%
After disposal, item protruding from trash bag or inappropriate waste container	2	0.5%
Restraining patient	1	0.3%
Device left on floor, table, bed or other inappropriate place	16	4.1%
Other, describe	37	9.5%

TYPE OF DEVICE CAUSING INJURY:		
Disposable syringe	197	51.3%
Prefilled cartridge syringe	5	1.3%
Blood gas syringe	2	0.5%
Syringe, other type	2	0.5%
Winged steel needle	70	18.2%
I.V. catheter (stylet)	27	7.0%
Vacuum tube blood collection holder/needle	39	10.2%
Unattached hypodermic needle	2	0.5%
Needle, unknown type	1	0.3%
Needle, describe	13	3.4%
Lancet	16	4.2%
Suture needle	3	0.8%
Scalpel, reusable	4	1.0%
Microtome blade	1	0.3%
Scalpel, disposable	1	0.3%
Pin	1	0.3%

SOURCE PATIENT IDENTIFIABLE?		
Yes	377	96.7%
No	9	2.3%
Unknown	3	0.8%
Not available	1	0.3%

INJURED WORKER ORIGINAL USER OF SHARP ITEM?		
Yes	317	81.1%
No	71	18.2%
Unknown	3	0.8%

SHARP ITEM CONTAMINATED?		
Yes	370	94.4%
No	2	0.5%
Unknown	20	5.1%

WAS SAFETY MECHANISM ACTIVATED?		
Yes, fully	44	12.1%
Yes, partially	61	16.8%
No	258	71.1%

IF YES, DID INJURY OCCUR:		
Before activation	182	56.9%
During activation	84	26.3%
After activation	54	16.9%

DEPTH OF INJURY:		
Superficial (little/no bleeding)	181	46.9%
Moderate (skin punctured, some bleeding)	191	49.5%
Severe (deep stick/cut, profuse bleeding)	14	3.6%

BODY PART INJURED:		
Arm	5	1.3%
Back	1	0.3%
Foot	1	0.3%
Front	2	0.5%
Hand, left	212	55.6%
Hand, right	156	40.9%
Leg	4	1.0%

GLOVES–Did sharp item penetrate:		
Single pair of gloves	286	74.9%
Double pair of gloves	10	2.6%
No gloves	86	22.5%

TYPE OF FACILITY:		
Non-teaching hospital	177	45.0%
Teaching hospital	216	55.0%

tional devices) has contributed to a marked decline in percutaneous injury rates overall.

A comparison of 2001 data for safety devices with the overall 2001 sharp-object injury data (safety and conventional devices combined) yields some interesting findings. Perhaps most striking is that, while nurses sustained 43.6% of PIs overall, in the safety device data they sustained 61%—probably because they handle the majority of safety devices in hospital settings. By contrast, while physicians sustained 14.7% of PIs in the overall data, they sustained only 2.3% in the safety device data. That is because physicians, in contrast to nurses, handle a relatively small proportion of safety devices (conventional, sharp-tip suture needles cause about 35% of injuries to physicians overall).

After nurses, phlebotomists have the highest proportion of PIs (13.5%) from safety devices, which is more than twice the fraction of injuries for phlebotomists in the overall 2001 data (5.9%). This may be attributed to the fact that most hospitals have made implementation of safety phlebotomy devices a priority, because blood-drawing and I.V. catheter placement are the highest-risk procedures for bloodborne pathogen transmission. With a significant increase in the use of safety devices for blood drawing, the number of PIs from safety phlebotomy devices will increase (although the overall risk of sustaining a PI while performing phlebotomy has markedly declined).

In the safety device data, 50.5% of PIs occurred in patient rooms, compared to 31.5% in the overall data; this probably reflects the relatively higher proportion of nurses using safer devices. The next most-frequent place of injury with safety devices was the emergency department (9.7%), which was close to the percentage in the overall data (9.4%). However, the low proportion of PIs from safety devices in the operating room setting (7.4%) contrasted sharply with the high proportion of injuries occurring in ORs in the overall data (28.8%). This likely reflects the generally low acceptance of safety devices in ORs. This is borne out when comparing data for suture needles: in the overall data, 17% of PIs involve suture needles, but in the safety device data, only 0.8% do.

The percentage of PIs that occurred while withdrawing a device from rubber or resistant material (as commonly occurs when transferring the contents of a syringe to a specimen tube) is slightly higher for safety devices (4.1%) compared to the overall data (3.6%). This suggests a misuse of safety syringes for blood drawing, which defeats the benefit of the safety feature.

In 56.9% of injuries from safety devices, the injury occurred before activation of the device, and in 26.3% of cases it occurred during activation.

Overall, we found that safety devices have resulted in important changes in the ways in which needlestick injuries occur. We also found that safety devices have resulted in marked reductions in PI rates, as described in the preceding article. However, we still need to be vigilant about how safety devices are used, because it is clear that they can be misused. Continuing efforts should focus on compliance and on using safety devices in the safest possible way.

Scalpel Blades: Reducing Injury Risk

By Jane Perry, M.A., Ginger Parker, M.B.A., and Janine Jagger, M.P.H., Ph.D.

Vol. 6, no. 4, 2003

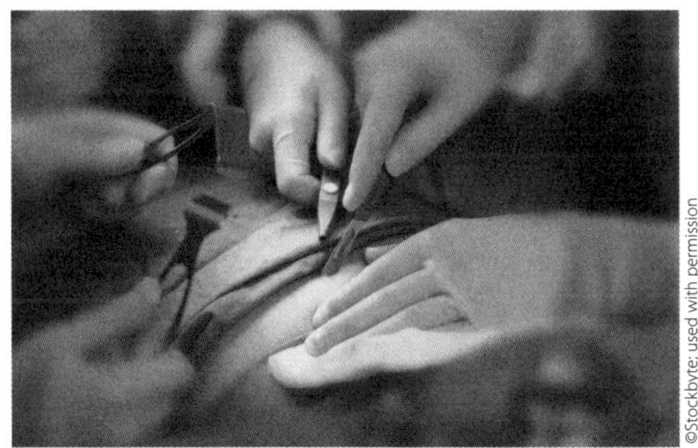

- *As a surgery attendant passed a scalpel to the surgeon, the surgeon simultaneously reached for it; they bumped hands and the attendant's left index finger was cut.*
- *A physician and nurse completed a C-section and were cleaning the patient. As the nurse reached back for a towel, she was cut by a scalpel held by the surgical technician.*
- *An attending surgeon was assisting the primary thoracic surgeon with a lobectomy. As the attending withdrew his left hand from the chest incision, he was cut by the scalpel held by the primary surgeon.*
- *Following an outpatient procedure a nurse was cleaning up the equipment, and used a hemostat to remove the blade from a reusable scalpel handle. The blade slipped and the nurse cut her middle right finger.*
- *A surgery attendant was reaching across a mayo stand to place a suture needle on a needle board and was stuck by a scalpel blade lying on the stand.*

These descriptions of scalpel blade injuries to healthcare workers, from the EPINet multihospital sharps injury database coordinated by the International Healthcare Worker Safety Center at the University of Virginia, illustrate how scalpel injuries can occur in surgical settings. In EPINet data from 1993 to 2001, reusable and disposable scalpels together ranked third as a cause of sharps injuries across all healthcare settings, accounting for 7% of injuries. In operating rooms (ORs) specifically, scalpels caused 18% of injuries—second only to suture needles, which caused 41% of injuries.

A detailed analysis of scalpel blade injuries in surgical settings, based on the most recent two years for which EPINet data are available (2000-2001; 133 scalpel injuries, 61 healthcare facilities contributing data), reveals that scalpel blades are more likely than needles to cause deep or otherwise severe injuries. While 39% of suture needle injuries were classified as moderate (skin cut, some bleeding) and 2% as severe (deep cut, profuse bleeding), these numbers were much higher for scalpels: 58% were classified as moderate, and 11% as severe. This suggests a higher probability of significant blood contact between patients and surgical personnel from injuries involving scalpel blades. Cases of HIV transmission to healthcare workers following scalpel injuries have been documented in both the United States and Italy; they involved, respectively, a pathologist performing an autopsy and a surgeon.[1] When a scalpel injury results in bleeding and the healthcare worker's hands are in or near the surgical site, there is a further risk of healthcare-worker-to-patient transmission of bloodborne pathogens such as HIV or hepatitis C.

Reusable vs. Disposable Scalpels

Reusable scalpels, which require removal of the blade in order to reuse the handle, caused more than twice as many injuries in this analysis as disposable scalpels, which eliminate this step (68% of scalpel injuries were caused by reusables, 32% by disposables). It appears, however, that use of reusable scalpels may be declining: in EPINet data from 1993-94 for percutaneous injuries in ORs, 21% of OR injuries were caused by reusable scalpels, while only 0.2% were caused by disposables; in 2000-2001 data, the fraction of injuries caused by reusables—9.5%—was much lower, while the proportion of injuries from disposables, 4.4%, had increased significantly (see **Figure 1**).

Who is Injured, and When

In EPINet data for 2000-2001, surgery attendants sustained the largest proportion of scalpel injuries (36%), followed by nurses (27%), physicians (18%), and non-lab technicians (9%) (**Figure 2**). In 76% of cases, the injured worker was *not* the original user of the scalpel. This is particularly significant because many injuries from scalpels occur when surgeons are passing them to nurses or other OR personnel. The largest proportion of scalpel injuries occurred "between steps of a multi-step procedure" (41%, 54/133) (**Figure 3**). Of these 54 cases, 12 workers specifically noted in the description area of the EPINet form that the injury occurred during passing (other "between steps" injuries may have involved passing as well, but were not noted as such by the injured worker). An additional 31% of scalpel injuries occurred during use, and 10% during disassembly.

When we separated injury data in the "disassembly" category for reusable and disposable scalpels, we found twice as many injuries during disassembly for reusable compared to disposable scalpels. (When workers indicate "during disassembly" for disposable scalpels, they may mean during activation of the safety shield. Injuries can occur during this

Figure 1. Injuries in Surgical Settings from Scalpel Blades: Reusable vs. Disposable

EPINet 1993–94 • 55 healthcare facilities
Reusable scalpels: 349 injuries, disposable scalpels: 3 injuries

EPINet 2000–2001 • 61 healthcare facilities
Reusable scalpels: 91 injuries, disposable scalpels: 42 injuries

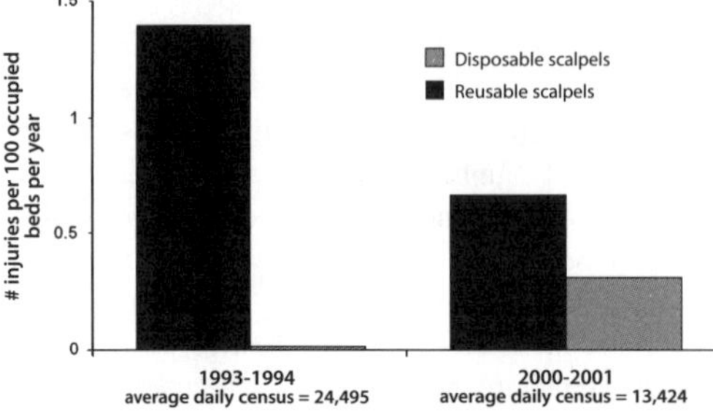

International Healthcare Worker Safety Center, University of Virginia

Figure 2. Job Category of Workers Sustaining Injuries from Scalpel Blades in Surgical Settings

EPINet • 61 healthcare facilities • 2000–2001 • 133 cases

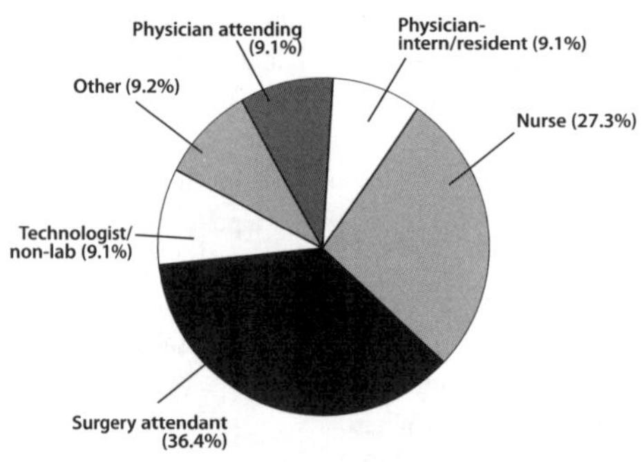

International Healthcare Worker Safety Center, University of Virginia

Figure 3. Mechanism of Scalpal Blade Injuries in Surgical Settings

EPINet • 61 healthcare facilities • 2000–2001 • 133 cases

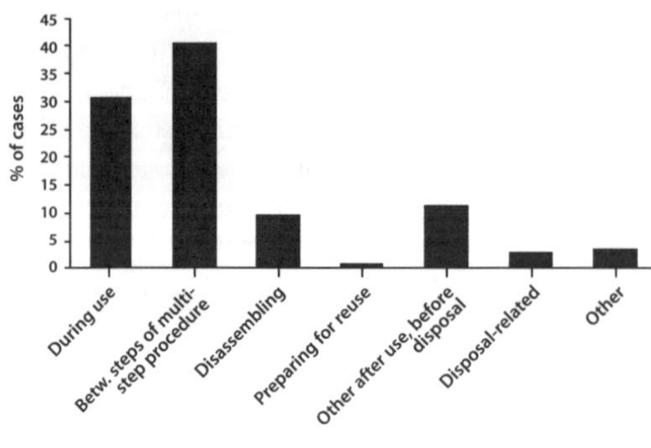

International Healthcare Worker Safety Center, University of Virginia

step if two hands are used to activate the shield instead of one. Activating the shield should be a one-handed operation, using the thumb to slide the shield over the blade or the blade into the handle. All parts of the hand thus remain behind the blade.)

An additional 14% of scalpel injuries occurred after use but before disposal (this category excludes disassembly injuries), or during or after disposal; these injuries could potentially be eliminated by using safety-designed scalpels where the blade can be covered between uses and after use. However, in 90% of cases workers reported that the scalpel injuring them was not a safety design.

The largest proportion of scalpel injuries, 94%, were to the hands; 3% were to the arms, and 2% to the feet. The right hand sustained 56% of injuries, and the left hand, 38%; this injury pattern may be related to hand-to-hand passing, in which the dominant hand (usually the right one) is used to receive instruments.

Preventing Scalpel Injuries

To reduce injuries from scalpel blades, a variety of prevention strategies are needed that target the different mechanisms of scalpel injuries.

Scalpels with shielded or retractable blades that can be placed in a protected position during passing and after use have the potential to prevent as many as 65% of scalpel injuries, if used consistently and correctly. However, it is important to document actual compliance in activating the safety feature, as well as prevention efficacy.

Disposable safety scalpels should be used preferentially to reusable ones, in order to reduce injuries associated with disassembly; there may be some procedures, however, for which reusable scalpels are still the only option. For example, some surgical procedures (such as those in deep cavities like the chest) require an "extended length" scalpel to reach the tissue being cut. Disposable safety scalpels are not yet available in these extended length sizes, although BD reports that it will soon have ones available.

Injuries that occur during cutting may be prevented by using alternative cutting methods when appropriate, such as blunt-tip scissors, blunt electrocautery devices, and laser devices. Other strategies include substituting round-tip for sharp-tip scalpel blades (blades with pointed tips cause more severe injuries than ones with round tips), and choosing endoscopic surgery instead of open surgery when possible. Manual tissue retraction (which puts fingers in closer proximity to the scalpel blade) should be avoided by using mechanical retraction devices.

Since the hands are the part of the body most frequently injured by scalpel blades, their protection requires special attention. No gloves provide total protection from cuts, but ones made of steel mesh, Kevlar, leather, or knitted cut-resistant yarn are resistant to lacerations and can be worn under latex or vinyl gloves[2]; these are particularly relevant for the non-dominant hand, where the added bulk interferes less with dexterity.

Products Designed to Reduce Risk of Scalpel Blade Injuries*

- ***Retracting and shielded scalpel blades*** *(e.g., BD Bard Parker Protected Disposable Scalpel, DeRoyal Retractable Safety Scalpel, Futura Safety Scalpel, Personna Safety Scalpel, Shackelford Retractable Safety Scalpel)*
- ***Rounded-tip scalpel blades***
- ***Disposable scalpels*** *(eliminate need for blade removal)*
- ***Ultrasonic scalpels*** *(e.g., Harmonic scalpel/Ethicon Endo-Surgery)*
- ***Products that facilitate no-hands passing*** *(e.g., BladeSafe Scalpel Transfer Device from Future Medical Devices; Devon Hands-Free-Transfer Magnetic Drapes from Kendall Devon)*
- ***Quick-release blade handles*** *to facilitate blade removal for reusable scalpels*
- ***Scalpel blade removal devices*** *(e.g., Qlicksmart Scalpel Blade Remover; Swann-Morton Surgical Blade Remover; BLDX-2260-K Scalpel Blade Exchanging System), for removing blades from reusable handles, to eliminate manual removal of blades.*

*This list is not exhaustive and does not necessarily represent all such devices currently on the market. Inclusion on this list is not, and should not be interpreted as, an endorsement by the International Healthcare Worker Safety Center.

No-hands-passing of sharp devices during surgery can help reduce passing-related injuries that result from collisions between hands and sharp instruments; this technique involves designating a neutral or "safe" zone, such as a mat, tray, or specially designed surgical drape, where instruments can be placed and picked up.[3]

Another important safety strategy is avoiding unnecessary speed and motion. Surgeon Mark Davis, in his book Advanced Precautions for Today's OR, notes: "When sharps are in use on the [surgical] field, there is rarely a need for excessive speed. The most technically proficient surgeons finish procedures faster than less experienced surgeons not by the use of rapid hand motion but by avoiding unnecessary and repetitive movements." According to Davis, good verbal communication is also important for injury prevention: "When a sharp is in use, avoid anticipating motion of anyone's hands; wait for the surgeon to request sponging. Do not assume anything. Often students or new graduates, eager to help, place their hands unnecessarily near sharps in use or perform needless maneuvers."[4]

By implementing and using these prevention strategies, OR managers and personnel can significantly reduce the

Using safety scalpels safely: Who should activate the protective feature during surgery?

Mark Davis, MD, FACOG, a surgeon who consults on exposure prevention issues in surgical settings, describes a current controversy relating to the use of safety scalpels:

"Who should activate and deactivate the safety feature of a safety scalpel? During simple unassisted procedures, such as insertion of a central line or newborn circumcision, the one using the scalpel clearly has that responsibility. But what about during assisted procedures in the operating room, where the scalpel must be passed between personnel? At a number of hospitals where I've consulted recently, I've heard that the OR staff wants the surgeon to uncover and cover the blade before and after each use. But many surgeons object, because during surgery their gloves may be wet and slippery, making such additional manipulation potentially hazardous. And the surgeons may have a point. If—and only if—a neutral or safe zone for no-hands-passing of the scalpel is established, it would be reasonable to allow the blade to remain exposed during no-hands-passing. Then the scrub person would activate the safety shield at the end of the procedure.

"OSHA says we need to involve the users of the devices in these kinds of decisions, as common sense would dictate. It's important that surgeons and OR staff weigh in on this and other device-related safety issues."

risk of scalpel injuries and bloodborne pathogen transmission in surgical settings.

Note: A version of EPINet is available that is designed specifically for surgical settings. Access/EPINet-OR can help facilities track sharps injuries in this unique environment; it is suitable for both inpatient and outpatient surgical settings as well as labor and delivery suites. To obtain a copy, send an e-mail to the Center's EPINet coordinator at: epinet@virginia.edu.

References

1. Ippolito G, Puro V, Heptonstall J, Jagger J, et al. Occupational human immunodeficiency virus infection in healthcare workers: worldwide cases through September 1997. *Clin Infect Dis.* 1999;28:365–383.
2. Diaz-Buxo JA. Cut resistant glove liner for medical use. *Surg Gynecol Obstet.* 1991;172:312–314.
3. Stringer B, Infante-Rivard C, Hanley JA. Effectiveness of the hands-free technique in reducing operating theatre injuries. *Occup Environ Med.* 2002:59(10):703–707.
4. Davis MS. *Advanced Precautions for Today's OR: The Operating Room Professional's Handbook for the Prevention of Sharps Injuries and Bloodborne Exposures.* Atlanta, Georgia: Sweinbinder Publications, 2001; pp. 70–71.

EPINet Data Report:
Injuries from Phlebotomy Needles

By Jane Perry, M.A., and Janine Jagger, M.P.H., Ph.D. | *Vol. 6, no. 4, 2003*

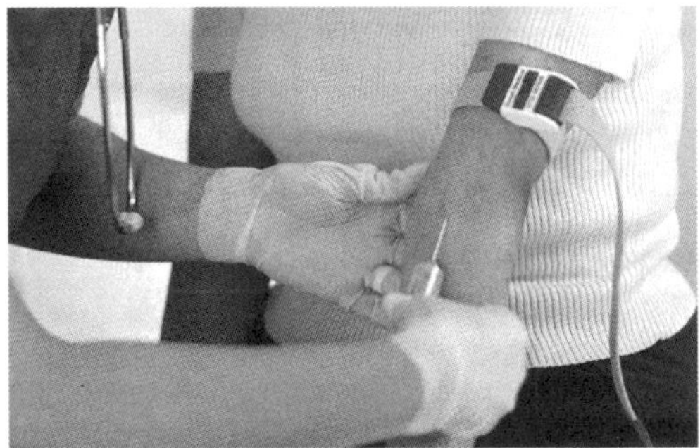

Over the last three years, the issue of reusing blood tube holders, and the safety concerns associated with this practice, have become a subject of debate, (see "Reuse of Blood Tube Holders, Redux"). To help shed light on the issue, the International Healthcare Worker Safety Center analyzed data on injuries from phlebotomy needles from the EPINet sharps injury database. The Center coordinates a network of healthcare facilities around the country that contribute data on a voluntary basis to its EPINet database.

Nine years of EPINet data (1993–2001) from 90 facilities indicate that approximately 5% of sharps injuries (1288/25,043) were caused by phlebotomy needles.[1] (On the EPINet form, the device is denoted as "vacuum tube blood collection holder/needle.") Of phlebotomy needles causing injury, 91% were used to draw venous blood, and 3% to draw arterial blood; thus, 94% of injuries from phlebotomy needles involved a blood-filled needle, the type that is highest risk for bloodborne pathogen transmission. For all other devices, the fraction of injuries involving blood-filled needles was 21%.

Figure 1 shows phlebotomy needle injuries by job category: nurses sustained the highest proportion of injuries (41.2%), followed by phlebotomists/venipuncture/IV teams (32.8%), clinical lab workers (7.4%), and non-surgical attendants (4.8%). The "other" category includes housekeepers, respiratory therapists, and paramedics.

Figure 2 shows the location of phlebotomy needle injuries: 50% occurred in patient rooms, followed by emergency departments (13.5%), outpatient clinics and offices (10.5%), intensive care/critical care units and venipuncture centers (4.4% each), clinical labs (4.3%), and home-care settings (3.8%).

Figure 3 shows the mechanism of injury (when the injury occurred in the use/disposal cycle) for phlebotomy needles. Almost one-fourth (23.4%) occurred during use, and 4.5% between steps of a multi-step procedure. The

Figure 1. Job Category of Workers Injured by Phlebotomy Needles

EPINet • 1993-2001: 90 healthcare facilities • 1,288 injuries

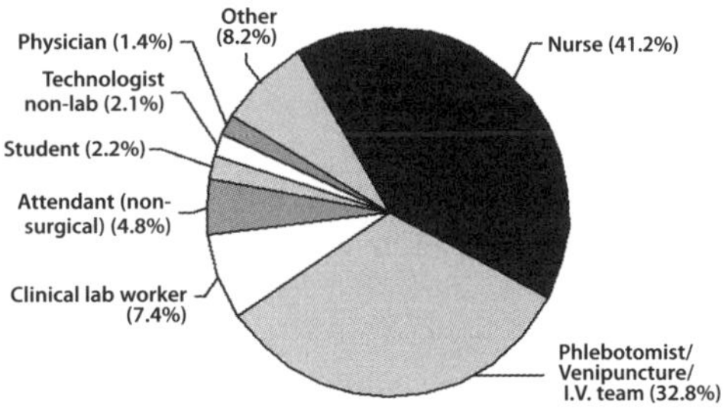

International Healthcare Worker Safety Center, University of Virginia

Figure 2. Location of Phlebotomy Needle Injuries

EPINet • 1993-2001 • 90 healthcare facilities • 1,288 injuries

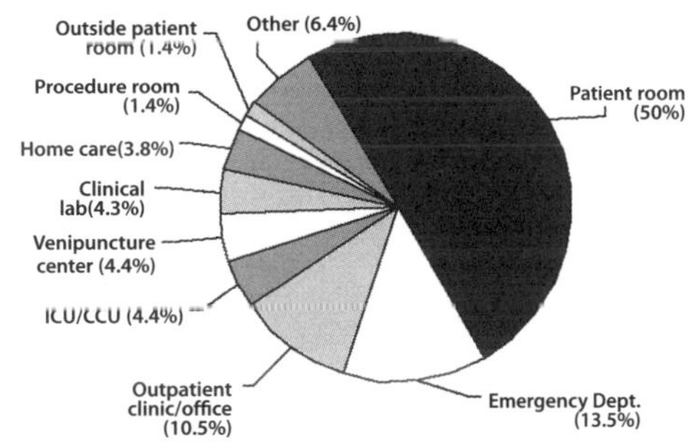

International Healthcare Worker Safety Center, University of Virginia

Figure 3. Mechanism of Injuries from Phlebotomy Needles

EPINet • 1993-2001: 90 healthcare facilities • 1,288 injuries

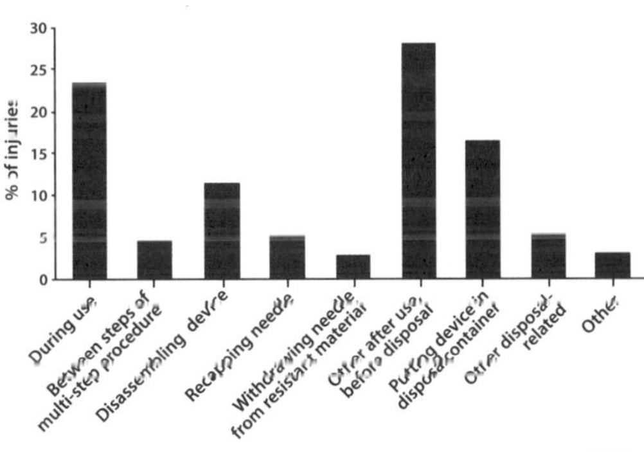

International Healthcare Worker Safety Center, University of Virginia

highest proportion of injuries, 28%, occurred after use but before disposal; for all other devices, 21.9% of injuries fell in this category. Of note, 11.4% of phlebotomy needle injuries occurred during disassembly of the device, more than double the proportion—5.2%—for all other devices. This is important because injuries that occur when the worker is removing a phlebotomy needle from a blood tube holder are most likely to be placed in the "disassembly" category or, alternatively, in the "other after use, before disposal" category.

Injuries related to disposal were also markedly higher for phlebotomy needles compared to all other devices: 21.9% for phlebotomy needles vs. 15.5% for all other devices. In particular, the fraction of injuries that occurred while putting the sharp in the disposal container were much higher for phlebotomy needles (16.6%) compared to all other devices (7.2%). One possible mechanism of these injuries, among others, is needle removal. Some phlebotomy systems are designed with a release mechanism: the worker places the tube holder, with needle attached, over the sharps container and activates the release mechanism on the holder to eject the needle into the container. Another mechanism of needle removal is by means of a notched opening in the disposal container, sometimes called an "unwinder": the worker places the needle in the unwinder and twists or pulls back the tube holder to release the needle. These two release mechanisms do not include an integrated needle safety feature, as specified in the bloodborne pathogens standard.

Needlesticks from the Back End of Phlebotomy Needles

On the EPINet Needlestick and Sharp-Object Injury Report form, the injured worker indicates the type of device that caused the injury; among the choices is "vacuum tube blood collection holder/needle." But the report form does not distinguish whether an injury from a double-ended phlebotomy needle involved the "front end" (the end that is inserted into the patient) or the "back end" (the end that punctures the blood tube). Elsewhere on the form, however, the worker is asked to describe the circumstances leading to the injury, and can give additional details about the exposure. These text descriptions have been entered into the EPINet database since 2000; there are now two years of descriptions (2000 and 2001) available.

Of the 146 injuries from phlebotomy needles in the EPINet database for 2000 and 2001, 114 included injury descriptions.[2] We reviewed all the phlebotomy needle injury descriptions, and found that 12, or 10.5% (12/114), indicated that the injury involved the back end of the needle.*

**Other injuries from phlebotomy needles in the EPINet database may have involved the back end of the needle, but unless the healthcare worker specifies in the description section which end of the needle caused the injury, it cannot be determined from the data.*

Of the 12 back-end injuries, 5 could be directly related to removing the needle from a blood tube holder (in the other 7 descriptions, needle removal was not mentioned). In addition to the 12 back-end injuries, another 7 descriptions of phlebotomy needle injuries indicated that the injury occurred while removing the needle from a blood tube holder (but did not specify which end of the needle caused the injury).

The texts for these 19 descriptions are provided below. Text in italics or brackets indicate editorial comments.

I. Descriptions: Needlesticks from the Back End of a Phlebotomy Needle, Associated with Removing Needle from Blood Tube Holder

(1) "After drawing blood ... disconnected needle from blood tube holder. Stuck self with covered end that pierces blood tube."

(2) "Did the venipuncture; unsuccessful attempt. Withdrew the needle, unscrewed the hub from the needle. Stuck self with the rubber end of the needle where it did not penetrate the patient."

(3) "I picked up the blood tube holder, twisted off the blood drawing piece and set it on the bed. In the dark I grabbed the blood drawing piece end to end between index finger and thumb. Rubber cover over needle was pushed back and needle penetrated through gloved thumb."

(4) "[I was] disposing of needle in the sharps container. Due to the spring-type action the needle (tube holder end) pushed back up poking me in the thumb. The phlebotomy needle did not fit in the unwinder in the sharps container to eject the needle into the container."

(5) "Venipuncture was performed [and] safety device [was] activated; was taking needle apart from holder and stuck finger." *[This appears to be a back end needlestick since the worker indicates that the safety feature—which would have shielded the front end of the needle—was activated. This is an interpretation.]*

II. Descriptions: Other Needlesticks from the Back End of a Phlebotomy Needle

(1) "Associate picked up what she thought was plastic lying on the floor with a tissue that was lying near it. Got stuck with needle used for inside blood tube holder."

(2) "After drawing blood was disposing of phlebotomy needle and holder; finger slid inside holder and stuck finger with needle inside of holder."

(3) "Stuck finger with phlebotomy needle after drawing blood. Rubber did not come back down over needle."

(4) "I was taking blood tube holder off of butterfly so that I could hook syringe to butterfly for next step of ... test and I punctured rubberized end of needle into finger on right hand."

(5) "Upon disposing of needle into sharps container she

was stuck by a phlebotomy needle that had not gone down into the sharps container. The rubber end of the phlebotomy needle is the end that stuck her."

(6) "Needlestick through glove from phlebotomy needle after central line blood draw. Adaptor stuck in end of central line lumen; outer cylinder came off, exposed the sharp."

(7) "Bottom end of butterfly needle stuck my thumb as I was disposing of it." *[Butterfly needles are sometimes connected to blood tube holders—with a single-ended phlebotomy needle attached—for the purpose of blood drawing.]*

III. Descriptions: Needlesticks from Phlebotomy Needles Associated With Needle Removal Problems

(1) "While removing needle from holder, stuck self with needle."

(2) "Disconnected blood tube holder from needle insert, needle poked thumb."

(3) "While removing needle from blood tube holder for disposal, needle pricked skin."

(4) "Needle was recapped and I was twisting the needle off the needle holder and the needle poked out the side of the cap."

(5) "After blood draw tried to engage safety device, device stuck so she tried to manually remove needle, stuck palm."

(6) "Associate finished drawing blood from a patient and went to put [needle] in a sharps container and realized there was no remover [i.e., needle unwinder]. She then set it on the counter behind her and got stuck."

(7) "After using phlebotomy needle/blood tube holder, was disposing of needle and it would not break off [i.e., would not disengage from holder]. Then attempted to put cap back on needle when punctured right index finger."

We looked separately at injury descriptions for safety-engineered phlebotomy needles in the 2000-2001 data; of 62 injuries, 51 included injury descriptions. Of these 51, 7 specified that the injury involved the back end of the needle (7/62=11%). These data indicate that, even with safety phlebotomy needles, the back end of the needle still poses an injury risk.

Conclusions

EPINet injury data on phlebotomy needles demonstrate that needlesticks from the back end of phlebotomy needles are not a rare event in healthcare settings. Although the rubber sheath covering the back end provides little or no protection to the healthcare worker, workers may be less likely to perceive the risk since the back end sharp is hidden by the sheath.

Because phlebotomy needles are high-risk for blood-borne pathogen transmission, it is especially important to utilize all possible engineering and work practice controls to minimize injury risk. The back end of a phlebotomy needle poses its own set of injury risks, in addition to those from the front end of the needle.

For optimal protection, both ends of the needle should be covered (or, in the case of the back end, *remain* covered) after use. As these devices are currently configured, the tube holder provides the only protection for the back end of the needle after use. Unless another shielding mechanism is provided to cover the back end of the needle, the needle and tube holder should be disposed of as a unit after blood drawing procedures in order to minimize high-risk injuries to healthcare workers.

References

1. EPINet Multihospital Needlestick and Sharp-Object Injury Data Report, 1993–2001. International Healthcare Worker Safety Center, University of Virginia Health System; 90 healthcare facilities contributing data. (Report run 5/22/03.)
2. EPINet Multihospital Needlestick and Sharp-Object Injury Data Report, 2000–2001: Injuries from vacuum tube blood collection needle/holder. International Healthcare Worker Safety Center, University of Virginia Health System; 61 healthcare facilities contributing data. (Report run 5/22/03.)

PART II

Occupationally Infected Healthcare Workers:

Interviews & Personal Accounts

L-r: Janine Jagger, Diane Mawyer, Lynda Arnold, Lisa Black, Julie Naunheim Hipps

Patricia Wetzel

A Physician Copes with Occupationally Acquired HIV

Vol. 1, no. 1, 1994

In September 1991, Dr. Patricia Wetzel became, according to the Centers for Disease Control and Prevention (CDC), the 29th U.S. healthcare worker with documented occupationally acquired HIV. She had just finished her residency in family medicine and was directing the in-patient AIDS unit at John Peter Smith Hospital in Fort Worth, Texas, when she sustained a needlestick injury while drawing blood from an HIV-positive patient. She subsequently seroconverted to the virus. She now works part-time for Caremark International's Corporate Human Resources Division, most recently as HIV Program Manager doing company-wide training on HIV in the workplace. Asymptomatic, she spends most of her time traveling in the U.S. and abroad, speaking on HIV and AIDS. During Bill Clinton's 1992 presidential campaign, he named her one of 53 "Faces of Hope," people he designated as "symbols of humanity." AEP was privileged to speak with her recently and discuss her experiences and views on preventing blood exposures in the healthcare workplace.

AEP: *Most articles on your work have focused on the personal side of your story, your transition from "AIDS doctor" to HIV patient, or your advocacy on behalf of women with HIV. How do you view your experience in terms of what other healthcare workers face in the workplace?*

PW: When you look at my needlestick and the immediate aftermath, my responses were very similar to those of other healthcare workers in the same situation. There's the initial panic, shock, and fear associated with the needlestick, and then the denial stage where you think, "Nothing bad will happen to me as a result."

AEP: *Was the denial a psychological response, or was it because people told you that the risk of becoming HIV-infected was low and there wasn't much to worry about?*

PW: A little of both. I certainly didn't want to believe that I could become infected. But there were moments when I was anxious and frightened and looking to talk about the situation with other people, and their reply was, "Oh, you're just being paranoid—you know the odds are slim."

AEP: *Did you have any kind of hospital training on exposure prevention?*

PW: No. I was three months out of my residency and my priority was getting our inpatient AIDS unit off the ground and trying to do the best we could with a minimal, bare-bones staff. As with a lot of people, the issue of occupational exposure was, quite honestly, something I had not given much thought to. The infection control department of the hospital hadn't focused attention on the dangers to a healthcare worker of working exclusively with an HIV-positive population, and our working group didn't have anything in place to deal specifically with the psychological aftermath of an occupational exposure.

AEP: *And nothing in place in terms of prevention?*

PW: No. I received no training specific to this during my residency, as I think is the case with many medical students. Even my knowledge of universal precautions was not from formalized teaching but just passed down through the grapevine in the way a lot of physicians learn things.

AEP: *Is that changing in medical schools now?*

PW: I'm aware of some medical schools that are doing a very good job of talking about universal precautions, OSHA regulations, and the whole topic of occupational exposures and protecting healthcare workers. There's still a huge gap in medical residencies in this area, however. It's assumed that residents know and understand about [universal precautions and occupational exposures], but again, too often there's no formalized training.

AEP: *Do you think residents are at higher risk for needlesticks based on the procedures they are performing and their level of training and experience?*

PW: I certainly have heard anecdotal reports from hospital infection control nurses who say that most of their needlesticks are coming from the inexperienced people on the medical staff—interns and residents who are learning procedures and providing frontline care. I do think that probably residents and interns—people in training—are at somewhat greater risk for sustaining needlesticks.

AEP: *But might that not reinforce the idea that if healthcare workers only have enough training, or are more careful, needlesticks won't occur? That needlesticks are really their fault?*

PW: Well, it may sound that way. There is something to be said for becoming facile with the equipment. On the other hand, I vigorously support the idea that needlesticks shouldn't have any kind of fault implication. They are, however, one of the hazards of learning the trade, which is why there needs to be more education so that medical students understand those risks and can be protected in the learning process.

AEP: *Do you believe that a specific safety device would have prevented your needlestick?*

PW: Yes. First of all, if there had been a sharps container at the patient's bedside, or if I had been provided with a portable sharps container, that would have been a possible intervention. More important would have been a sheathing device on the needle. These kinds of devices and equipment were certainly available at the time.

AEP: *Are they now routinely used by the hospital where your injury occurred?*

PW: What the hospital instituted after my seroconversion was to make sheathing devices available for phlebotomy needles and syringes in the areas they deemed "high risk."

AEP: *How did they determine what was "high risk"?*

PW: They decided that the AIDS unit and the emergency room were the "high risk" areas.

AEP: *That's interesting. Our data [from EPINet] don't show that emergency rooms are at higher risk for needlestick injuries than other areas. Anecdotal notions of what is "high risk" can be very misleading.*

PW: Right.

AEP: *You were practicing universal precautions when your needlestick occurred. What could be done at the federal level to promote healthcare worker safety beyond recommending universal precautions?*

PW: I think we are far behind where we should be in promoting healthcare worker safety and creating a greater level of awareness. Since local efforts seem to be inadequate, I think this issue needs to be promoted from the federal level. It's ironic: for tuberculosis, OSHA came out and mandated certain procedures and precautions. But when it comes to [HIV], they make a rather superficial comment about "Where available, use a safety device." That's just not good enough. In a perfect world, employers would voluntarily comply, but there just doesn't seem to be enough incentive for them to do that right now. We have enough data at this point to suggest in what situations safety devices should be required, and I think there needs to be stronger language coming from the federal government.

AEP: *That's an important point. OSHA has mandated that "engineering controls" such as effective safety devices be the preferred method of compliance with the OSHA bloodborne pathogens standard. The difficulty has been in getting OSHA to specify what an "engineering control" is. Do you agree that getting at that issue is crucial?*

PW: I would agree, absolutely. I'd also like to emphasize another crucial issue: getting accurate numbers from the CDC on healthcare workers who are occupationally infected with HIV. Right now we have what I believe are gross underestimates, and the low numbers have allowed hospital administrators to brush aside the issue. I estimate that at least 50 or 60 healthcare workers have contacted me in the last 18 months who are occupationally infected with HIV and haven't reported their infections or seroconversions—they are totally undocumented.

AEP: *Have you had much dialogue with government officials on this issue?*

PW: Well, I didn't become actively involved in the general issue of healthcare worker safety until more recently; I was focusing on women and HIV. I've always done some lecturing on occupational exposures—as an infected healthcare worker, not as an expert in the field—but now that seems to be the most in-demand topic for my talks to healthcare worker groups. I met recently with representatives from the CDC; I haven't yet talked to anyone at OSHA, the FDA, or NIH.

AEP: *When you speak to healthcare worker groups on the topic of occupational exposures, what kinds of concerns do they have?*

PW: I get many comments from audiences about their appreciation for helping them see that this is an issue. There is still a great deal of denial among healthcare workers that [infection from exposures] can happen to them.

AEP: *So one of your tasks is to increase awareness at a very basic level?*

PW: Exactly. I am the tangible evidence that HIV from needlesticks can happen—I'm not just the infection control person saying this *might* happen. It forces them to really acknowledge the issue. There are also questions about what they can do to protect themselves—so many healthcare workers don't have any idea about available safety devices that their hospitals could provide. Many people want to know about the reporting issues—why accidents need to be reported and how they should be documented. And there are questions about what to do if you have a needlestick with a bad outcome: if you become infected, then what? Some of those concerns are related to workers' compensation.

AEP: *Do you think that the workers' compensation system is adequately providing for the needs of healthcare workers who are exposed to or infected by HIV or other bloodborne pathogens?*

PW: Well actually, in my case I didn't *have* workers' compensation. This is very important: in the state of Texas an employer is not *required* to provide workers' compensation. If the employer has chosen not to provide workers' compensation and an employee has an on-the-job injury, then the employee's only recourse to recover compensation for his damages or injury is to file a lawsuit. If you *are* covered by workers' compensation, then you cannot pursue litigation. I don't know what the laws are in other states, but when I talk to physicians in Texas and tell them that my group didn't have workers' comp, their reaction is: "What do you mean, don't they *have* to have it?"

AEP: *So you think a lot of healthcare workers just assume that their group has workers' compensation?*

PW: Yes. People need to be aware of what kind of compensation benefits they have. It's especially important for physicians, who are generally not employees of hospitals but are employed in small group professional associations (PAs). In terms of how workers' comp has been for other employees who have had a blood exposure and subsequently seroconverted, you'd probably have to look at that on a state-by-state basis.

AEP: *When you talk to groups around the country, what sense do you get of the barriers to a safer workplace that healthcare workers face?*

PW: Basic lack of awareness of and appreciation for the risks to healthcare workers in the workplace—both on the part of healthcare workers themselves and of hospital administrators. There's just not enough education provided all the way through the training process. In many instances there is no access to any kind of safety equipment.

AEP: *And with healthcare costs going up and hospitals facing pressure to cut costs, they may be reluctant to invest in safer devices.*

PW: Right. They think it will be a money-saver if they cut the training programs and don't buy the more expensive safety devices—but that's a very short-sighted approach. Still, some hospitals seem to be getting the message: I was lecturing recently in Tampa and heard about a 450-bed hospital that had gone totally needleless because the administration finally realized that their needlesticks were costing them a lot of money.

AEP: *What kinds of groups invite you to speak? Who is interested in hearing your message?*

PW: I've spoken to a number of large state physicians' associations such as the Florida Medical Association and the California Medical Association. It's interesting, though: when I'm invited to speak at a hospital and it's an open invitation to the staff rather than a mandatory lecture, generally there are many more non-physicians than physicians, with nurses attending in the greatest numbers.

AEP: *Some healthcare workers—perhaps especially physicians—seem to have the attitude that exposure to infectious diseases is just one of the risks you take by being in the healthcare profession and that it's a sign of weakness to try to protect yourself. How widespread do you think that attitude is?*

PW: Well, I know that when I visit my husband in the emergency room and I see the trauma team of surgeons come down to handle a gunshot wound, many of them have a rather cavalier attitude about putting on protective equipment. I do think this attitude is held by many in the health care profession; if there was more education about exposure prevention during medical training perhaps it wouldn't be so widespread.

AEP: *What do you think industry should do to promote healthcare worker safety?*

PW: I'd like to see unsafe equipment pulled off the market where safety devices are available. I don't see why there should be an unsafe vacuum tube phlebotomy set on the market, for example, when there's an alternative that's much safer. I'd also like to see some compromise on the *cost* of safety devices. If you lower the cost, more people will buy them, the companies will make just as much money, and everybody will be happy. Industry can also help by putting pressure on federal agencies to draw attention to the issue of healthcare worker safety.

AEP: *Do you think hospitals and industry are working together effectively in experimenting with the first generation of safety devices, to find out what's working and what's not?*

PW: I think so; but what I'd like to see happen in conjunction with that is more information-sharing among the hospitals themselves. There's not enough communication at a national level about which safety devices have been found to be effective; there's a lot of duplication of effort.

AEP: *Is there anything else you'd like to add for our readers?*

PW: As healthcare workers, we often get caught up in our everyday work and don't take time to think about ourselves. I often say that denial is what allows us to do our job—if you didn't put a psychological barrier between yourself and your patients you'd think you had every disease you saw. Still, that kind of denial can do a great disservice to us as healthcare workers; we need to let that barrier drop and realize that there is much that we can do to protect ourselves. Not only do we need to be aware, we also need to take specific actions to protect ourselves.

"Jane Doe"

HIV Infection from a Needle Used to Access an I.V. Port

Vol. 1, no. 2, 1995

"Jane Doe" is the pseudonym of a nurse at San Francisco General Hospital who was infected with HIV from a needlestick she sustained in July 1987. In the following interview she discusses, among other things, her reasons for maintaining her confidentiality. Asymptomatic, Jane Doe continues to do bedside nursing at San Francisco General, and helps educate her co-workers about needlestick risk and prevention.

AEP: *We would like to talk first about how your needlestick occurred. We understand that it was from a needle on an I.V. line, and we think this is important, because many healthcare workers seem to think you can't get infected that way. Can you describe what happened?*

JD: At the time of my accident, I was working on a medical unit of San Francisco General, a large, urban, public hospital. I was in the 12th hour of a 12-hour shift; it was 6:00 a.m. I was caring for a patient who was being given an antibiotic through an I.V. line with a heparin lock; at that time the heparin lock was capped with a rubber port. The antibiotic was finished and I was going to disconnect the I.V. tubing from the heparin lock and flush it. I withdrew the needle from the rubber port, and there I was—standing with an unsheathed needle.

I really didn't have any good choices in terms of how to dispose of the needle. I was working in the smallest patient room on my unit; it was crowded with furniture and presented an obstacle course. I was on the side of the bed farthest from the bathroom where the sharps container was located. I could have taken the whole assembly down and gone around the bed to the bathroom. I could have recapped the needle. I could have disconnected the needle from the tubing and then walked it around the bed. What I decided to do was to try to insert the needle into the rubber port at the bottom of the I.V. bag. Once I had it in a safe place, I thought I could take the whole set-up into the bathroom and dispose of the needle in the sharps container. But as I attempted to insert the needle into the rubber port, the needle slipped and instead pierced my finger.

AEP: *What could have prevented your needlestick from happening?*

JD: A sharps container immediately at the bedside, or a safety needle device. One that was available at the time was the Click Lock [by ICU Medical]. Then, when the needle was withdrawn from the rubber port, it would have been in a sheath inaccessible to my fingers. Of course, another thing that would have prevented my needlestick would have been for the hospital to have adopted a needleless I.V. system; since my needlestick, the institution has moved toward gradual implementation of a needleless system.

AEP: *Given such changes, do you feel that healthcare workers at your institution are at less risk for needlesticks now than they were when you were injured?*

JD: Yes, definitely. It's still not perfect; equipment design could use improvement. Also, the education needed to make people aware of the new devices and teach them how to use them isn't necessarily where it should be.

AEP: *So you think the use of these new devices isn't as consistent as it should be?*

JD: Yes. You'll find pockets of healthcare workers throughout the hospital who aren't aware that a particular safety device is available, or who may have noticed it but didn't know what it was used for; or they know what it's used for but they haven't had a good in-service, so they're not really comfortable with it; or they have a good understanding of its use but don't feel motivated to use it. You're always dealing with resistance to change, the reluctance to learn the techniques involved in handling a new device—it takes a lot of education to overcome that. As nurses, we're used to having to work quickly; there's a rhythm to what we do—our hands and minds have memorized certain routines. Having to stop and think about what you're doing, and maybe feel awkward, creates a resistance. Of course, people will learn new equipment in order to enhance patient care, or when it's mandated by the institution. But you wonder why healthcare workers are so resistant to learning something new when it's for their *own* protection. I think there is some denial; you know—"The chances are really small, it's not going to happen to me." I have found over the years that I have been reticent to tell people how my accident occurred because I think it's easy for people to say, "Oh, I would never do it that way, so therefore I don't have to worry."

AEP: *Your accident occurred at the end of a 12-hour shift, and you were tired; do you think you would have done something differently at another point in your shift?*

JD: Yes I do, because at the time my accident occurred I remember being really rushed, as well as tired. It was 6:00 a.m., which, as any nurse can tell you, is a really busy time. You have to do meds, dressing changes, there are patients waking up with various needs to attend to, and you're trying to organize your charting so you can report to the next shift.

What might I have done differently? Although neither of these is good practice, I might have stuck the needle in the mattress and then come back to deal with it thoroughly, or I might have recapped the needle. I recapped all the time

> **"We need to be protected from the needles."**
>
> *"I have a very strong response to the suggestion that healthcare workers just need to slow down and think—they just need to prioritize as they work, and then accidents will be less likely to happen. I think that's a myth; it puts the onus and it puts the blame on healthcare workers, in a situation where our workload generally increases without any supports. We are asked to take care of more patients. We're told there's a hiring freeze. The budget is down. We need to spread ourselves a little more thinly. And then in the midst of that, we're asked to slow down and think. And we're asked to do that while we're using sharp instruments.*
>
> *"I think that institutions in general place a lot of responsibility on healthcare workers in settings where they have very little control. We can alter our behavior. We can put on our gloves—and may I say here that I have never seen a puncture-resistant glove. We can follow rules that are laid down for us, but a needle is a needle and accidents happen. We need to be protected from the needles."*
>
> —Jane Doe, testifying at OSHA hearing on the Proposed Rule on Occupational Exposure to Bloodborne Pathogens, 1/16/90

when the sharps container was in the bathroom, even though it was against hospital policy, because to me that was safer than walking with it uncapped. The sharps containers were ultimately moved out of the bathrooms and into the patient rooms, but it took three years. And then, most of the sharps containers were mounted either at the foot of the bed or at the entrance to the room, rather than at the head of the patient's bed.

AEP: *So instead of having to walk to the bathroom with the needle, in some cases you would have to walk to the door of the room?*

JD: Yes, but that's about to change. This is at an institution that I actually believe cares, and that handles a higher volume of AIDS cases than most. But like all institutions, mine gets tangled up in bureaucracies; one committee doesn't know if it's another committee's responsibility—those kinds of things make change happen slowly.

AEP: *With respect to budget issues: most hospitals have gone through financial difficulties. Do you think in general that safety efforts are being slowed down or compromised because of that? Is there an incompatibility between healthcare workers' push for safer devices and administrators who are trying to save money?*

JD: I have seen a situation in which the hospital wanted to limit use of a safety I.V. catheter to the E.R.; it wasn't until after a major fight, including a high-level union grievance, that the device was finally introduced hospital-wide. One of the arguments against hospital-wide use was that "only 8% of our needlesticks are related to I.V. insertion."

AEP: *Our EPINet data shows that for most hospitals, I.V. catheter stylets cause fewer than 10% of injuries—in fact, most devices, except for disposable syringes, cause a very small proportion of the total number of needlesticks. So you can use the same argument for almost every device.*

JD: They didn't cite budget as one of their issues, but the reality was that in order to implement the safety device they were going to have to do intensive training across the board—doctors, nurses, medical students—and that was going to cost big money. But the Critikon ProtectIV safety catheter was finally introduced hospital-wide, and needlesticks in that category have really diminished.

AEP: *What is your opinion of hospitals that haven't converted to needleless or protected needle I.V. systems?*

JD: I think they are totally irresponsible. In economic terms, needlesticks are expensive, and then if you have a seroconversion it's going to cost even more. I'm asymptomatic and I'm costing the city and county of San Francisco over $5,000 a year ($30,000 in these last 7-1/2 years). So I don't think these institutions are really being honest about their cost/benefit ratio. In terms of the moral imperative, I have a really hard time responding, because it seems so obvious to me. Sometimes I wonder how many infected healthcare workers it takes for administrators to get it. I really respond strongly to language like "statistically insignificant"; I want to say, "I'm a statistic, *and* I'm significant—and I'm only one person. Do you need more?" Certainly there are a lot more healthcare workers who are occupationally infected—many more than are reported by the CDC.

AEP: *What do you recommend healthcare workers do who work in institutions that haven't gone to needleless or protected needle systems?*

JD: There's a lot they can do. You know, in my testimony to OSHA I said that I think institutions generally place a lot of responsibility on healthcare workers in settings where they have very little control. Well, we *don't* have a lot of control, but we do have *some*. Healthcare workers need to find out about the health and safety infrastructure in their institutions, and look for the hot points where their input is needed—like the product evaluation committee. Prior to my accident I had no idea that a product evaluation committee existed; at the time there were no representatives from the rank-and-file frontline workers, and now there are. Healthcare workers can start a needlestick prevention committee, or be active in management-labor negotiations.

AEP: *What activities does San Francisco General's needlestick prevention committee support?*

JD: They look at equipment that's coming out; they do pilot studies; they survey staff on problems with new equipment—what they like, what they don't; they plan and implement education.

AEP: *On a different subject: Why were you so concerned about protecting your confidentiality when your needlestick occurred?*

JD: The emotional reason was that I had enough to deal with inside myself without having to cope with other people's responses. But the primary reason was to protect my job security, both at present and in the future. While I didn't see any immediate threat, I didn't know what might be coming down the road. For instance, there was a bill in the Senate that would have mandated HIV testing for all healthcare workers who do "invasive procedures"—but that term was never defined. Does it include putting in an I.V.? My love is bedside nursing, and I feel comfortable that my scope of practice does not put my patients at risk. Do I really think that mandatory testing for healthcare workers could happen in my lifetime? I doubt it. Do I work in an institution that is very firm about my being allowed to continue working there? Yes. But I've chosen to maintain my confidentiality because I don't know what will evolve on the state and national scenes. I'm ambivalent about it, because there are losses; I could make more of a difference if I were "out." I do think that the greater the number of healthcare workers who go public with their infections, the more powerful the issue will be.

AEP: *Why was it so difficult for you to get an agreement with your institution about confidentiality?*

JD: San Francisco General is part of a city and county governmental system. The institution itself was very committed to maintaining my confidentiality—they were very clear when they decided to hold a press conference announcing my needlestick that they were not going to reveal anything about my gender, age, or worker category. But the city and county have their own worker's compensation board, and they would not guarantee my confidentiality. What my attorney and I were asking for was some kind of protocol for processing my claim and handling my files that would keep the number of people who knew my identity to a minimum. The board kept saying they would do what we asked, but when we'd ask for a written plan, they wouldn't give it to us. We went around and around on this.

AEP: *Was it bureaucratic inertia? Why were they so resistant?*

JD: Well, this is where you come back to how much the budget issue comes into play. They weren't saying it was about money; but my situation was going to set a precedent, and perhaps they were afraid that it was going to open some sort of floodgate. They would come up with the most outlandish responses; for instance, rather than having me file a needlestick injury report, they wanted me to file an "injured finger" claim which would have allowed them to handle the matter according to established procedures. But any number of people in their office would be privy to my paperwork, and if that finger claim was suddenly being reimbursed for a t-cell count—well, it wouldn't take a rocket scientist to figure out what had happened. Finally, a sympathetic reporter published a very straightforward article about the fact that, 18 months after my needlestick, I still hadn't gotten any compensation and was paying out-of-pocket for my expenses. As a result, the mayor of San Francisco, probably out of embarrassment, told the board to give me what I wanted. In the end I got a guarantee that a maximum of four people at the worker's comp office would know my true identity; in addition, the conditions of my settlement were incorporated into our 1989 contract negotiations, so that what I got is now available to any other healthcare worker at my hospital who might be in my position. This was very gratifying to me.

AEP: *From your experience, what do you recommend other healthcare workers do to protect their confidentiality if they sustain a needlestick?*

JD: I think healthcare workers need to find out what they can do *before* a needlestick happens, because afterwards you're dealing with too much stress. Find out who the worker's comp carrier is, and how the process would be handled, and whether information would be kept confidential. Does the employee health department have a protocol for dealing with needlesticks? Is a healthcare worker's seroconversion documented in employee health records? If so, under a pseudonym or under a real name? If under a real name, is that name kept locked up, with only one person having access to it? If the healthcare worker was tested at the institution, where is the documentation of tests and test results stored?

AEP: *Do you feel you've been fairly and adequately treated by the worker's compensation system?*

JD: My benefits cover my medical and mental health costs. Worker's comp is paying me what state disability would pay me—I can't get both. And that's it. You decide for yourself if that's enough. In an age of shrinking health coverage, I do feel lucky that I have 100% coverage of my medical costs.

AEP: *A lot of people have worker's comp benefits that don't cover 100%, not nearly 100%; and some don't have worker's comp at all.*

JD: In that respect, California has generous worker's comp laws, and I am grateful for that. But one of the issues for someone like me, getting worker's comp, is that I cannot sue my employer. I can only file a third-party suit—i.e., I could have gone after the manufacturer of the needle—and that is governed by a statue of limitations. I'm not sure what I would have done if I had the chance, but at some level I do feel locked in. I don't have a lot of recourse. I don't have major complaints about the system; but trying to discuss compensation intellectually for something like this is a challenge, because there is no compensation. Perhaps the biggest compensation I could have would be for needlesticks to be obsolete in my lifetime.

AEP: *How can a healthcare worker tell if their hospital is doing enough in the area of needlestick prevention?*

JD: If a healthcare worker comes upon equipment or devices that they're not really familiar with, they should stop and say, do I know what this is for? If you notice that devices are being used inconsistently—some staff are using one piece of equipment for a procedure, some another—then probably not enough is being done in terms of staff development and education. You need to go to the staff educator and tell them what you're seeing. If a needlestick occurs, bring the device to the staff educator, or the health and safety person, or the product evaluation committee, and explain to them—however many times it takes—how the accident happened, and what could have been different to prevent it. Be very, very specific. Because those people are probably dealing with the manufacturers. Or healthcare workers can go directly to the product reps and say, this isn't working, this is a bad design. Somehow healthcare workers need to get better versed in being evaluators; we need to learn the manufacturers' language and they need to learn ours so we can understand each other. When a needlestick happens, people are frightened, they shut down—I've seen co-workers of mine say, well I'm not going to worry about it; if I seroconvert, then I seroconvert. That might be someone's personal reaction, but you will also be doing everyone a big favor if you go to people who can make a difference and say, this is what happened, this is how it happened.

AEP: *That's an important point—a healthcare worker makes a contribution to the safety of other healthcare workers by reporting an exposure incident and providing detail, because that's the foundation of prevention.*

Do you think hospitals should provide a special level of protection for healthcare workers on AIDS or HIV units?

JD: Well, I wouldn't want to see a two-tiered system develop where people are more strict about enforcing universal precautions on an AIDS unit than elsewhere; that could feed into the false sense of security that some people have when they don't know for sure whether a patient is HIV-positive, and into their personal biases about what an HIV-positive person looks like. But I do think that an AIDS unit should have, at the very least, an active working group that is reviewing and piloting the newest equipment. Of course there are risks in piloting; but you can do focus groups outside the clinical area, study the equipment and have some kind of criteria for then trying it on patients. We certainly shouldn't withhold any safety equipment from a group where, if a needlestick occurs, you know it will most likely be with an HIV-contaminated needle. And perhaps the cost/benefit analysis should be looked at in a different way in the case of an AIDS unit. Certainly the level of attention should be very high.

AEP: *In closing, is there a message you'd like to convey to our readers?*

JD: It can be frustrating and tedious dealing with bureaucratic inertia, but working to bring about institutional change can also be very rewarding. Find a channel in which to make a change—you can really get creative. If you're mechanically inclined, you can work with the product evaluation committee; if you're good with process, you can work with administrative committees. But I'd like to emphasize that it can be empowering to get involved with the system.

Marie Jasmin

Providing Homecare for an HIV Patient, A French Nurse is Stuck—and Infected

Vol. 1, no. 6, 1995

Moi, Marie Jasmin...

I, Marie Jasmin, am a nurse infected by HIV. I was infected while working as a home healthcare nurse with l'Assistance Publique of Paris. On the 14th of September, 1990, my life fell apart. On that day I was sent to the home of an AIDS patient who had just been released from intensive care, where he was treated for septic shock.

I find him very tired, covered with cutaneous lesions and hematomas—he has Kaposi's sarcoma. I listen to him as I prepare my equipment: the cardboard container for disposing of waste such as soiled gauze and used needles, which are collected from patients' homes once each week. This container is on the floor about one meter from the patient's bed. I slip on a pair of gloves.

The veins are difficult to find, but after some palpation I succeed in entering a vein. I draw several tubes of blood. When I withdraw the needle, I place the vacuum tube holder, with its needle still attached, on the bed in the leftover packaging from the gloves. I was not given (and I did not know about) the small box used by other units of l'Assistance Publique for immediately discarding needles.

After applying pressure to the puncture site to stem the bleeding—the patient bled a lot—and putting on a sterile bandage, I pick up the glove wrapping, with the needle in it. I take one step forward to put everything in the cardboard container on the floor. My right hand moves past my left hand and I feel the needle stick into the palm of my left hand. The pain is brutal.

Immediately, I rip off the glove and scream for someone to bring bleach. I go into the kitchen. The patient's wife comes in and applies bleach to my palm. I tighten a tourniquet around my wrist to induce bleeding. My hand turns blue. It doesn't matter; I'm too scared. Trembling, I finish caring for the patient. I keep a bandage, saturated with bleach, on my palm.

It takes me one hour to get to the hospital that employs me. I see the patient's physician. He reassures me that it's nothing and gives me two tablets of Retrovir to take immediately. On the unit I meet the patient's nurse, a woman I know: "What's wrong, Marie?" "I stuck myself with a needle from your patient." "What happened?" I explain to her about the cardboard container and how I handled the needle. She is floored. She takes me to her office and shows me a little red plastic box that she uses on her unit—a disposal container—to discard used needles. It's the first time that I have seen one. Since then the vision of this red box has pursued me. It could have saved my life.

I go to the lab to have my blood drawn. My HIV test results are negative. Six weeks later, during a bout of fever, extreme fatigue, and swollen lymph glands, I am retested. They send me to a professor for my results, who announces, "You are infected. You are seropositive." I can't breath; I ache; I scream. They send me to a psychologist. He tells me, "They killed you." I think of my children who are four and nine.

The next day my husband, who was out of town, called to announce that he was bankrupt. I told him that I was infected. After that he slept on the couch. We separated. I paid his taxes and raised my children for four years on my salary as nurse.

I do not receive treatment because Retrovir makes me vomit and that frightens my daughter. I wanted to be reassigned to a position as clinical coordinator, but I was forced to leave my patients for a job in a storeroom distributing bandages and cleaning products. It was like a prison. I was accused of having been clumsy.

For four years I suffered. I kept quiet, carefully managing my budget so that I could feed my children. Sometimes I wanted to die.

I learned about a group, the Association for the Protection of healthcare workers and their Patients (U.P.S.P.). I went to one of their meetings. I spoke, and I felt relieved. I was invited by the U.P.S.P. to give public testimony in France's national assembly hall on June 24, 1994. I also testified on French television before 10 million viewers, without concealing my identity, on December 16, 1994. This testimony greatly assisted legal efforts to obtain compensation for all French healthcare workers occupationally infected with HIV. Then I met with Professor Luc Montagnier,* honorary president of U.P.S.P.

We finally obtained an audience with Madame Simone Veil, France's Minister of Health and Social Affairs, on January 6, 1995. This warm-hearted woman trembled as I recounted the history of my infection. We embraced each other. On March 13, 1995, after another meeting with her, and in front of national television cameras, she announced the government's decision to provide a lump-sum compensation for healthcare workers occupationally infected with HIV. I received my compensation on September 7, 1995, along with other infected healthcare workers like me.

I have fought along with U.P.S.P. to obtain this reparation, which can never give me back my life. I no longer work. My CD4 count is 30. I am sad. But I'm still fighting to prevent what happened to me from happening to other healthcare workers.

Paris, 10-2-95

**Professor Luc Montagnier discovered the AIDS virus.*

Marie Stevens

My Struggle for Compensation

An HCV-Infected Healthcare Worker Speaks

Vol. 2, no. 1, 1995

In the March/April 1995 issue of AEP (Vol. 1, no. 3) we published a review and analysis of the complex workers' compensation system in the United States, and its implications for occupationally infected healthcare workers. Here we present an account of a hepatitis C-infected phlebotomist, Marie Stevens, and her struggle to win workers' compensation benefits. Her case illustrates many of the hurdles and unanticipated barriers confronting healthcare workers when seeking compensation, as well as characteristics of hepatitis C (HCV) infection. The perspective and opinions expressed in the following are Marie's.

I was exposed to hepatitis C on January 2, 1990, while working as a phlebotomist on the dialysis and transplant floor of a small private hospital in Nebraska. I had been employed there for four months. On the day of the injury I left my phlebotomy tray at a nurses' station while I went to do a line draw. When I came back, there was a contaminated sharp lying on the tray, still attached to a blood tube holder. I knew it was contaminated because the seal was broken. When I saw it lying there, I thought, "Someone is going to get hurt." I went to dispose of it in the new sharps container we were using; it had a feature that removed the needle mechanically so you didn't have to touch it. I decided to try it out. The device jammed, and the needle broke off in my hand and stuck me. It was my first needlestick.

The injury was very deep and it bled. Of course, I had that initial panic—"Oh God, I'm dead." I immediately induced bleeding at the puncture site and applied disinfectant. I finished drawing blood on some patients because we were really backed up and their labwork was needed immediately; then I went down and notified the phlebotomy supervisor and she gave me an incident report to fill out. She asked if I knew who the source patient was, and I told her I didn't because the needle was left on my tray. I was then sent to the employee health nurse.

At the time of my needlestick, the hospital did not routinely test for hepatitis after an exposure; they said it was too expensive. But I had been tested for hepatitis B four months earlier, as part of my preemployment screening. The results indicated that I had a prior infection with hepatitis B. An HCV test had just been developed but was not widely available at the time. Therefore there was no baseline test to establish that I was HCV-negative at the time of the exposure. The nurse drew blood only for HIV tests. I did not receive the HIV test results until March 1990; they were negative.

About a month after the needlestick, I started experiencing severe fatigue, to the extent that my job performance began to suffer. I did everything I could to lighten my load. I took time off from nursing school—I wanted to become a trauma flight nurse—and quit my other part-time job. But the fatigue just got worse. I was so sick during the first few months after the needlestick that I would sleep 20 hours on my day off, but I still couldn't catch up. Eventually I made some serious work-related errors, and was fired in March 1991. I later learned that the appearance of my symptoms was consistent with the incubation period for HCV. [*Editor's note: The incubation period for HCV is 15-150 days.*]

I did not learn that I was infected with HCV until more than two years after my needlestick. My blood was screened in February 1992, prior to surgery for an unrelated condition. The results showed I was positive for HCV. I then had ELISA and RIBA tests and a liver biopsy to confirm the diagnosis. Because it took so long to diagnose my infection, the statute of limitations for filing a workers' compensation claim almost ran out.

After I had a confirmed diagnosis, it took me a couple of months to backtrack and figure out how I got infected. I had no knowledge of hepatitis C or routes of transmission, and had received no training about it when I worked as a phlebotomist. So I educated myself about the disease, and went through my past history with several experts, including my treating physician, to figure out how I got it. It became clear that the most likely source of infection was my needlestick, particularly since I had been working at the time primarily with dialysis and transplant patients, a population at high risk for hepatitis C.

My main caregiver was a physician at the University of Nebraska Medical Center, who runs one of the top three transplant centers in the country. At first, when he looked at my medical records and saw a document that incorrectly stated that I had received a transfusion, he was reluctant to attribute my HCV to the needlestick, since transfusions are one of the most common transmission routes. When I told him I never had a transfusion, he wanted to know how the record got in there. That began a long complicated process of trying to get my medical records corrected so there was no question of the HCV being due to a transfusion. I was able to satisfy my physician that the transfusion statement was an error, and he eventually wrote a letter supporting my claim that I got HCV from the needlestick. But the document which stated that I had a transfusion remained in the medical records that were submitted to the workers' compensation court as evidence and was never corrected; that circumstance created a major obstacle in my case.

Once I realized that I might have a workers' compensation claim—after my diagnosis had been confirmed and after I had traced the infection to the needlestick, which in itself involved a lot of time and research—I then had to find an attorney to represent me before the workers' compensation court. I had no support or help whatsoever in putting my case together. I gathered most of the supporting evidence myself. I finally filed a claim in January 1993, but my case was not heard by the Nebraska Workers' Compensation Court until May 1994. The court rejected the claim in late June 1994.

At the court hearing, the only physician present was a defense witness, who was the state epidemiologist. He did not treat patients and had no particular expertise in HCV. The best answer he could give to the question of whether my infection was occupationally acquired was "I really don't know." The judge took that to mean that I hadn't met the burden of proof, even though I had a letter from my treating physician, one of the leading experts in the field, stating that there was "reasonable medical probability" that my infection was due to the needlestick. But that was disregarded by the court, which chose to listen to the defense witness, and the court finally concluded that there was insufficient evidence that my illness was the result of an occupational exposure.

In my view, the attorney for the hospital used disinformation as his main strategy. If he could muddy the waters enough and make me appear not to be a credible witness, then the court would rule against me. And that's exactly what happened. During the deposition, I was asked when I lost my virginity, if I was promiscuous, and how many people I had slept with. In court they referred to my two miscarriages as "numerous abortions." They referred to my prior marriages as "unprotected sex with multiple partners." They did everything they could to smear my reputation.

I requested a review by a three-judge panel of the workers' compensation court and in December 1994 the review panel upheld the previous decision. I appealed the decision to the Nebraska Court of Appeals in March 1995, which then upheld the lower court's decision. I filed a motion for rehearing by the Court of Appeals, and that was overruled. Finally, in November 1995, I filed a petition to have my case reviewed by the Nebraska Supreme Court. That request was denied. So after a three-year struggle, it appears that I have exhausted all avenues of recourse in my quest for workers' compensation.

In terms of my medical options, I was told there was no treatment other than an experimental drug called interferon. When I tried to obtain the drug through the pharmacy, I was told they didn't carry it because it wasn't "cost effective." Finally, I visited the University of Nebraska Medical Center because they were conducting an interferon study there, and that's how I found my current physicians, who are liver specialists. They put me on interferon for a 90-day trial period, but I had a very severe reaction to it, including terrible migraines. My husband was out of town six days a week at the time with his job, and I was alone and couldn't move around much. It was a long 90 days; at the end I was so weak I couldn't stand, and they couldn't find a blood pressure with a doppler. At that point, my doctors elected to take me out of the study and off the interferon. They were reluctant to do it, but obviously a fatality would have been worse. That was the end of my medical options.

My liver enzymes continue to be only slightly elevated, but I've learned that lab results don't necessarily reflect the severity of the symptoms. There are some people who can continue to work with HCV and lead fairly normal lives, while others can't get out of bed, even though their enzyme levels aren't very high. It's different for each patient, which makes it difficult from a legal standpoint, because courts look at lab results like mine and say, "They're not that bad."

I'm not on a transplant list. I don't qualify financially because my workers' compensation claim was denied, and my current insurance considers the HCV to be a pre-existing condition. It's ironic that I can't qualify for a transplant at the hospital where I took care of so many transplant patients, and where I became sick in the first place. Even if I did qualify financially, I probably wouldn't qualify medically until I'm more endstage, because livers are the most difficult organ to get. Also, I have a number of drug allergies, so I would have a very hard time with the anti-rejection medications. Still, I would like to have that option; right now I don't.

Since I lost my position as a phlebotomist in March 1991, I've held only one job. I worked as a cab driver for a few months. I am now totally incapacitated by HCV. I sleep at least 12 hours a day, and can only be up and moving about for two or three hours at most. I applied for and received social security disability, and my medical expenses are now covered by Medicare. Besides the emotional costs of HCV—my previous marriage ended in divorce, partly as a result of the disease—I've lost almost five years of wages and spent several thousand dollars of my own money pursuing my case.

Changes are needed in the present workers' compensation system in Nebraska. The main issue is what the limit should be on the burden of proof required of a healthcare worker. If you have a letter from your attending physician stating that there is reasonable medical certainty that the infection was occupationally acquired, that should be the end of it, the worker should be compensated. If a hospital contests a claim, they should be required to provide absolute evidence of another source of transmission. I also think if an employee reports an exposure, and the hospital doesn't test for a specific pathogen, the hospital should not be allowed to refute a later claim for an infection with that pathogen.

I think, in general, healthcare workers are well informed about the risks posed by their jobs, but I do think phlebotomists need better information about the particular risks associated with blood-drawing. No one told me when I started phlebotomy that it was one of the highest risk jobs. We were treated like general unskilled labor, and paid accordingly.

Lynda Arnold

A Nurse with a Mission

Vol. 2, no. 2, 1996

Lynda Arnold received her nursing degree in May 1992. In September 1992 she sustained a needlestick that resulted in her seroconversion to HIV within six months. On February 28, 1996, at a press conference in Philadelphia, Lynda announced her decision to launch a national campaign asking for a written commitment from every hospital in the U.S. to provide protective catheters and blood-drawing devices to their employees, devices that could have prevented her infection. She tells us about her experience and what led her to undertake this campaign.

The Needlestick

On September 9, 1992, I was working as a registered nurse in the intensive care unit (ICU) at Community Hospital in Lancaster, Pennsylvania. An aide brought us a very ill patient from the outpatient clinic and told us to use blood and body fluid precautions. That was her way of letting us know that the patient was probably HIV positive. We didn't know why he was admitted, since we hadn't gotten his chart yet. But we were putting on a cardiac monitor and getting an I.V. line going, which was standard procedure in the ICU.

I had my gloves on and was starting the I.V. line, using a 20-gauge I.V. catheter. As I withdrew the needle from the patient, he jerked his arm, hitting my right hand and forcing the needle into the palm of my left hand. Soon after the needlestick, I found out he was a known AIDS patient with a history of abusive behavior toward the staff. He'd been admitted for pneumonia, and died two weeks later.

My immediate concern was to finish the procedure. If I lost the I.V. line, then someone else would have to stick the patient. So I finished putting the heplock on, and disposed of the I.V. catheter. Then I went to the sink and took off my glove. The needlestick was moderate, but the needle had caused a kind of jagged tear as it went in, and the wound was bleeding. After I washed my hands with disinfectant, I went out to the desk and told the day-shift supervisor that I'd been stuck, and asked her what I was supposed to do. It was my first stick.

I had a little training in nursing school about needlesticks. We were made aware of the risk, and also that there were safety products on the market to prevent needlesticks. I later learned that a shielded I.V. catheter would have prevented my injury.

After the needlestick, everyone on my unit was very reassuring, telling me nothing would happen. Some told me about their own needlesticks. The doctor in the emergency room was also reassuring; he was more concerned about my risk of getting hepatitis B than HIV, because I hadn't yet completed my HBV vaccination series. He told me that the risk of HIV infection was one in 250. Of course, no one ever thinks they are going to be the one in 250 who becomes infected. Even though the doctor didn't seem that worried about my exposure, the other E.R. personnel looked concerned, even shaken. I remember seeing the fear in their eyes, reading their thoughts—"It was bound to happen." They were familiar with the source patient and his history of aggressive behavior toward the staff, including biting and spitting. They seemed to have a sense of foreboding, and that scared me.

Although my identity wasn't revealed, word of my needlestick traveled quickly throughout the hospital. A couple of days later, I was at home talking to a friend who worked on another floor of the hospital, and she said, "Did you hear about that nurse who got stuck in the ICU?" And I said, "Yes, that was me."

Seroconversion

I was tested for HIV on the day of the needlestick, and the results were negative. At around three weeks I developed a fever, rash, and sore throat. The doctor I was referred to told me not to worry; again, he was more concerned about hepatitis. I did finish my hepatitis B vaccination series, and all the hepatitis tests came back negative. I wasn't taking AZT. At the time there wasn't any conclusive data about its benefits as a prophylactic drug. A few months before, another hospital employee had chosen to take AZT after an at-risk needlestick, and had gotten very sick as a result. So employee health didn't really encourage me to take it.

Around six weeks, the fever and sore throat reappeared,

although they weren't as severe as before. But I also developed abdominal pains, with nausea and vomiting, which was diagnosed as appendicitis. I underwent surgery and it turned out that I didn't have appendicitis, but rather moderate to severe abdominal lymphadenopathy. I was retested for HIV, and was still negative. At my three-month test I again tested negative, and I truly thought that the nightmare was behind me. I was told that if you tested negative at three months there was a 90% chance you didn't have it.

After I went in to get my blood drawn for my six-month test, I began to feel that something was wrong. With every other test, I'd gotten the results back within a couple of days. This time I waited and waited, and didn't hear anything. My phone calls were not returned. This went on for about 10 days.

I later learned that follow-up testing was being performed, but also that hospital administrators were meeting to try to figure out what to do with me. I think they realized later that the process took too long; the waiting was very hard on me.

Finally, on April 7, 1993, I got a call asking me to come in for my results. This scared me, because before I was given the results over the phone. As soon as I walked in the door of the employee health office, I knew. I looked at the nurse and she had tears in her eyes, and I started crying. It was all over.

I was the first employee in my hospital to seroconvert to HIV from a needlestick, and I was only the second on record in the state of Pennsylvania.

Aftermath

Once I was diagnosed as HIV-positive, I was referred to an infectious diseases specialist. There was a lot of lab work to do. I asked for repeat HIV tests to confirm the diagnosis; I still couldn't believe I had it. Then they did lymph site subset panels to test my t-cells. The usual advice is to start AZT therapy when you're below 500, and mine were at 490. So I started AZT right away. I didn't react to it very well. After six months I started getting terrible headaches and nausea, and went off it. Since then, I've been on DDI, DDC, and D4T [nucleoside reverse transcriptase inhibitors], but have had serious adverse reactions to all of them and had to stop taking them. I haven't taken any drugs for over a year.

Although I do not have full-blown AIDS, I have been symptomatic almost from the beginning of my infection 3-1/2 years ago. I have had chronic fatigue and headaches, night sweats, insomnia, swollen lymph glands, yeast infections, and fungal infections of my feet. I've been hospitalized twice, once for meningitis and then for pancreatitis brought on by the DDI. I had surgery for cervical dysplasia, and recently underwent a liver biopsy because my enzyme levels were elevated.

After seroconverting, I never returned to patient care. That was my choice, because the hospital had ruled that, based on CDC guidelines, I wasn't a threat to patients and could remain in the intensive care unit. But along with the t-cell test I had an anergy panel, which determines the level of your immune response. I was completely anergic, which means that something was wrong with my immune system early on. I decided to take an administrative job because I was fearful of being exposed to infectious diseases, particularly tuberculosis, in the ICU. The hospital supported my decision.

I stayed in the administrative job for about a year but found it unfulfilling, and also stressful having to hide my illness from others. I tried cutting back to part-time hours, but finally, in April 1994, citing fatigue and abdominal pain from the lymphadenopathy, I resigned and went on permanent disability.

To a certain extent I regret the decision to leave patient care. I miss it every day. I had so much training and was just starting my nursing career. I was thinking of becoming a nurse practitioner, and planned on getting a master's degree. My goals at the time were completely focused on my career. I became engaged shortly before I found out I was HIV-positive, but did not feel any urgency about getting married. After I seroconverted, my feelings changed. I wanted to get married while I was still healthy. I didn't want to go down the aisle in a wheelchair.

Everyone in my immediate circle of family and friends was shocked; they had never known anyone with HIV before. They were also angry, and wondered how this could have happened. There had been so much emphasis on AIDS education and awareness in 1992, and their feeling was, what about the healthcare workers? How come Lynda got stuck if there was so much effort at prevention?

My fiance (now husband) got an HIV test, which was negative. The hospital requested it for workers' compensation, and the CDC wanted it for surveillance purposes. My husband is still HIV-negative.

My husband and I very much wanted to have a child, but we felt strongly that we did not want to risk having a baby born with HIV. Through my experience in the HIV community, I had seen the pain and suffering HIV can cause in children. We agreed that even a 1% risk was too high. So we decided to adopt. I didn't think having HIV would be a barrier to adoption, and for us it wasn't. It turns out, however, that we are the only "serodiscordant" couple in the country that has officially adopted a child—that's where one partner is HIV positive and the other is not. When we went to the adoption agency, we had to provide a lot of information about life with HIV and how we would deal with various contingencies. Finally, we came home with a newborn baby boy who is now 9-1/2 months old.

I do believe that honesty is the best policy. It's the key to everything. One of my favorite quotes is, "Truth is not a secret to be hidden. It's a gift to be used." It's become a kind of motto for me as I launch my public campaign.

When I first learned I was HIV positive I was worried about the word getting out and someone burning down my house or throwing rocks at my car. But I was assured that things like that didn't happen anymore, and that the hospital would do everything it could to protect my identity. After that I really didn't give much thought to confidentiality. Certainly, after I left the intensive care unit and went to the administrative job, many people surmised what had happened. In a small hospital it's hard to keep things like that a secret. But my co-workers and the administration tried to protect my confidentiality. Finally, when I filed a lawsuit against the product manufacturer in March 1994, I made a decision to go public with my injury and illness. I was tired of the shame and secrecy. I needed to speak out; people were taking care of me, supporting me, protecting me, but that wasn't going to make it better for anyone else. That's why I chose to use my actual name when I filed the suit. Our local newspaper published the headline "Nurse Gets AIDS Virus Through Needle" and included my name, age, and the hospital where I worked. From then on the silence was broken.

There was no explosion of media coverage, but people from the hospital and in the community felt free to offer their support. They expressed compassion and solidarity. My fears about adverse reactions turned out to be groundless. I was sure that I had done the right thing and it was a relief to be out in the open. I wanted to make a difference, to try to ensure that what happened to me didn't happen to someone else.

I sought the advice of Patti Wetzel soon after I learned I was infected. I sent her a letter, she wrote back, and since then we've had several conversations. She's been a role model for me. She was the only occupationally infected healthcare worker I knew of who was speaking publicly about her disease. I wanted to know why she had made the decision to go public, and whether she felt she was making a difference. I am very interested in the idea of some kind of national support network for occupationally infected healthcare workers.

Compensation

I have received complete compensation for my injury and illness, 100% care. I've gone to the doctors of my choice, seen specialists, gotten psychological counseling. All of my labwork and hospitalizations have been covered. I also received back pay for the period when I was hospitalized before I seroconverted. I now receive about 2/3 of my former salary in workers' compensation benefits.

The key to getting compensation is documentation. Without the documentation, what would they have had to go on? I was a 23-year-old single heterosexual female. I could have gotten HIV from another route. But my case was thoroughly documented. Of course I realize that for other occupationally infected healthcare workers, the path hasn't been so smooth, and my heart aches for them. I'm one of the few who has gotten full coverage. I've come away from this experience luckier than most. I have full medical care; I have family and friends and an employer who support me; and I have the right to go back to my job. Most people with HIV don't have that.

I don't have anything negative to say about the workers' compensation system. I did have to answer every embarrassing question you can think of. But I didn't want there to be any doubts about my case. The issue was not whether I would get compensated, but how. How would my confidentiality be protected and an HIV infection be handled under workers' compensation? I was asked whether I wanted the paperwork to refer to my illness in vague terms like "blood virus." I said I wanted it written up as "HIV infection from needlestick that occurred on 9/9/92." I didn't want any grounds for confusion in the records, because my case would set a precedent for how the next occupationally infected healthcare worker in Pennsylvania was treated under workers' compensation.

I believe that healthcare workers in patient care who have an occupationally acquired disease should be allowed to stay at their job as long as they don't pose a risk to patients. I do think their status needs to be evaluated on a six-month basis. But healthcare workers need to be protected, and they need to have a choice. In my case, because of the course of my disease, I chose not to work. But that decision should be up to the healthcare worker. By the same token, however, nurses and other healthcare workers should not be able to refuse treatment to an HIV-positive patient.

When I do public speaking—and I've spoken to over 6,000 people locally in the last year—inevitably there are people who come forward and say, "My uncle was a surgeon who was infected on the job, and he died of AIDS," or "My sister was a nurse who was infected on the job, and she's never gotten any compensation for it because her injury wasn't documented." I hear these stories, not in the medical community, but from the general public, and that's just in my local area. It makes you wonder about the true number of occupational infections, and how many healthcare workers are afraid to reveal their infections and seek compensation.

Ignorance and Denial

What is the greatest barrier to a safer workplace? Ignorance and denial, on the part of hospital administrators, product evaluation committees, purchasing departments, and healthcare workers themselves. Healthcare workers need to be made aware, not only that there are occupational risks, but also that those risks can be reduced. They need to have the power to influence purchasing decisions. They are the consumers and users of products; if there is a safer product available, they should have a right to demand it.

Are hospitals and healthcare workers reacting appropriately to the risk of occupational exposure? It depends. There are three major hospitals in the Lancaster area. Since March 1993, my hospital has been using shielded I.V. catheters in most units. But neither of the other two hospitals in the

Lancaster area currently use these kinds of safety devices.

Since my needlestick, my hospital has made greater efforts to educate staff about exposure prevention and risk reduction. The impetus for that has come not only from the administration but from the nurses themselves. Their attitude is, "We support Lynda, but we're not going to be the next ones." At the time of my injury, the hospital had a needlestick prevention committee, but there is greater employee participation now. There is also more participation in the nursing standards and protocols committee.

Appeal to the Nation's Hospitals

Since leaving my job at the hospital in April 1994, I have done extensive HIV education in the Lancaster County area, mostly with public schools, colleges, and community groups. I chose not to address the medical community while my court case was pending.

I pursued a legal remedy because I wanted to increase awareness and make hospitals and manufacturers sit up and take notice. But a motion for summary judgement by the manufacturer was granted, which meant that I never got my day in court. The case was then settled out of court. Since I didn't have an opportunity to speak out in court, I am now launching a public appeal.

My immediate goal is to promote a safer workplace for healthcare workers. I am making a public appeal to the nation's hospitals and medical centers, as well as to medical device manufacturers, to unite and work together to fully implement safety I.V. catheters and blood-drawing devices. This will include asking hospitals across the country to sign a pledge agreeing to convert to these kinds of safety devices within a year's time. Research has shown that I.V. catheters and blood-drawing devices are associated with the highest-risk injuries. It makes sense that hospitals should focus their efforts on these device categories. Of course, it's of special importance to me since I was infected via an I.V. catheter.

I don't want this campaign to be seen as one woman's platform. This is for all healthcare workers, and for their children who may someday become healthcare workers themselves. It is also for all those HIV-positive people who have made such enormous efforts to further HIV awareness and education. This is just another step in that process. Implementing safer devices in the healthcare workplace will help not only healthcare workers, but also HIV-positive people who are their patients, by reducing the fear and discrimination they sometimes experience in the healthcare environment.

Some injuries will continue to happen, but that doesn't mean that the risk of infection can't be reduced. Healthcare workers, whether they are nurses, physicians, aides, or maintenance workers, have the right, and the responsibility, to demand safer products and a safer work place. And hospital administrators, purchasing departments, and manufacturers need to provide better access to safety products.

There is so much debate about the cost of implementing safety devices—so many people who say that they are too expensive, especially with hospital budgets and staff being cut. But all they need to do is pick up the phone and call Community Hospital of Lancaster; my infection alone will cost hundreds of thousands of dollars. Even without an infection, however, the cost of following up occupational exposures is high, including the medical treatment, lab work, counseling, lost work time, and the fear and stress that take their toll. Hospital administrators need to look at the total picture.

My hospital is a small, private, financially challenged facility. If it can make the switch and find the money for safety devices, then any hospital can do it. It's impressive when you hear about large hospitals implementing safety devices, but it's more impressive when small hospitals make that kind of commitment. My hospital did it because the administration understands the economic impact a needlestick injury can have. Of course, I wish my hospital had implemented safety devices before my injury. But I'm glad they have them now, because there won't be another case like mine. And for that I am grateful.

Donna Cieniawa

Just Another Needlestick ...

Vol. 3, no. 1, 1997

Donna Cieniawa's occupational exposure occurred in a busy emergency department one Saturday night. But unlike Patti Wetzel, Lynda Arnold and others, she did not become infected with a bloodborne pathogen. In that sense, her experience is like that of most healthcare workers who sustain needlesticks. But her experience also shows that, even when a healthcare worker does not become infected, the long wait for final results can be traumatic. Donna's experience underscores the importance of providing professional and compassionate post-exposure follow-up care.

At the time of my needlestick, I was working as a staff nurse in the emergency department [ED] of a 110-bed private hospital. I had done clinical nursing for 25 years and loved my work. I also had a part-time administrative position as the hospital's trauma coordinator, logging in data for a state-mandated trauma registry. The hospital where I worked was the smaller of two health care facilities in a city of approximately 110,000. We frequently saw HIV-positive and AIDS patients and, with a major city less than 50 miles away, also saw our share of the transient population.

On Saturday, June 29, 1996, I was working a day shift; it was late afternoon, and the ED had been busy all day. Even though I was preoccupied with my own patients, I was aware that a baby had been crying intermittently for hours in one of the cubicles. One of the nurses told me that they were trying to draw blood from the baby and it was a particularly difficult stick.

I finished taking care of my patient, then answered a call from pediatrics asking for a report on the baby. I offered to relieve the baby's nurse so she could give the report. When I entered the cubicle, I was faced with several potential hazards. First, it was crowded—the baby's father was there plus an attending physician, a resident, and the nurse. The nurse handed me an uncapped butterfly needle and said, "They want to keep this." I took the needle and tried to hold down the baby, but she was fighting so I got permission to dispose of the needle in order to free up my other hand. I then noticed a second butterfly needle with a heparin lock attached lying on the bed; blood was visible in its tubing. I picked up that needle as well and put both of them in a disposal container on the wall. I held down the baby while the doctors tried to draw blood, but after a few minutes they gave up. I wrapped the baby and handed her to her dad. As I was walking away, he said, "Watch out, there's a needle on the floor." I looked down and saw an uncapped phlebotomy needle. I picked it up and disposed of it.

I removed the I.V. tray from the baby's cubicle, set it down in another room, and started to convert a patient's I.V. line into a heparin lock. I disconnected the I.V. tubing and walked over to the tray to get the necessary equipment. As I reached in, I felt a searing pain. I pulled my hand out and with it came a needle attached to a phlebotomy holder.

The needle was jammed into my left middle finger, right down to the bone. It had been sticking directly upwards, and there was blood on the needle.

I removed the needle from the holder, placed it in a disposal container, and went to the sink to scrub the puncture site with soap and water. Then I went to the front desk to find out which patient the needle was used on.

I finally pulled aside the baby's nurse and told her I'd been stuck; we concluded the needle had been used on the baby. I began to ask questions about the baby, and found out she had sickle-cell anemia and had had multiple hospital admissions for pneumonia and other infections. I remembered that the baby's mother was very thin. Now I wondered if she might be HIV positive—and whether her baby might be, too. With all I had seen as an ED nurse, I knew you could never be sure.

I notified my nursing supervisor and reported the incident to the ED doctor, according to our needlestick protocol. I filled out the blood and body fluid exposure form and had blood drawn for baseline HIV and HCV tests and an HBV profile. (I was already immunized for hepatitis B, but the following week my test results indicated I needed a booster shot, which I received.)

I had been stuck two years earlier while trying to resuscitate a patient; blood was drawn for testing soon after the stick, once the patient had died, but the three days it took to get the test results back seemed like forever. So I had already experienced the emotional consequences of a needlestick injury, and dreaded going through it again.

An infectious diseases (ID) physician, who was in charge of the hospital's needlestick protocol, was contacted. I told the ID doctor about my exposure, but he did not seem to treat it seriously; he quoted statistics to me, such as the chances of seroconverting to HIV are 3 in 1,000. He didn't ask any questions about the baby or her health history. He simply decided that my risk was minimal. I asked him, "How do you know that? If this needle was used on an HIV-positive patient, then it seems to me that my risk is significant." I told him I wanted to reduce to a minimum my risk of seroconverting to HIV or hepatitis C if the needle was contaminated, and asked what he recommended. His response was, "Well, you could take ZDV [zidovudine] if you want." I remember thinking, "If I *want*? This isn't like taking aspirin for a headache." I asked him for information about side effects, and after briefly summarizing them, he said, "I guess you could take ZDV for three weeks."

The ED doctor wrote in my chart the information from the ID physician about ZDV dosage, and that was it. The question of testing the baby was never even raised. The attitude was, "We filled out the paperwork, we drew your blood—you're done."

At that point, I called my husband, an ED physician at another hospital who frequently treated blood and body fluid exposures among the staff. When I told him what the ID doctor had said, he responded, "No, it's not 'you can go on ZDV if you want'—you *will* go on ZDV, immediately, and you also need 3TC [lamivudine]." He said he would consult with an ID doctor at another hospital and call me right back. The other ID doctor confirmed that I needed to go on ZDV and 3TC right away. CDC recommendations on postexposure prophylaxis (PEP) had been published three weeks earlier in *MMWR* and she based her advice on those recommendations. Her reasoning was that, even if we could test the baby right away, we would not get the results back for a few days. I could always go off the drugs if the baby tested negative; the critical thing was to start prophylaxis within one to two hours of my exposure. We considered adding a protease inhibitor, but agreed that ZDV in combination with 3TC should provide sufficient protection. But I worried that the stress of waiting for a negative test at six months would aggravate the symptoms of a disease I already had—multiple sclerosis.

I took the first dose of ZDV within two hours of the needlestick. Our hospital pharmacy did not have 3TC, and did not offer to get it for me. It took my husband and the ID physician he consulted several hours of telephoning to arrange for me to pick it up from another hospital about 40 minutes away. I left work to get the 3TC, and took my first dose about five hours after my exposure.

The next day I called my nurse manager at home to tell her what had happened because she had been off the night before, and I asked for her help in getting the baby tested. I also wanted to discuss ED safety with her. After my previous needlestick, I recommended that we introduce needles with safety features, but nothing was ever done. The only safety equipment we had in the ED was needleless I.V. connectors.

I was not prepared for her response. "You're being absolutely ridiculous," she said. "How many times have all of us been stuck in this job? Only a minuscule number of people become HIV positive. It almost never happens." When I told her I was not sure if I could continue to work in the ED for fear of another exposure, she said, "If you want to leave there are plenty of people who would be happy to have your job. I'll just need a letter of resignation." I was shocked; I had a good professional relationship with her, and she had given me responsibilities above and beyond those of other staff nurses, including scheduling and writing policies and procedures.

She did agree, however, that the baby should have been tested at the time of the stick, and promised to take care of it the next morning. She also said she would speak to the head of the residency program about the careless handling of needles that I had witnessed in the baby's cubicle. I didn't hear from her again until late Monday afternoon; by that time, her attitude had changed. And my battle to get the child tested had begun.

My nurse manager discussed the needlestick with the residency head; based on that conversation, she said the needle was sterile and hadn't been used on the baby. The physicians who worked on the baby claimed to have used only butterfly needles, not a phlebotomy needle, on the baby. And the residency head said it would be too much of an ordeal for the baby to have her tested. My husband responded, "What about the ordeal my wife is going through?" Recently, I learned that our state law allows HIV testing, without source patient consent, on previously obtained blood samples. The hospital already had samples of the baby's blood, but no one seemed to know about this law. I also later learned that the hospital's needlestick protocol required that a committee, including the head of Infectious Diseases, the Employee Health physician, and a nurse epidemiologist, be convened when questions arose concerning testing a source patient. That protocol was not carried out.

During an exam by the Employee Health physician, she told me I had reported the third needlestick from the ED in the last three months, and that before my stick the infection control nurse had offered to do an in-service on needlestick prevention, but had been refused. The Employee Health physician later called ID doctors at two large healthcare facilities in our area regarding their PEP protocol, and both confirmed that combination therapy with ZDV and 3TC was the appropriate treatment.

After three days, the ZDV and 3TC made me so nauseous

and tired I couldn't work. I checked with my ID physician, and she said she would do blood work weekly to monitor side effects. She recommended that I continue with the medications for four weeks and that I take a medical leave of absence from work.

About three weeks after my needlestick, my lab tests indicated that my white blood cell count had dropped to abnormally low levels, and my ID doctor, in consultation with my husband, decided that I should discontinue the medications.

I continued on medical leave after I stopped taking the drugs because I still felt very sick. I was also experiencing some MS symptoms, such as blurred vision. I remained on leave for six weeks.

My six-week HIV test in mid-August was negative, and I went back to work. However, I decided not to go back to bedside nursing. If I was stuck again, I didn't think I could deal with another round of chemoprophylaxis and the accompanying physical and emotional stress. And nothing had changed in the ED. There had been no in-services on needlestick prevention and no evaluations of safety devices.

Before returning to work, I spoke to the hospital's risk manager and told her that none of my concerns had been addressed, and that I was willing to do anything other than clinical nursing. I had taught in several areas of nursing in the past and was qualified to do staff development, in-service education, or quality assurance. But nothing was offered.

So when I went back to work I was limited to the trauma registry job, and lost about 12 to 16 hours of work per week as a result. My nurse manager was antagonistic towards me, berating my job performance in front of my co-workers and criticizing my reaction to the needlestick. She also tried to take away my trauma position. My work environment became increasingly hostile.

One of the most frightening aspects of my post-exposure experience was that in August, shortly after I went back to work, I started to develop a rash and enlarged lymph nodes in my groin. I realized that these could be symptoms of early HIV infection, and was more anxious than ever to have the baby tested.

After three months of struggling with the hospital over having the baby tested, I met with the senior vice president and told her all my concerns: the hospital's refusal to test the child, the struggle to get appropriate medical care, the nurse manager's harassment, the ID doctor's failure to discuss the CDC's recommendations with me, the carelessness with needles among the baby's treating physicians. I also told her that my workers' compensation claim had been denied, I had received no reimbursement for my six-week leave of absence, and human resources had not taken care of the medical bills that were starting to pile up. The vice president seemed appalled and said she would be the liaison for my workers' compensation claim, and would try to resolve the other issues as well. Ultimately, however, I was only partially reimbursed for my leave of absence, and my medical bills were not paid until months later.

She also promised that the baby would be tested, as quickly as possible. However, for the next three months, I only heard a series of excuses about why it hadn't been done.

The one bright spot during those months was that I happened to see a piece on television about Lynda Arnold. [*Editor's note: Lynda is a nurse who contracted HIV from a needlestick injury*]. I contacted her and she was a great help. Talking to her made me feel I was not alone. At the same time, it brought home the reality that healthcare workers can and do seroconvert to HIV.

When I finally got my six-month HIV test results at the end of December and learned I was negative, I felt deeply grateful, knowing there are many healthcare workers, like Lynda, who have not been as fortunate and are now living with HIV or hepatitis. The baby wasn't tested for HIV and HCV until ten days before my six-month test; she was also negative.

I finally transferred to the hospital's case management department, and continue to do trauma registry work there; I have also gotten additional hours doing quality assurance. But I worked in the ED for most of my 25 years as a clinical nurse, and that is where my heart will always be.

I am disappointed with my hospital and how I was treated by my fellow healthcare professionals—the ignorance they showed about needlestick injuries, the way they stuck their heads in the sand, their lack of compassion. We are in the business of caring for people, yet my hospital didn't show much concern for me.

I'm more fortunate than many healthcare workers; because I am married to a physician, we were able to quickly find out what the appropriate treatment was for my exposure. Also, my family didn't depend on my income to survive; that gave me the opportunity to speak out and fight the system to some degree. For many, this would not have been possible.

What most healthcare workers don't fully understand are the physical and emotional consequences of an at-risk needlestick injury. Whether you test negative or positive in the end, the six months of uncertainty until the final HIV test are an ordeal.

As healthcare workers, we need to talk more about occupational exposures. We need to have open discussions about the reality of living through the six months after an exposure and what it is like to take the PEP drugs. We need to reach out to co-workers who have been exposed, and not be afraid to be supportive and kind. It will make our job of caring for patients so much easier if we show care and compassion for one another.

Lisa Black

One Unnecessary Needle = HIV + HCV

Vol. 4, no. 3, 1999

My name is Lisa Black. I am a twenty-eight year old registered nurse living in Reno, Nevada. From a very young age, I knew I wanted to be a nurse. After completing high school in 1988, I immediately enrolled in nursing school at the University of Nevada, Reno. While a student there I married and had my first child. Though it was grueling to be a full-time student as well as the mother of an infant, my determination to realize my goal of being a nurse never wavered. I graduated from nursing school with high honors in 1993.

On October 18, 1997, I was working the night shift as a per-diem staff nurse on a combination medical-surgical-telemetry unit in a small, privately owned hospital in Sparks, Nevada. That night I was assigned to eight acutely ill patients, one of whom was in the terminal stages of wasting due to advanced AIDS.

While checking on my patients, I noticed that the one with AIDS—I'll call him Mr. Jones—had blood backed up in his intravenous (I.V.) line tubing, which had become occluded. In order to prevent having to re-establish his I.V. access, I needed to irrigate the line quickly. As I assembled my supplies, I noted that Mr. Jones' I.V. line was not equipped with the needleless I.V. access system that the hospital had made available in order to prevent needlestick injuries. Because the I.V. line was occluded, I was unable to change over to the needleless system.

I filled a syringe with saline irrigation solution and inserted the needle into the rubber port on the patient's I.V. line. Using a push-pull method, I attempted to aspirate the blood clot from the occluded catheter and then flush the solution through the line. During this procedure, Mr. Jones startled and his arms jerked, causing the needle that had been inserted through the rubber stopper of his I.V. line to dislodge and puncture my left palm.

My first reaction was sheer panic, but this soon yielded to robot-like action. The staff of our unit had recently attended an in-service on needlestick injuries and post-exposure chemoprophylaxis. Replaying the in-service in my head, I quickly expressed as much blood from the wound as possible and scrubbed it with betadine. By the time I left the patient's room and walked to the nurse's station, I was in tears. I was sent immediately to the emergency department and had blood drawn for baseline tests for HIV, HBV and HCV, which eventually came back negative. I was started on a regimen of AZT, 3TC and Crixivan within two hours of my exposure. I was instructed to take some of the pills with food, others on an empty stomach, and continue this regimen around the clock for one month, returning for frequent blood counts to monitor my response to these potentially toxic chemicals. Although the drugs made me ill, I adhered to the regimen faithfully. I thought, "I can stand this for one month if it prevents me from becoming HIV positive." As though to reinforce this conviction, I was assigned to care for Mr. Jones again ten nights later, when he finally lost his battle with AIDS and died.

The weeks after my needlestick were among the most tumultuous of my life. I experienced severe fatigue and nausea from the HIV medications, difficulty with tasks that required concentration, and trouble with short-term memory. My hospital provided little support. Eventually I made several on-the-job errors that resulted in my being fired. At the time, I didn't attribute my poor memory and difficulty paying attention directly to my needlestick. I could not understand why I was so scatter-brained. Now, in retrospect, it makes sense to me.

I started work at another hospital, but I still felt tired and unable to concentrate. I was often irritable and short-tempered and, in the end, I was let go from that job as well. Being fired from two nursing positions in such a short time devastated me. Prior to this, I had been employed by the same hospital for seven years. I did not understand what was going wrong with my life and my career, but I knew I needed to take some time out to regroup.

During this period, I did not share my needlestick experience with anyone except my immediate family and a few very close friends. I felt scared and alone. I feared judgement from my colleagues about my competency as a nurse, having been fired from two jobs, to say nothing about being possibly infected with HIV.

While I was taking some time to get myself together, my three-month post-exposure blood work came back negative. I felt like the weight of the world had been lifted from my shoulders. The literature I read indicated that close to 90% of seroconversions took place within the first twelve weeks of a blood exposure.

I was offered a position on an oncology unit at yet another hospital, and though I thought frequently of my needlestick, I believed that the period of acute risk was behind me. I was very successful at this new job, made many friends, and discovered a love of oncology nursing.

In June 1998, however, I again started to feel fatigued. I had swollen lymph nodes in my neck and groin and regularly ran low-grade fevers. After two weeks of this, I started to worry that I'd gotten mononucleosis or something of the kind, and made a mental note to see my physician. It did not cross my mind that these symptoms were consistent with acute HIV infection.

On July 9, 1998, I woke up with a very high fever and an excruciating headache. I was admitted to the hospital under airborne isolation precautions with a tentative diagnosis of acute meningitis, which later was ruled a viral meningitis syndrome.

During my stay in the hospital, I was again tested for HIV, and again I tested negative. My lymphadenopathy persisted, however, and my white blood cell and platelet counts remained dangerously low—symptoms my physicians were at a loss to explain.

After I was discharged, I spent many hours researching my symptoms on the internet and in medical journals. At a follow-up appointment with my physician, I showed him literature outlining the symptoms of early HIV infection, which were ominously consistent with the symptoms I'd been experiencing. He ordered a test for an HIV-RNA by PCR examination, which was done by a molecular biology laboratory in Virginia.

On July 27, 1998, nine months and nine days after my needlestick injury, I learned that I had, indeed, been infected with HIV. Though my ELISA antibody test was still negative, the PCR test was positive, with a viral load of 250,415 HIV viral particles per milliliter of blood.

There are no words to describe the horror of the moment when I learned I was HIV positive. Although I could not have asked more of my physician or his staff, their kind words could not change the finality of my diagnosis. It was the beginning of a new journey in my life one I would never willingly have chosen. In the end, I was sent home with a prescription for Valium and a referral to an infectious diseases specialist.

By October 1998, I was starting to adjust to the fact that I was now one of the statistics—a person with "occupationally acquired HIV infection." I had gotten used to the routine of taking upwards of sixteen pills every day to keep the virus at bay. But there was more bad news ahead. During some follow-up blood work, my liver enzymes were found to be severely elevated. Subsequent testing revealed that I was seropositive for hepatitis C virus (HCV). The source patient had been tested for HCV at the time of my exposure and was found to be negative. Since I had no risk factors for HIV or HCV other than being a healthcare worker, the only possible conclusion was that the source patient had been in the seroconversion window period at the time of my needlestick, and that I was simultaneously infected with both pathogens from my injury.

Fortunately, my needlestick was well documented. The exposure was reported to the proper supervisory personnel and the necessary baseline tests were performed. While my experience with worker's compensation has not been without stumbling blocks, the claim for my occupational HIV and HCV infections was accepted and my mounting medical bills are fully covered. I feel fortunate, since many occupationally infected healthcare workers have been denied benefits, often because of problems related to inadequate documentation.

I am unable to work at this time due to a deficiency in the number of infection-fighting white blood cells in my body—a side effect of the bone marrow-suppressing HIV drugs I must take. This deficiency puts me at risk for contracting infections; therefore I must limit my contact with those who have communicable diseases. Recently I was hospitalized with a severe systemic infection that led to my being admitted to the intensive care unit for septic shock.

Although I am receiving disability payments, my benefits, which would have been 66-2/3% of my previous salary, have been reduced by 25% because I was not wearing gloves at the time of my injury. This was done even though there is no clear indication from the medical literature that wearing latex gloves would have prevented my infection. As a divorced mother with two children, the disability benefit, which comes to $18,749 annually, is barely enough to live on.

I recently started a regimen to treat my HCV infection, which consists of three self-administered injections of interferon per week, as well as oral ribavirin, which brings my daily pill total to 22. Current medical literature indicates a 75% chance that this treatment will result in long-term remission of my hepatitis C. I say "remission" because you are not "cured" of hepatitis C—you can only hope for a long (and possibly permanent) period without clinical signs of the disease. Though I am hopeful that this treatment will be successful, it comes at a price. Interferon causes me to have periodic fevers, chills, muscle aches, and further increases my already-high fatigue level. The ribavirin causes severe nausea, as well as changes in my skin tone and hair texture. But I can live with these side effects for the chance to suppress the progress of this potentially lethal disease.

Telling my family about my infection with HIV and HCV was the most difficult thing I had to do. My mother is a nurse and knows all too well the ramifications of my diagnosis. I have yet to fully explain my illness to my own daughters, ages four and eight. I dread the time when they will realize that I might not be there to share the important moments in their lives.

On the other hand, with all of the research and advances in treating HIV, I may very well may be alive and well when my children are grown. But I have to accept that there are no guarantees, and make plans for a time when I may not be part of my children's lives.

This brings me to the reason I am telling my story. I cannot turn back the clock and undo the events of October 18, 1997. Nothing will give me back my life as it was before I was infected with HIV and HCV. But I would like my experience to be used to prevent similar tragedies from happening to other healthcare workers, and to educate hospital administrators and the general public about the real dangers of occupational blood exposures.

Although I certainly have questioned the meaning of all of this pain and turmoil in my life, I am not bitter. I believe that all things happen for a reason and that God will not hand us more than we are able to endure.

Dealing with illness and being forced to face my own mortality has given me a completely new perspective on life. I no longer take for granted the time I enjoy with my children, family and friends. I have discovered that this life I have been given is a precious gift, and strive to live better and to love better every day.

Karen Daley

Demanding Safety

Vol. 4, no. 4, 1999

The president of the Massachusetts Nurses Association tells about her needlestick and subsequent infection with HIV and hepatitis C—and her mission to get needle safety legislation passed in her state

On April 6, 1999, I testified before the Massachusetts state legislature in support of House Bill 969, "An Act Relative to Needlestick Injury Prevention." This bill holds special significance for me: I am co-infected with HIV and HCV from a needlestick injury I sustained on the job.

I have been a practicing nurse for over 25 years. I love clinical nursing and have felt privileged to care directly for thousands of patients over the years. My passion for nursing and the critical role I believe nurses play within the healthcare system—not only as direct care providers, but also as patient advocates, teachers, and interdisciplinary team members—led to my current role as president of the Massachusetts Nurses Association (MNA).

For 22 years I worked in the emergency department of Brigham & Women's Hospital (BWH), a 650-bed acute-care hospital in Boston. BWH is a teaching hospital for Harvard University and part of a healthcare system that includes Massachusetts General Hospital and Dana Farber Cancer Institute.

The stick

My needlestick occurred on July 25, 1998; it was a typically busy day in the emergency department (ED). After drawing blood from an elderly patient with mild dementia, I turned to dispose of the butterfly needle in the sharps container, which was above my eye level. Suddenly, I felt a stinging pain. I had been stuck on the index finger of my right hand by another needle that had become wedged in the container's hinged opening. The needlestick was deep and bled profusely through the puncture site in my glove.

I immediately removed my glove and washed the area, squeezing blood from the wound. My first instinct was to go back to triage as if nothing had happened. I was angry that I had been stuck; the needle box had been the subject of complaints in our unit. I also didn't want to deal with what would follow once I reported my exposure: the lab tests, the occupational health visits, the anxiety of waiting for results and thinking about the possible consequences of the exposure. But a colleague who witnessed my needlestick convinced me to report.

Follow-up

I filled out an incident report and saw the nurse practitioner in our "fast track" area who offered me post-exposure prophylaxis (PEP) for HIV as part of our occupational health protocol. I was already hepatitis B-antibody positive from a blood exposure in the ED in the late 1970s. (My HBV infection resulted in only mild symptoms that resolved quickly, and I have had no symptoms since then.)

I followed up with Occupational Health within 48 hours for lab work and counseling. The OH nurse was attentive and thorough in the information she provided. I returned later that week for my lab results, which confirmed that I was negative for HIV and HCV at the time of my needlestick.

In the days and weeks that followed, I remember how upset I was that the injury had occurred, and I had numerous conversations with colleagues and friends about it. That was my way of dealing with the anxiety, I guess. I figured it would be a full year before I knew if I was infected. As a result of my experiences during the first couple of months after the needlestick, and because of a growing awareness of the emotional toll and preventable nature of most sharps injuries, I asked the staff at the MNA to file a needlestick prevention bill with the state legislature.

By the time I had my six-month follow-up labwork on December 20, I had already experienced a wide range of symptoms; they began four or five weeks after my exposure and included unexplained weight loss, severe waves of nausea and mild abdominal pain, excessive hair loss, and insomnia. I experienced as many as 15 episodes per day of chills and heat and profuse sweating, and discrete neurological episodes lasting 25-30 minutes, in which I lost my ability to concentrate. I went to see my primary care doctor twice following the onset of these symptoms.

From nurse to patient

The results of my six-month HIV and HCV tests came back on December 23, two days before Christmas. I was positive for both viruses. Repeat tests were performed, and the results were confirmed on December 30. I started the new year with the knowledge that I was infected with HIV and HCV.

I can't describe to you how drastically one moment—the moment of my needlestick—has changed my life. Since January, I have had to come to terms with the fact that I am infected with not one, but two potentially life-threatening viruses. Up until this time, I had enjoyed good health, with little need to seek medical care beyond an annual visit to my primary care provider. Now I am truly a patient.

I have a new physician who specializes in infectious diseases and HIV management. I also see a hepatologist who is managing my HCV. Despite the confidence I feel in the quality of my medical care, both of my doctors admit that knowledge regarding treatment of hepatitis C is limited, in part because of the disproportionate resources directed toward HIV and AIDS in recent years. Experience in treating co-infections is even more limited. What they do know is that treating a person co-infected with HIV and HCV can be more difficult. This means I could develop liver cirrhosis or face the onset of AIDS in five or ten years rather than 20 or 30. In the meantime, I take a daily regimen of potent antiviral drugs that have caused nausea, weight loss, extreme fatigue, a low white blood cell count and a drug rash. To say these side effects interfere with my normal day-to-day routine is a gross understatement.

I am no longer a practicing healthcare provider. Because of a reluctance to be around needles, and based on health issues, I made the decision not to return to clinical practice. This change has cost me more than I can ever describe in words. I have been forced to abruptly leave colleagues with whom I've worked for many years, who are like family to me.

Since leaving my job, I have received regular payments from workers' compensation amounting to two-thirds of my former salary. Drug prescriptions, which cost almost $3,000 a month right now, are fully covered by insurance and are filled at my hospital. Over the long run, I'm not sure if workers' compensation will be adequate for my needs, but for now, my medical expenses are covered. I also received a lump-sum payment under an HIV insurance plan for Harvard employees; to my knowledge it was the first time such a payout was awarded.

My needlestick injury, which was entirely preventable, was related to two factors. First, if the needle in the container had been a "safety" needle with a protective needle shield, I could not have been stuck by it. Second, the needle became wedged in the window flap part of the box. As a result of my injury, a needlestick prevention committee has been formed to expedite the adoption of safer needle devices. Also, the emergency department at BWH has replaced the old needle boxes with ones that don't have a flap or lid.

Massachusetts legislation

The needlestick prevention legislation filed by the MNA requires that the Massachusetts Department of Public Health (DPH) compile a list of recommended safe needle products; institutions and providers regulated by DPH would be required to use safe needle technology, with exceptions granted by the commissioner on a case-by-case basis. Healthcare facilities would also be required to establish a needlestick prevention plan and to collect data on each exposure incident; this data would then be reported in aggregate form to the DPH. The circumstances of my injury have made me very concerned about the safety of some disposal container designs. Aggregate data from exposure logs, centrally collected, could provide evidence not only of the effectiveness of safety needle products, but also of the safety and appropriateness of particular disposal container designs.

I am hopeful about the chances of this legislation being passed. We have a long way to go to get it through the House and Senate with the current language intact, but there appears to be bipartisan support for the intent and direction of the bill. The major hurdle we face is educating legislators about the importance of keeping the language strong so adoption of safety needle technology is no longer voluntary. We anticipate that there will be efforts to weaken the language. If we are successful, we hope to see safer needle technology and significant reductions in the number of needlesticks throughout the state within the next couple of years. So far, the legislation has been reported favorably out of the Joint Health Care committee, and is now with the Ways and Means Committee. [Editor's note: Massachusetts passed a needlestick prevention bill, H.B. 5394, on 8/17/00.]

Demanding safety

As president of the MNA, and as part of my role as an advocate for RNs, I want to increase nurses' awareness of the very real health and safety risks we face in the workplace. I think we all have a tendency to minimize or dismiss those risks, thinking that the worst will never happen to us. While in some respects that makes it easier to do our daily work, it also keeps us, individually and collectively, from advocating for safer work environments—something we should *demand* from our employers.

We need to get involved with our hospitals' safety and product selection committees, where many of the basic decisions are made that affect those who provide frontline care. We all fight complacency about the unsafe conditions we sometimes work under, particularly given our workload and the demands placed on us in trying to provide care for our patients. And we allow employers to be just as complacent about situations that can threaten our health and lives. It is an attitude we can no longer afford.

Diane Mawyer

The High Cost of Hepatitis C

My battle with occupationally acquired HCV

Vol. 5, no. 2, 2000

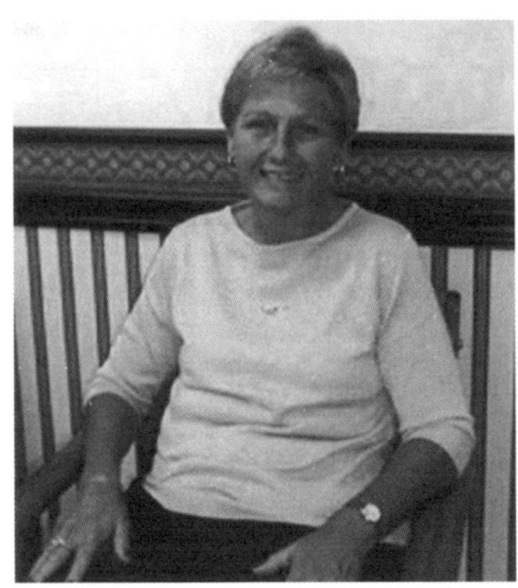

Diane Mawyer is a 46-year-old nurse who lives in Afton, Virginia, a small town in the Blue Ridge mountains near Charlottesville, home of the University of Virginia. The mother of one daughter who is currently an undergraduate at U.Va., Diane earned her L.P.N. in 1976 and R.N. in 1978.

Diane is infected with hepatitis C (HCV) from an occupational blood exposure. The International Healthcare Worker Safety Center estimates there are as many as 1,000 new occupationally acquired HCV infections each year in the U.S. Diagnosed with hepatitis C in 1993, Diane has been critically ill numerous times from HCV-related complications since her illness was discovered. Remarkably, she has survived not one but two liver transplants, and one kidney transplant (performed at the same time as her second liver transplant). Diane shares her story to promote understanding about occupationally acquired HCV and the risks that healthcare workers face.

From 1976 to 1981, I worked in the OR at the University of Virginia Hospital. I was first a staff nurse, rotating through all the surgeries, and then, during the last year, I was head nurse in eye surgery. This was before the bloodborne pathogens standard, and universal precautions were not yet in force. Typically, scrub nurses wore gloves, but circulating nurses did not. My time was divided about 50–50 between working as a scrub and a circulating nurse.

In 1981, I left the hospital to take a job supervising the local Red Cross blood bank. I worked there for the next 13 years. The blood bank was very busy, serving the Charlottesville-Albemarle County region, with a population of 100,000 and a major medical center, as well as 30 other counties in Virginia and West Virginia. We drew blood from approximately 40,000 blood donors a year. Although I was in a supervisory position, I performed blood drawing duties every day; up until 1985, when universal precautions were issued and glove-wearing became standard practice, blood drawing was performed with bare hands. During this time, I had daily contact with blood from as many as 20 donors a day. Screening for HIV didn't start until around 1985, and for HCV until 1989.

Segment sampling was a particularly hazardous procedure. Blood banks need donor blood samples in order to perform various typing and screening tests. Once the blood is drawn and the needle is removed from the donor, the blood-filled tubing attached to the blood bag is heat-sealed at intervals, creating a series of separate segments, which are later used for cross-matching with patients. I found that during the sealing part of segment sampling the tubing frequently burst—probably two or three times a day—splashing my hands and cuticles, and sometimes my face, with donors' blood. I must have been exposed to hundreds of patients' blood this way. [*Ed. note: The American Association of Blood Banks indicates that approximately 2–10% of donor blood is rejected due to positive tests for HIV, HTLV-1, HBV, HCV or syphilis; although there are false positives among the rejected blood, these figures show that an exposure to donor blood is clearly a potential hazard.*]

Starting in 1989 and for the next several years, I was plagued with chronic fatigue, joint and muscle aches and pains, edema in my legs, and frequent nausea. I generally felt unwell most of the time. Then in 1992, during a gastric stapling operation, the doctors discovered that I had cirrhosis of the liver. I was given an HCV antibody test as well as a battery of other tests, but no cause for the cirrhosis could be found. In the fall of 1993, during a preoperative work-up for a hernia repair, my liver enzymes were found to be very high, and about that time I developed jaundice. This prompted further testing for HCV in November 1993. In December I got the results back: I was found to be infected with HCV by both the RIBA and PCR tests. I never tested positive with an HCV antibody test. By the time my HCV infection was confirmed, I had extensive liver damage and needed a transplant.

In the spring of 1994, shortly after I was diagnosed with HCV, I filed a workers' compensation claim, since my physician believed that the only possible route of infection was occupational—I had none of the other risk factors. Those at risk for HCV include injection drug users, healthcare workers who handle blood products, and people who have had transfusions or multiple sexual partners. During my 18-year career as a nurse, I had had at least six documented needlesticks (two in the OR, four at the blood bank) and innumerable blood exposures. However, before 1989 there were no diagnostic tests for hepatitis C available—there was no way to find out if a source patient was HCV-positive, or to run a baseline test to establish that an exposed healthcare worker was HCV-negative prior to an exposure.

For six weeks after my diagnosis I received interferon treatment, but it didn't really help and suppressed my platelet count. Since I was extremely ill, my doctor advised me to stop working, and I left the blood bank at the end of January 1994. In early March I was listed for a liver transplant and, amazingly, received a liver only three months later, in late May. After the transplant, I spent six weeks in the hospital because of complications—I had a severe rejection, followed by pneumonia, and then renal failure, for which I received dialysis for a couple of weeks. Finally I started to improve and was released from the hospital.

My workers' compensation claim was denied in the fall of 1994. However, Red Cross offered me a settlement, which I accepted. When I retired from my job at the Red Cross, I began receiving social security disability.

For about a year after the transplant I felt better, but then my kidneys started to deteriorate. I discovered that the immunosuppressant drugs that transplant patients must take can often cause damage to the kidneys, and that, along with liver damage, HCV can also cause kidney damage. Over the next several years I lost about 80% of my kidney function. During 1997 and 1998, I experienced a massive recurrence of symptoms related to HCV and cirrhosis, and was hospitalized 11 times over a 12-month period for severe encephalopathy (deteriorated brain function) from ammonia and other toxins that my liver was unable to get rid of. During these hospitalizations, I was in and out of comas, and I remember little from this period.

In the two years leading up to my second transplant, my main symptoms, in addition to the encephalopathy, were pleural effusion, which required two thoracoscopies to drain the fluid from my chest wall; severe edema in my legs; severe ascites—a form of edema in the abdomen—which made me so bloated I looked nine months pregnant; seizures; peripheral nerve damage in my feet and legs; bleeding disorders; and recurrent infections. I was listed for another transplant in October 1997, but instead of three months, this time I had to wait two years. Finally, on June 18, 1999, I had a double transplant: I received a new liver and a new kidney (my existing kidneys were not removed, so I now have three). Since the second transplant, I have felt much better—better than I have in a long time.

I have often thought about the needlesticks and blood exposures I experienced as a nurse, and wondered which one infected me with hepatitis C. Of my two documented exposures in the OR, one involved a scalpel blade, the other a needle. The four needlesticks I sustained at the blood bank are particularly vivid. In December 1982, while I was inserting a 16-gauge needle used to draw donor blood into a vacuum tube, the needle slipped and punctured my finger. In 1985, after I had filled a glass capillary tube (used to determine hematocrit) with donor blood, the tube broke in my hand while I was pressing the end into clay. [*Editor's note: The FDA, in conjunction with the CDC and OSHA, issued a safety alert in 1999 about the hazards of glass capillary tubes. There is at least one documented case of occupational HIV infection from this kind of injury—a physician at Johns Hopkins who subsequently died of AIDS.*]

I had two needlesticks in 1986. The first was like the one in 1982—I was attempting to fill a vacuum tube with donor blood using a 16-gauge needle, my hand slipped, and the needle punctured my finger. The other one happened while I was disposing of a contaminated butterfly needle. While carrying the needle to a sharps container, I lost hold of it and, trying to catch it, my finger was stuck. This needlestick involved blood from an HIV-positive patient—significant because it is common for HIV-infected patients to have HCV as well. Three of these needlesticks involved hollow-bore, blood-filled needles—the highest-risk device category for transmission of bloodborne pathogens.

My first transplant cost approximately $250,000; if you add to that the cost for my second transplant and my numerous hospitalizations, I believe my total treatment costs add up to over a half-million dollars at this point. My healthcare expenses have been covered by Medicare and private health insurance, although I have had to pay the deductibles. These have totaled close to $60,000 over the last six years, which I have paid for out of my own pocket.

My health is now better than it has been in years, but I have liver biopsies every three months to monitor the condition of my liver. So far my doctors have seen little, if any, damage. But who knows how long that will last? My long-term prognosis is a big question mark. With a disease like cancer, once you reach the five-year mark after treatment usually you are in the clear. But hepatitis is a lifelong illness that never goes away. The least health problem requires immediate medical attention, because it can quickly become something serious. My goal right now is to maintain my health until new and better treatments for HCV become available. I don't make long-range plans; my challenge is to stay positive and live life to the fullest each day.

The impact of my illness on me and my family has been profound. My daughter, in particular, has had to live with the fear throughout her teenage years that her mother might die at any time. She had to grow up too quickly—often she was the one caring for me, instead of vice versa. My husband has been my rock; he learned how to run the house when I was too sick to do it. For me, one of the hardest parts about this disease was losing my professional identity. I was devastated when I was told by my doctor that I had to quit my job. All I ever wanted to be was a nurse. As supervisor at the blood bank, I truly believed in what I was doing, and that our work made a big difference in others' lives. I used to do a lot of traveling, going to meetings and conferences—I loved that. I loved the sense of community with the volunteers, donors, and my co-workers, and seeing so many people on a daily basis.

Ironically, when I worked at the blood bank I was responsible for blood donor safety—assessing the suitability, history and health of potential donors. I spent 13 years of my nursing career trying to provide the safest possible blood supply for patients. The irony, of course, is that I myself became infected from a bloodborne pathogen.

I think the blood exposures and sharps injuries I sustained over my career speak to the broad spectrum of occupational risks that healthcare workers face in the course of doing their job. Fortunately, there are many, many different kinds of safety devices on the market now that weren't available in the 1980s when most of my exposures occurred. When universal precautions first came out, most healthcare workers thought they were ridiculous—they grossly underestimated the risk of becoming infected with a bloodborne pathogen. What I would say to healthcare workers now is: don't underestimate the risk. Don't cut corners. Use whatever safety devices are available and if they're not available, seek them out—take a stand. Constantly look for the safest way to do a job or procedure. If necessary, get procedures revised—the best suggestions for improvements in technique come from frontline healthcare workers.

I strongly support needle safety legislation at both the state and federal level. I think it is needed to put pressure on facilities to adopt the safety devices that are available and to expand the market for those devices. And I very much hope to see needle safety legislation passed in Virginia by the end of the next legislative session.

Hepatitis C: CDC Findings & Recommendations

- HCV infection is the most common chronic bloodborne infection in the U.S., affecting approximately 4 million people.
- HCV infection often occurs with no symptoms or only mild symptoms. But unlike HBV, chronic infection develops in 75% to 85% of patients, with active liver disease developing in 70%. Of the patients with active liver disease, 10% to 20% develop cirrhosis, and 1% to 5% develop liver cancer.
- Although the prevalence of HCV infection among healthcare workers is similar to that in the general population (1% to 2%), healthcare workers clearly have an increased occupational risk for HCV infection. In a study that evaluated risk factors for infection, a history of unintentional needlestick injury was independently associated with HCV infection.
- The number of healthcare workers who have acquired HCV occupationally is not known. However, of the total acute HCV infections that have occurred annually (ranging from 100,000 in 1991 to 36,000 in 1996), 2% to 4% have been in healthcare workers exposed to blood in the workplace.
- There is no vaccine against hepatitis C, and currently no treatment after an exposure that will prevent infection. Immune globulin and antiviral agents are not recommended for postexposure prophylaxis. For these reasons, following recommended infection control practices is imperative.
- As soon as possible after an exposure to HCV, healthcare workers should be given baseline HCV antibody and liver enzyme tests, with follow-up tests at 4-6 months.

Sources

- National Institute for Occupational Safety and Health. NIOSH Alert: Preventing Needlestick Injuries in Health Care Settings (November 1999). DHHS Publication No. 2000-108.
- Centers for Disease Control and Prevention. Recommendations for prevention and control of hepatitis C virus (HCV) infection and HCV-related chronic disease. MMWR 1998;47 (No. RR-19):1-39.

Julie Naunheim Hipps

When Home is Where the Risk Is

By Jane Perry, M.A.

Vol. 5, no. 3, 2000

Julie Naunheim-Hipps, a nurse in St. Louis, Missouri, is a 43-year-old mother of four. Julie worked hard to become a nurse. It was a second career for her; she had returned to school in her thirties to get her associate nursing degree. While taking classes towards her B.S.N. she worked at Jewish Hospital, which is now part of BJC Health Systems, the largest healthcare provider in St. Louis and a nationally known research institution. She worked in the medicine and oncology units, first as a secretary, then as a patient technician, then, after obtaining her R.N., as a nurse.

Julie loved her work. "I felt that at the end of the day I had made a contribution to helping others, by providing compassionate and loving care. That's the reason we become nurses—because we want to make a difference in peoples lives." She had developed expertise in oncology nursing and infusion therapy, and found her job challenging. However, she was working night shifts and wanted a regular daytime schedule. So, in November 1998, she began working for a home health agency affiliated with the BJC system.

While caring for patients in home-care settings, Julie noticed important differences from hospital-based nursing. "First of all, there is much more paperwork, which is very time-consuming," she says. "Also, you have little or no control over the environment in which you're providing care. You might walk into a house and find it neat as a pin, or a total mess. A number of our patients were very poor and lived in run-down housing in high-crime areas. Sometimes they were victims of violence, abuse or neglect right in their own homes." She found that disposal of sharps and other contaminated waste was often a problem: "Although the home health agency provided the patient with a sharps container at the beginning of treatment, it was the patient's responsibility to call the agency when replacements were needed. This didn't always happen. You might walk in and find an overflowing sharps container, or no container at all."

The day of Julie's needlestick, October 29, 1999, was, she says, "a day I will never forget. It changed me and my family forever." She began, as usual, by getting her work assignments for the day and calling her patients to schedule visits. When she arrived at the first patient's home, she found it cluttered, untidy and chaotic, with a small child running around. She needed to draw a blood sample, change the dressing on the patient's peripherally inserted central catheter (PICC) line, and replace the I.V. medication bag. She washed her hands and put on gloves. It was difficult to find a place to work; she crouched down on the floor to arrange her supplies on a barrier pad that was provided by the agency for this purpose.

Matthew Hipps

Julie prepared the vacuum tube phlebotomy set. She had a blood collection tube holder into which she screwed a single-ended, rubber-sleeved needle (which pierces the rubber stopper on the blood tube). Instead of a phlebotomy needle on the other end, the device had a luer fitting, which she connected directly to the I.V. line. She slid a blood collection tube into the holder and over the rubber-sleeved needle, and watched the tube fill with blood. When it was full, she disengaged the phlebotomy set from the I.V. line, removed the tube of blood, and put it in a cooler for transport.

Julie had been instructed to save and reuse the blood tube holders (although only on the same patient), so she unscrewed the adapter with the rubber-sleeved needle from the holder and set the holder aside. Then she needed to dispose of the adapter/needle. The disposal container was on top of the refrigerator to keep it out of reach of the patient's child. As Julie rose to go to the container with the device in her hand, she inadvertently squeezed the rubber sleeve covering the sharp. The needle that had pierced the blood tube now pierced her left thumb.

She cried out, and saw the blood oozing through her glove. As she stripped it off, the patient asked what happened. Julie told her the needle had pierced her thumb. It was then the patient told Julie she had hepatitis C.

Julie ran to the bathroom and squeezed the blood from her thumb under running water. She washed her hands with soap and cleaned the wound with alcohol. Then she called her supervisor to report the event. The supervisor was reassuring, but told her to come to the office immediately, and arranged to cover her other patients.

Another nurse was called in to draw blood from the patient for postexposure testing for HIV and hepatitis B and C. Julie, scared and shaking, drove back to the office, where she had her own blood drawn for baseline testing. She was started on an HIV postexposure prophylaxis regimen of Epivir and Retrovir. After five days, she learned that the patient was HIV-negative, but positive for hepatitis C. All of Julie's baseline tests were negative. She stopped the HIV PEP drugs, and was told to come back in four weeks for follow-up testing.

Her supervisors assured her that the risk was very small and she did not need to worry. She had experienced one other needlestick as a nurse, which did not involve a blood-filled needle. She had reported it, and her follow-up tests were negative.

After her needlestick, Julie returned to work, but three weeks later she started having difficulty concentrating and became unusually tired. She thought it might be related to her thyroid condition or to the recent loss of her father.

She asked her physician to do some blood tests. When the results came back, he told her everything looked fine except for slightly elevated liver enzymes. He suggested that she take some time off. Julie had blood drawn for her four-week postexposure follow-up tests, and then left for Texas to visit her sister and get some much-needed rest. Although she was terrified at the thought of contracting HCV, there was nothing to do but wait for the results.

A week later, on December 10, Julie called the employee health department at BJC. She was told that her hepatitis C test was positive. Like the patient, she had type 1b HCV, which is the most difficult to treat. Her liver enzymes were more than ten times the normal level. She could not believe what she was hearing, that this could happen to her. Her experience as a nurse had not prepared her for this.

She returned home and saw a hepatologist for treatment. Blood was drawn for more tests and a liver biopsy was performed. At this point, Julie says, "It was only a couple of weeks before Christmas. I had lost my father, my health and my financial security in one month's time—and possibly my career as a bedside nurse. I had neither the physical strength nor the money to buy Christmas gifts. What would happen to my career as a nurse? Would I be able to attend my son's high school graduation? Was I going to live or die?"

Hepatitis C has been called the "silent epidemic" because many of those who carry the virus do not become symptomatic until years after they are infected. Julie's case, however, is remarkable for how quickly she became symptomatic and was diagnosed with acute HCV infection. During the first few months, she was too sick to get out of bed; routine tasks became "mountains to climb." Now, nine months after her diagnosis, she is undergoing combination therapy with ribavirin and interferon. With combination therapy, there is up to a 30% chance that her type 1b HCV can be suppressed to undetectable levels and her liver enzymes normalized over a sustained period. However, the side effects of the drugs, all of which Julie has experienced, include flu-like body aches and pains, nausea, extreme fatigue, and depression. And though the drugs can suppress the virus, they do not eliminate it. She could have a major recurrence of symptoms at any time, especially after therapy is complete.

Because the HCV treatment has compromised her immune system, and because she was unable to mount an immune response to hepatitis B vaccine, Julie chose not to continue with bedside nursing. If she were occupationally infected with hepatitis B, the combined impact of HBV and HCV could be fatal. She is hoping to find a research position, preferably related to hepatitis C. She started receiving workers' compensation almost immediately after she was diagnosed, and she feels that her employer has been very concerned and expeditious about her care.

Since her diagnosis, Julie has become an advocate for needlestick prevention and safety, and is working with the Missouri Nurses Association, Missouri state representatives Joan Berry and Harry Kennedy, and others to pass needle safety legislation in Missouri. "I have excellent role models," Julie says, "including Lisa Black, Lynda Arnold, Karen Daley, Diane Mawyer and other occupationally infected healthcare workers who have been willing to share their experiences to help others. They have provided tremendous emotional support. I no longer feel alone."

Julie supports the widespread implementation of sharps with safety features that protect healthcare workers from injuries and potentially life-threatening bloodborne pathogens. Ironically, however, a safety device was not even necessary to prevent her injury, given the device configuration she was using. It would merely have taken a change in her employer's policy—requiring that blood tube holders, which cost only pennies, be disposed of after each use rather than reused. (Julie notes that the lifetime cost of treating her illness "is estimated to be in excess of one million dollars in medicine, lost wages, and possible liver transplants.") If the policy had been different, she would not have detached the adapter from the holder, thus exposing the rubber-sleeved needle, and herself to injury.

In fact, many experts on the bloodborne pathogens standard believe that the practice of removing adapters from blood tube holders constitutes a violation of section (d)(2)(vii)(A) of the standard: "*Contaminated needles and other contaminated sharps shall not be recapped or removed unless the employer can demonstrate that no alternative is feasible or that such action is required by a specific medical procedure*" (emphasis added). California OSHA cited a healthcare facility for this practice in December 1999. The citation referenced the above section of the standard and stated: "The employer failed to ensure that contaminated sharps were not removed from devices. In the emergency room, STAR lab, and outpatient surgery areas, employees removed contaminated phlebotomy needles from tube holders."

There are three key prevention messages that emerge from Julie's experience. **First, healthcare workers need to beware of the "back-end" of a needle adapter**. Although most healthcare workers do not notice it, inside the rubber sleeve is a sharp that can cause a high-risk injury with a hollow-bore, blood-filled needle. **Second, employers should not tell workers to reuse blood tube holders**, which requires them to remove a blood-filled needle and is both a hazardous practice and a potential violation of the bloodborne pathogens standard. Julie would like to see companies stop marketing reusable blood tube holders altogether.

Finally, in unpredictable home healthcare settings, safety blood-drawing equipment—particularly phlebotomy, butterfly and I.V. catheter needles—**and needleless I.V. systems should be routinely used**. OSHA cannot regulate private homes, but it does require that home healthcare agencies, like all healthcare employers, have an exposure control plan in place that provides employees with the safest devices and equipment to prevent sharps injuries.

Julie's message is simple: "My injury was 100% preventable—as were those of so many others. The nursing profession has lost too many dedicated nurses to infections resulting from needlestick injuries. It has to stop."

What Julie's experience tells us:
Key Prevention Messages

1. **Healthcare workers need to beware of the "back-end" of a needle adapter.** The needle inside the rubber sleeve can cause a high-risk injury.

2. **Employers should not tell workers to reuse blood tube holders,** because it requires that they remove a blood-filled needle. This is both a hazardous practice and a potential violation of the bloodborne pathogens standard.

3. **Safety blood-drawing equipment**—particularly phlebotomy, butterfly and I.V. catheter needles—**and needleless I.V. systems should be routinely used in home healthcare settings.**

"Dr. Jones"

A Surgeon, a Suture Needle—and Hepatitis C

By Jane Perry, M.A., and Janine Jagger, M.P.H., Ph.D.

Vol. 5, no. 6, 2001

AEP interviewed a surgeon infected with hepatitis C, presumably from an occupational exposure. Although "Dr. Jones" (not his real name) wished to remain anonymous for this article, his willingness to come forward and speak to us was unusual. His experience sheds light on the personal and professional realities confronting an infected healthcare worker, and the factors that determined his professional future.

Dr. Jones is chief of plastic and reconstructive surgery at an academic medical center; he is married and the father of four children.

AEP: *When did you start performing surgery?*

Dr. Jones: During my residency, starting in 1979. I went through five years of residency training, then started my own practice in 1984.

AEP: *It's historically significant because measures to prevent occupational exposures were implemented gradually during the late 1980s and early 1990s, and a test for hepatitis C wasn't available until 1989.*

Dr. Jones: It's hard to imagine, but back in 1979-80, at the beginning of my residency, I remember coming out of the operating room literally soaked in blood, from the waist down—pants, underwear, socks, shoes. And I'd sit down and have a cup of coffee in my blood-soaked clothes, and then I'd go back in to fight again. I remember going into someone's chest in the emergency room with just sterile gloves—no gown or goggles—because he had been shot in the heart and had gone into cardiac arrest. And when you did put on sterile gloves, it was to protect patients from you, not the other way around. You were bathed in blood regularly, got stabbed with instruments. It was no big deal. You just changed the instrument and got back to work. There was little protection.

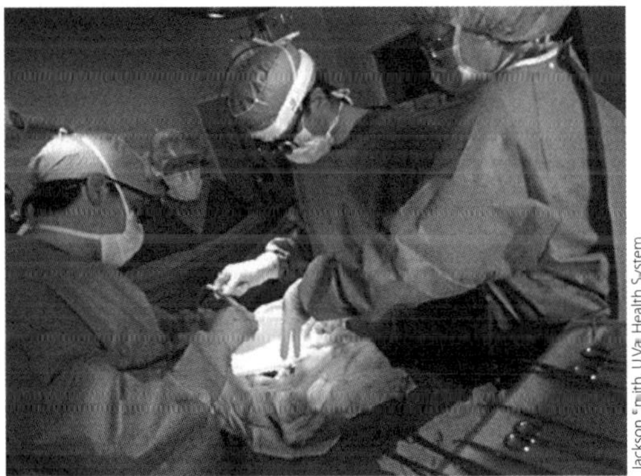

Jackson Smith, U.Va Health System

AEP: *So you didn't have a documented exposure that you could correlate to your HCV infection?*

Dr. Jones: That's right; it was rare to report exposures back then. But I suspect that my infection occurred sometime during my residency. There was one patient in particular I'll always remember. I was using a large retention needle, and really got harpooned. These are very large suture needles, maybe 3-1/2 inches long, that are used for big closures that are subject to a lot of stress. I remember getting cut really badly, and the patient having non-A non-B hepatitis. As I look back, that could have been the event that led to my hepatitis C infection. But that was before a test for hepatitis C was available.

AEP: *How did you find out the patient had non-A non-B hepatitis?*

Dr. Jones: In the course of treating the patient we discovered she had elevated liver enzymes, and everyone was worried that she had hepatitis B. We tested her for that, but she was negative. We concluded that she had some form of hepatitis, but not A or B.

I remember getting very sick in 1981, during my residency. I was out of commission for about a month, and my liver enzymes were elevated. At one point they thought I had mononucleosis, but I tested negative for it, and was also negative for hepatitis B. Then, after about a month, I recovered and went back to work.

AEP: *Did you have any other risk factors for hepatitis C, outside of work?*

Dr. Jones: I never had surgery requiring a blood transfusion. I had a septum fixed and wisdom teeth removed, and that was it for surgery. I didn't have tatoos, had never been on drugs. I had no other risk factors.

AEP: *Do you recollect how much time elapsed between the exposure event you described and your first illness?*

Dr. Jones: It was about a year. I had horribly elevated liver enzymes, but not jaundice. After that illness, I finished my residency and went into private practice. I did not give it any more thought until ten years later, when I got very sick again—that was 1992. Again they thought I might have mono; they checked my liver enzymes, which were about twice the normal level. At that point a marker for hepatitis C had finally been discovered, and that was when I tested positive for HCV. A liver biopsy was performed, and it showed moderate inflammation with signs of chronic, persistent HCV infection.

I was having a hard time practicing. I would do a couple of cases then would have to go home. I was sweating all the time and lost weight; I was nauseous, had constant diarrhea and no appetite. That's when my doctor offered me treatment with interferon. He said it was grueling, but thought I should go ahead and try it. I took it for three months while I was still practicing. After the treatment, my liver enzymes were back to normal.

About six months later I got ill again. This time, the tests showed that my liver enzymes were normal, but I was feeling terrible. So I reduced my work schedule and took it easy for a while. After about a year, I felt fine, but my liver enzymes had gone back up. It made no sense. I had another liver biopsy and it showed a little fibrosis this time, so they wanted to put me back on interferon. I was told I probably would not be able to practice this time while I was taking the drugs; so I left my private practice for about three months and went to another part of the country. I was tired of people asking questions, wondering if I had AIDS or liver cancer or some other disease. I did not want to reveal that I had hepatitis C. But eventually it got out after I left.

AEP: *How did that happen?*

Dr. Jones: I was getting my own medicine from the hospital, and people started to talk—"Oh, Dr. Jones is ill." By the time I got back, everyone knew I had hepatitis C. When you are a physician and you are gone for three months, rumors start to fly—people automatically assume you are a drug abuser, an alcoholic, or deathly ill.

When I got back, I was still on interferon and planned to practice three days a week, on the days when I wasn't taking it. But my colleagues told me, "Look, we don't feel comfortable referring patients to you because you have hepatitis C, and we could get sued for that, because we knowingly sent our patients to you." They said if I told my patients that I had HCV, they would feel comfortable referring people. Well, I tried that a few times, and it just did not work. As soon as patients find out you have hepatitis C, they suddenly change their minds about having surgery.

At that time—around 1992–93—I had two lawsuits filed against me. One was for a scar from an abdominoplasty operation I performed. In the course of the litigation, the patient's lawyer found out I had hepatitis C and added a claim for emotional distress, saying that I did not properly inform my patient about my condition and that she was now deathly afraid she might have been infected. Soon after, another patient sued me for a silicone breast implant. The patient claimed the implant ruptured—this was during the "silicone wars." And again, attached to that suit was a claim for lack of informed consent about my hepatitis C. One of the lawsuits cited a case in which a surgeon who was a recovered alcoholic and a member of AA was successfully sued in court, on the basis that the patient had a right to know that the surgeon was a recovered alcoholic. The decision was upheld by the state supreme court.

Because of all this uproar, one of the hospitals where I had surgical privileges formed a committee to discuss the issue of my hepatitis C and informed consent. Their final recommendation was that I inform my patients about my HCV status. I was on friendly terms with the people at this hospital; they were providing my interferon free of charge, because the drug was considered experimental at the time and insurance didn't cover it. At that time, it cost between $400–$600 a month. I was on and off it for eight years—I kept failing treatment, with continued symptoms and elevated liver enzymes.

The lawsuits, which were ongoing for about six years, were eventually dropped. The patient who brought the silicone breast implant suit ended up suing the implant manufacturer instead. As far as I know, neither of the patients developed hepatitis C. But, of course, I still had to pay my attorney's fees.

At that point I really had to reconsider my career path. No patients wanted to see me, I had been sued twice, my colleagues did not want to refer patients to me, and the hospital where I worked required me to tell my patients that I had hepatitis C.

But I struggled on for the next couple of years, trying to maintain my practice. Things just became worse and worse, however, and by 1995 I could not afford to stay in practice because of lack of patients, and because I was ill. So I finally gave up. You can only fight the system so much.

Within a month of the time I closed my plastic surgery practice, I was called by the chairman of the surgery department at a nearby state university. He understood my situation and invited me to join the faculty, strictly in a teaching role. I would oversee and instruct the residents under me, and they would perform the actual surgery. I provided the knowledge, the resources, and the direction.

For me, it was a great opportunity: it meant I could still be involved in medicine in some capacity, and use my experience to teach others. I definitely wanted to give it a try. So, in March of 1995, I joined the faculty at the university. It turned out to be a great situation. I would go to the operating room with the upper-level residents under my supervision, and diagram on the chalkboard the operation they were going to perform. I would tell them what to do, and they would do it. During the operation I had a laser pen and a pointer to aid in giving them directions. Occasionally I would scrub in, and would push and pull, using a blunt hemostat, to show them where to go, but they did all the cutting and sewing. I was double-gloved, of course, which helped ease my own fear of infecting patients. The hospital was very comfortable with this arrangement. It did not have an explicit policy about infected surgeons, and I did not sign any written statement. It was a gentlemen's agreement. I was very grateful to have a job, and they were happy to have me, so it turned out to be a great partnership.

So that has been my life for the last six years. I rarely wield a knife, unless there is an urgent need to do so.

AEP: *Sounds like a great example of the medical community taking care of its own.*

Dr. Jones: Absolutely.

AEP: Another alternative for someone in your position

might be conducting training for endoscopic procedures, since the training is performed on animals.

Dr. Jones: Yes, that would be a possibility. Another is something called "Zeus," a computerized virtual-reality surgery program where you operate with hand-held devices in another room using the computer, and the actual surgery is done by a robot. There are usually scrub techs or residents in the room with the patient who place the instruments to get the robot going, and the surgical staff is in the other room operating.

AEP: *That sounds like a new horizon for you.*

Dr. Jones: It is about ten years away from being perfected. So for now, I teach others. My biggest fear in the operating room is getting stuck by one of the residents, who are relatively new to surgery. I only have one type of hepatitis C, and there are several variants of it. I have been stuck by residents a couple of times—one time with a needle-tip Bovie coagulator. The device was plugged in when I got stuck, so I got an electrical shock at the same time. I assume any pathogens would have been killed by the electricity, so I'm not worried about another infection. I was also stuck with a "rake," a retractor with sharp little claws on it. A resident was using it to pull a piece of tissue out of the way; it went through the tissue and stuck my finger.

AEP: *Of course, with sharps exposures we always want to know what the device was and whether it needed to be sharp, particularly in the surgical setting. In both those situations, with the Bovie and the rake, a sharp-edged device was not necessary, correct?*

Dr. Jones: Yes—there are blunt Bovies and blunt retractors. I think the sharp versions should be removed from the operating room altogether. There are many unnecessarily sharp instruments that endanger healthcare workers that ought to be removed from the OR.

AEP: *The number-one sharp device to get out of the OR is the sharp suture needle—there are very few cases where this device is really necessary. Blunt suture needles can be substituted for the suturing of less-dense internal tissues; for cutaneous closures, you can use staples or tissue adhesives or adhesive strips.*

Dr. Jones: You are talking to a plastic surgeon now, and if I told you I was going to close your eyelids with staples I think you might get a little upset.

AEP: *I'm sure you have to be selective about the use of staples. What about tissue adhesives, what is your opinion of them?*

Dr. Jones: I use them frequently, but you can't use them around the eyes, the mouth or inside the nostrils.

AEP: *It would be useful to develop an inventory of surgical procedures, with a list of all the sharp items that can be eliminated from those procedures.*

Dr. Jones: Absolutely. I would have sharp towel clips removed from every operating room in the country right now. Those are ridiculous—who needs them?

My objective now is to stop the transmission of these diseases to healthcare personnel. Being infected with hepatitis C has ruined my life. My life as I knew it, with all the training I underwent to be a plastic surgeon, is over. The crazy thing is that it could have been prevented so easily. It is all a question of awareness. Today, everybody is aware of the risk of infection. We are heading in the right direction of being more aware and better protected—but we need to go much further.

AEP: *Getting the attention of surgeons on this issue has been very difficult.*

Dr. Jones: Well, they need to realize the potential consequences of an infection. Your life is ruined, and the financial consequences are devastating. After I got sick, I was able to get disability, but I never got workers' compensation because I never filed a claim. I didn't know exactly when I got infected, because there was no test for hepatitis C, and I was in training at the hospital where I had that major exposure—I did not want to rock the boat. And, of course, now it is too late. But at least I had the disability.

Recently, though, my disability has been cut off. I am paid for my teaching, but it is very little compared to my former income. It is enough for food and shelter, but certainly not enough to put my four kids through college or lead the lifestyle that I trained and struggled so hard to have.

AEP: *What happened with your disability?*

Dr. Jones: The disability company said to me, "Having

Between a rock and a hard place:

"The disability company said to me, 'Having hepatitis C is not sufficient reason to claim disability anymore. Because, according to the CDC, having hepatitis C should not put restrictions on your practice.'

But in courts of law, issues of informed consent are settled in favor of patient-plaintiffs. When it comes to getting hospital privileges, neither courts nor hospital administrators are going by the CDC position regarding HCV-infected physicians—but disability carriers are. They are saying an HCV-infected physician should be able to practice without restriction.

That leaves infected surgeons between a rock and a hard place: Should they inform patients of their serostatus, as the AMA advises, and risk losing their practice—and their disability—or practice without restriction, but risk being sued by patients for lack of informed consent?"

—"Dr. Jones," HCV-infected surgeon

hepatitis C is not sufficient reason to claim disability anymore. Because, according to the CDC, having hepatitis C should not put restrictions on your practice."

AEP: *That is an unintended use of the CDC position.*

Dr. Jones: It's frightening. According to my disability company, which is one of the major carriers in the country, having hepatitis C or HIV is not a reason to restrict your practice. That is a point of view that will scare any surgeon in private practice. If the carrier ends up being successful in denying me disability on this basis—obviously I am going to have to go to court with them—you are going to have more infected surgeons hiding in the closet than ever before. You know, if I had kept my mouth shut, I could have kept practicing.

"My life as I knew it, with all the training I underwent to be a plastic surgeon, is over. The crazy thing is that it could have been prevented so easily ... We are heading in the right direction of being more aware and better protected—but we need to go much further."

—"Dr. Jones"

Regarding the issue of informed consent, I think that if someone is physically able to work and is totally asymptomatic, they should be able to practice. But I also think that patients need to be informed of their surgeon's sero-status, based on ethical and legal considerations. Surgeons should do everything in their power to make their surgical practice as safe as possible, by using protective equipment such as mesh gloves and blunt suture needles. And infected surgeons should avoid performing exposure-prone procedures.

AEP: *Your experience was that once you informed your patients, you did not have any more patients.*

Dr. Jones: That is true, and it is a difficult decision to make. But I feel it is still the patient's right to know. I think most of the time you are going to find people like me saying, "I better make sure I *never* expose a patient to this illness." Of course, one option is to switch specialties to something like family medicine or radiology. Or, if I wanted to continue practicing plastic surgery, I could limit myself to teaching and to procedures like laser resurfacing, laser photocoagulation, children's port wine stains, chemical peels, and microdermabrasions.

Safety Checklist for the OR

- Are blunt suture needles, stapling devices, adhesive strips or tissue adhesives used whenever clinically feasible in order to reduce the use of sharp suture needles?
- Are scalpel blades with safety features used, such as round-tipped scalpel blades and retracting-blade and shielded-blade scalpels?
- Are alternative cutting methods used when appropriate, such as blunt electrocautery devices and laser devices?
- Is manual tissue retraction avoided by using mechanical retraction devices?
- Has all equipment that is unnecessarily sharp been eliminated?
  (Example: towel clips have been identified as a cause of injury in the operating room, yet blunt towel clips are available that do not cause injury and are adequate for securing surgical towels and drapes. Other examples of devices that do not always need to have sharp points include surgical scissors, surgical wire, and pick-ups.)
- Is double gloving employed in the surgical setting?
- Do circulating nurses, as well as personnel close to the surgical site, wear eye protection such as goggles or faceshields that have a seal above the eyes to prevent fluid from running down into the eyes?

Figure 1. **Potentially Preventable Suture Needle Injuries**
6 hospitals, 15 months, 197 suture needle injuries

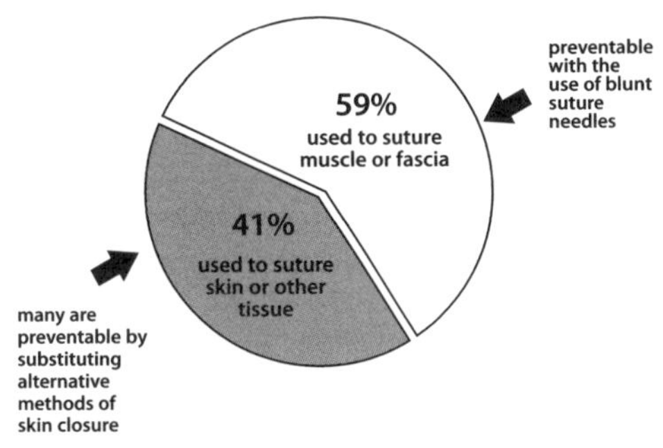

From: Jagger J, Bentley M, Tereskerz P. A study of patterns and prevention of blood exposures in OR personnel. AORN Journal *1998; 67(5):979-996.*

AEP: *Is there a need to inform patients who have these kind of procedures of your HCV status?*

Dr. Jones: I don't think so, if the procedures are non-invasive. Of course, there are some aspects of this issue on which I have not reached a conclusion. For instance, if I were a general surgeon and I had hepatitis C, I would just restrict my practice to non-exposure-prone procedures. But if I were a heart or orthopedic surgeon with HCV, I would probably just quit, because most of their procedures are exposure-prone.

AEP: *Have you thought about switching specialties?*

Dr. Jones: Oh yes. I've thought about going into psychiatry or radiology. But I've kind of carved out a niche for myself in a totally different avenue of plastic surgery. I run a very large clinic where I direct wound therapy and treatment. It doesn't involve my doing surgery at all—just seeing patients and taking care of wounds, particularly diabetics. That's where most of my research is now. It is very rewarding; if a diabetic is able to keep a leg, he or she is extremely grateful.

AEP: *Are you still on the interferon?*

Dr. Jones: No, I have been off interferon for about a year and a half now. I have hepatitis C "fatigue attacks," where one day I am doing great, and then the next day I am totally wiped out and cannot move. It feels like the flu and lasts from 24 to 48 hours. I get them about once or twice a month—they are well documented in the literature on hepatitis C. I also get night sweats and myalgia. But my prognosis is good; I've been infected for probably 20 years, and I have no evidence of cirrhosis. I'll probably continue to live like this, hopefully for a normal lifespan. My biggest fear, of course, is liver cancer. I may be in sustained remission right now, because I don't have any detectable virus—but that doesn't mean you're not infectious. Nobody can tell you whether you are cured or not.

AEP: *What kind of effect has your illness had on your family?*

Dr. Jones: Everybody works now—my kids are 20, 18, 16 and 11. Everybody has a job, including the 11-year-old. It was quite a change in lifestyle for them—from having everything to suddenly having very little security.

AEP: *Do you have any policy recommendations?*

Dr. Jones: I do not understand why we don't have a national policy on the issue of HCV-infected surgeons and invasive procedures. I think surgeons have a right to know if their patient is infected, and I think patients have a right to know if their physician is infected. At my hospital, of the two surgeons besides myself who are occupationally infected with hepatitis C, one has limited herself to teaching and the other is doing what I do, having the residents perform the surgeries. So there are three surgeons who are all adhering to the same self-imposed restrictions, because we do not want to give the disease to anyone else. If you have this disease and you have been through the hell, you would not want to pass it along to anyone else, not even your worst enemy—much less a patient.

Why don't surgeons report?

"As an occupational health nurse, I know that blood exposures are under-reported, particularly from the OR. Recently our surgeons appear to be even more reluctant to report blood exposures and have baseline testing performed. OR nurses helped me understand why when they invited me to speak to them about hepatitis C. Stories are circulating in the OR that a well-known surgeon withdrew from his practice due to HCV infection. Fear of inability to practice surgery, related to HCV, may contribute to even greater resistance among surgeons to report their exposures.

All of this made me think about what I could do, since HCV is treatable. I need to make sure that testing results are confidential, educate healthcare workers about progress in treating HCV, and utilize this as an opportunity to actively involve surgeons in exposure prevention efforts."

—Occupational health nurse
Seattle, Washington

Vanessa Burkhart

A Needlestick in the ER

A patient turned aggressive—and Vanessa Burkhart was infected with HCV

By Jane Perry, M.A.

Vol. 6, no. 2, 2002

In 1999, Vanessa Burkhart was a 39-year-old emergency nurse, supremely confident in her clinical skills. She had worked in a variety of settings during her thirteen years as a nurse—including medical/surgical, orthopedics and home health—but emergency nursing had always been her passion. In nursing school, when she had to choose two areas for her clinical training, she wrote down "ER and ER." She loved the excitement and challenge of having to think quickly on her feet, the constant opportunities to expand her clinical knowledge and skills, the adrenline rush of trauma situations. Emergency nursing was the perfect fit for her: it demanded assertiveness, self-assurance—someone who was driven. Vanessa was all three.

Vanessa grew up in Arkansas and earned an associate's degree in nursing from Arkansas State University in 1986. She had a natural toughness and resiliency that had already seen her through some major crises—including her 21-year-old daughter's automobile accident four years earlier that had left her with brain damage and in need of care. Vanessa also had a 17-year-old son who was a senior in high school, and she had recently remarried.

She had worked in the emergency departments (EDs) of several healthcare facilities, including Barnes-Jewish in St. Louis, a top-rated hospital and a teaching facility for Washington University School of Medicine. With her breadth and depth of experience, she thought she had seen just about everything in the ED.

In 1995, wanting to be closer to home and family, she accepted a position in the ED of a medium-sized hospital in southeast Missouri, near the Missouri-Arkansas border. The hospital had a mental health unit and thus saw a higher-than-average number of patients with depression and suicidal tendencies; there was also a high incidence of drug use in the local population.

On the evening of December 17, 1999, Vanessa was working as a charge nurse on the night shift. About 11:30 p.m., the police brought in a woman who had attempted suicide by taking an overdose of pills, which she washed down with alcohol. "We called such patients 'pseudo-suicidal,'" Vanessa recalls. "Right before they take the pills, they call 911." Vanessa took charge of her. The patient seemed fairly cooperative; close at hand was another nurse as well as the two policemen who brought her in.

Vanessa explains what happened next: "I started an I.V. in the patient's left forearm using an 18-gauge catheter needle; it wasn't a safety design. With this type, you had to pull the needle out a little from the catheter after insertion to visualize blood return. As I did that, the patient suddenly tried to hit me with her right fist. Letting go of the needle, I blocked her—that becomes a reflex when you work in the ED. In the scuffle, the needle came all the way out of the catheter. The patient then hit the holding end of the needle, pushing it towards my right hand, which was on the stretcher supporting my weight as I leaned over to grab her with my left hand."

"She immediately calmed; I let go of her arm and moved to pick up the needle on the side of the stretcher. The catheter was still positioned in her arm with blood dripping out of it. We reached for the needle at the same time; I thought she was trying to be helpful. I told her to leave it there or one of us might get stuck. She didn't pick it up—instead, she used the base of her hand to jam it into my finger. It went through the medial side of my right third finger; when I felt the initial prick of the needle, I must have moved just enough to rotate my finger, allowing the needle to pass through the joint complex. It came out the lateral side about 3 cm."

"It didn't dawn on me how badly it hurt until I looked down and saw the needle sticking through my finger—it felt like a railroad spike had gone through it. I finally managed to dislodge it; meanwhile, blood was still dripping from the catheter in the patient's arm. Because she was so inebriated, she bled more easily. But she had calmed down again and was alert enough to put her finger over the catheter end to stop the bleeding, because she didn't want to get stuck for another I.V."

Agitated patients who try to attack healthcare personnel are not uncommon in the ED—although they don't usually use needles as weapons. But what happened next was totally unexpected. As the patient sat calmly twirling her hair, she said, "You're going to have to get your blood tested, because I have hepatitis. I might have HIV, too—I've been living on the streets for a while." Vanessa was both flabbergasted and furious. "One of my first thoughts was, I could be dead in five years. Of course, I wanted to have her tested immediately, but we had to get her permission. The patient refused at first, but after a few other members of the ED staff talked with her, she finally said yes." One week later, Vanessa learned that the patient was indeed positive for hepatitis C, but negative for HIV.

After the needlestick, Vanessa called her supervisor and filled out an exposure report. Three days later, the infection control/employee health nurse was notified. Baseline tests were not performed immediately after the incident. Five weeks prior to the needlestick, Vanessa had been scratched by a suicidal patient's pierced earring; she was tested at that time for HIV, hepatitis B virus and hepatitis C virus (HCV), and was negative for all three. Her caregivers did not think retesting was necessary. However, her workers' compensation physician decided otherwise, and her blood was drawn for baseline tests on January 2, 2000. HCV and HIV antibody tests by enzyme immunoassay were performed, along with an alanine aminotransferase (ALT) test for liver enzyme levels. She was notified three days later that her baseline tests for both HIV and HCV were negative; her ALT levels, however, were elevated.

> *"According to CDC statistics, I had less than a 2% chance of seroconverting. But I have learned that I should never play the odds."*

Although the patient tested negative for HIV, Vanessa was encouraged to have follow-up HIV tests in case the patient was in a window period of infection. Her first follow-up test was performed at 15 weeks, and was negative. Her blood was drawn for a second test in May 2002, and she is currently awaiting the results.

Vanessa was carefully monitored and received excellent post-exposure care; she wishes, however, that she'd had the opportunity to take interferon and ribavirin prophylactically, immediately after her exposure, even though this is not recommended by the CDC. "I would have done it in a heartbeat," she says, "especially given the level of risk involved in my exposure."

Vanessa was seen by the workers' compensation physician soon after her exposure. He was not concerned about her elevated ALT levels, since she had been taking therapeutic doses of ibuprofen for arthritis, which could potentially affect her liver enzymes. She discontinued the ibuprofen and her daily allergy medicine; however, her ALT levels continued to rise.

By mid-March, her liver enzymes were twice the normal level, and Vanessa began to realize that she might be infected. She insisted on being seen by a hepatologist; her physician arranged an appointment at Barnes.

On April 3, at 15 weeks post-exposure, Vanessa had blood drawn for a qualitative HCV RNA test by PCR. Three days later, on April 6, 2000, she received the news that her HCV test was positive. "The infection control nurse was very distraught about giving me the news. She was also afraid one of the lab personnel might say something to me before she had a chance. It's such a small hospital and small town. She took me aside in an ED exam room and showed me the piece of paper with the test results. At first I didn't understand—I thought she was showing me results from the source patient. I already knew *she* had hepatitis. But the nurse said, 'Vanessa, you're not paying attention. It's *you*.'"

Vanessa was surprised at how emotional she became. "As my liver enzymes shot up over the previous months, the clinician in me recognized that infection was a real possibility. But another part of me refused to believe it. Besides, the statistics were on my side: According to the CDC, I had less than a 2% chance of seroconverting. But I have learned that I should never play the odds."

After Vanessa got the news, her coworkers were very supportive. "There were offers to drive me home and stay with me until my husband got home from work, but what I needed more than anything was to be alone. I had to find a way to tell my husband. I had to tell him not just that I had a life-threatening disease, but that he, too, was potentially in danger of being infected."

Vanessa went on leave in mid-April. She was having joint pain and swelling and some problems with memory loss. After she learned she had seroconverted, she had a liver biopsy. The results were positive for HCV but did not show signs of liver damage. Lab work was also performed to determine her HCV genotype, but it was inconclusive.

Vanessa started combination therapy with interferon and ribavirin on May 8, 2000, 20 weeks after her exposure, and underwent treatment for one year. During the first three months of therapy, she experienced severe nausea, fever, and myalgia; her hair, which had been down to her waist, began to thin and she had to cut it short. She suffered from malaise and depression, common side effects of interferon. She was able to do very little during this period; she could hardly care for herself, much less her husband, son and daughter. "There were days I felt like I was going to die; it was like the worst case of flu you ever had, only it went on for a year. If I crawled out of bed, it was just to crawl over to the couch." The treatment, she notes, felt worse than the disease: "For every ribavirin pill or shot of interferon I took, I had to take two more pills of something else to counteract the side effects."

> ***During her leave of absence she had come to a heart-wrenching conclusion: she would have to give up her career as an emergency nurse.***

Although the combination treatment was physically grueling, an HCV RNA quantitative test (PCR) performed in November 2000, 6-1/2 months after the start of therapy, indicated that her viral load was less than 1,000 viral particles per ml of blood, an "undetectable" level. Her most recent test, in December 2001, showed that she continued to have undetectable levels of HCV (a "sustained response"), and she is no longer considered a transmission risk. This result was, of course, a tremendous relief to Vanessa—especially since she had already decided that if her viral load was over 1,000, she would not undergo another course of combination therapy. "The treatment destroyed my thyroid; I will have to take thyroid medication for the rest of my life. I just didn't feel I could cope with another year of those side effects."

In October 2000, 10 months after her needlestick and while still undergoing combination therapy, Vanessa returned to work. However, during her leave of absence she had come to a heart-wrenching conclusion: she would have to give up her career as an emergency nurse. "It was one of the hardest decisions I ever had to make. Being an ER nurse was a huge part of who I was. Until my needlestick, I didn't imagine doing anything else." But she no longer felt she could perform at her highest level. The work was physically demanding, and she suffered from severe fatigue at times. And she continued to experience memory problems. "ER nursing requires the ability to think quickly on your feet and react instinctively in life and death situations. I could no longer rely on my ability to do that in all circumstances." She also wasn't sure she could be a very compassionate caregiver for patients who were drug or alcohol abusers, especially ones who became aggressive; she still felt a lot of anger towards the patient who had attacked her. So when Vanessa came back to work, she had a new position at the hospital: clinical nurse educator.

The transition from ED nurse to nurse educator was difficult at times. Vanessa especially remembers the day a co-worker said to her, "When are you going to go back to being a *real* nurse?" However, while it wasn't the career she envisioned for herself, Vanessa found that she enjoyed her new role. Over the last year and a half, her responsibilities have included all nursing education, skills and competency checks, and providing in-services on new procedures or techniques. She does not provide any patient care or do patient education.

"My passion is teaching others how to protect themselves from needlesticks and blood exposures, and creating a safer work environment," Vanessa says. "If I had been using a safety I.V. catheter, my needlestick and HCV infection would have been prevented. After my stick, my hospital completely converted over to safety devices. When we are doing training on a new device, I will sometimes literally drag people away from work to come to the training. I make sure they understand that this is important, that it has to be a priority. And I'm not afraid to use my own story to help educate others in the hospital. I tell them, 'Remember me? I used to work in the ER, and now I'm here to tell you why you *are* going to use safety I.V. catheters. You're going to use them because you don't want to end up like me. You do not want to live with a life-threatening disease.'"

Her hospital administration has been very supportive: she has a flexible schedule and is able to leave work if she is too fatigued to put in a full day. She can also work from home. In addition, she has not had to fight for workers' compensation. Until recently, she was able to get all the medical treatment she needed. However, the workers' comp provider has refused to pay for her thyroid medication.

After Vanessa completed her combination therapy in May 2000, she accepted a position as the hospital's coordinator of clinical education. The competency checks she performs, and the education she does on compliance with standards set by the Joint Commission for Accreditation of Healthcare Organizations (JCAHO) are, she believes, of utmost importance for healthcare facilities today—yet she also finds the work frustrating at times. "Because nurses are already stretched to the maximum, it is difficult for them to find time to attend educational sessions and mandatory in-services or skills fairs. For them, it means either staying late, coming in early, or coming in on your day off. It can be hard to get their attention."

Vanessa is doing educational work on a larger scale, too: she has started speaking at national conferences about her needlestick and subsequent HCV infection. "It makes me feel good to use this terrible experience to help others. When I share my story, it brings home to my audience the reality of the risk we face as nurses in a way no amount of statistics or data can."

"I really try to shake them up, help them understand that safety has to be a high priority for all of us. We must change our attitudes. Safety devices are not an annoyance devised by employee health or infection control people; they are our first line of defense."

When Vanessa is asked about risks in the ED, she focuses

on mindset. "When you get really good at your job, you can get complacent. And ER nurses are usually very good at what they do. We like being specialty nurses; we like the sense of control that gives us. But sometimes we can get an attitude—OK, here's my first overdose for the night, where's my second and third? When I worked in the ER, we had the yearly bloodborne pathogens update, and I recognized the risk, but I think we all tend to have the attitude, 'It won't happen to me.'"

Living with HCV has changed Vanessa's life in a number of ways, both small and large. "We observe strict precautions at home. We cover our toothbrushes, and I change them every 30 days. We never share combs or brushes, and my daughter would never dream of borrowing a pair of my pierced earrings. I probably go overboard, but I don't want to take any chances." On a deeper level, she finds her attitude towards life has changed dramatically: "The fear of liver cancer, liver failure—a variety of potential ills related to the hepatitis—is always with you. You do learn to live with it to some extent, but it changes your perspective. Many things that I took for granted before, I don't take for granted now. Before my needlestick and seroconversion, my all-consuming thought was making enough money to provide for my daughter in case something happened to me—I worked constantly. But now that something *has* happened to me, I have been forced to slow down; I have a more relaxed attitude. My priorities are different. I treasure the time I have with my daughter, with my son who is in college, and with my husband who has stood by me throughout this ordeal. I talk to my mother, who I am very close to, almost every day on the phone, and I make sure I see her at least once a week."

She will always miss emergency nursing. "I can't begin to explain the feeling of a clinical save. They were the driving force behind my commitment to the ER. I remember each and every one of the patients whose lives I helped save in all the years I worked in ERs. Would I still be working as an ER nurse if I hadn't had that needlestick? In a heartbeat, without a doubt, without hesitation—yes."

"Would I still be working as an ER nurse if I hadn't had that needlestick? In a heartbeat, without a doubt, without hesitation—yes."

Stephen Derrig

On-the-Job Exposure to HIV

A firefighter/paramedic, occupationally infected with HIV, shares hard-won lessons

By Jane Perry, M.A., and Janine Jagger, M.P.H., Ph.D.

Vol. 6, no. 2, 2002

The following is a revised version of the article that appeared in AEP. The original article, "HIV Infection in a Firefighter/Paramedic," was published before Steve went public with his HIV infection, and referred to him as "John Smith." This version brings the information about Steve and his family up-to-date.

In 2000, Steve Derrig was a 32-year-old firefighter and paramedic living in Akron, Ohio. That was the year he made a shattering discovery: he was infected with HIV.

In his nine years as an emergency worker, Steve had frequent on-the-job exposures to patients' blood and body fluids (BBF), and he had no other risk factors for HIV. At the time of his diagnosis, he had been married for 10 years; he and his wife, Melissa, have two children.

In Akron, firefighters are trained as paramedics and respond to both fire and medical emergency calls. Steve estimates he spent approximately 75% of his time responding to paramedic calls. He typically worked in inner city neighborhoods and treated numerous stabbings and shootings.

In November 1999, he developed an afebrile dry cough with shortness of breath. His condition worsened over the next month, and he went to see a doctor in early January of 2000. The doctor diagnosed pleurisy and prescribed an anti-inflammatory, but Steve's health continued to deteriorate. He was then put on a course of prednisone, but that, too, had no effect. He had numerous chest x-rays and two bronchoscopies, all of which were normal. In late January, Steve was admitted to the hospital for extensive testing and was seen by a leading pulmonologist, who diagnosed asthma and put Steve on inhalers. A month later, he could not walk even a short distance without gasping for air, and had lost 20 pounds.

On March 7, 2000, Steve was admitted to the emergency room in critical condition. A cardiologist in the ER, who had previously worked as a paramedic, thought Steve should have an HIV test; no other doctor had suggested this. Two days later, as Steve was about to be put on a ventilator and have a lung biopsy, his HIV test results came back positive: he had pneumocystis carinii pneumonia (PCP) and full-blown AIDS. His CD4 (t-cell) count was below 200 cells per cubic ml of blood, and his viral load was over 300,000 copies per ml of blood. (A few weeks later, he was also tested for hepatitis C; the result was negative. He had already been vaccinated for hepatitis B.) His caregivers couldn't say for sure when he was infected.

Immediately after Steve's diagnosis, his wife was tested for HIV as well. The result was negative, and she remained HIV-negative at a one-year follow-up test.

Steve routinely wore gloves on the job, and sometimes safety glasses, but rarely, if ever, put on fluid-resistant gowns or goggles. And he says that is typical for most paramedics he knows.

Steve does not recall a needlestick injury that might have caused his infection, but during his nine years as a paramedic he sustained massive BBF exposures on a number of occasions. He had only one exposure on record—a patient with tuberculosis who coughed in his face—but he says that during most of his paramedic runs, which he estimates at around 5,000, he sustained some kind of BBF exposure. He believes he might have been infected in 1995, when he had a severe case of poison ivy with open sores and blisters on his forearms. He took prednisone, an immunosuppressant, to control the inflammation. Steve's supervisor took him off the medical unit for a day because of his sores.

Steve remembers treating a patient in 1995 who was stabbed in the chest in a park known for the high-risk groups that congregated there; the patient had massive bleeding. On another occasion, Steve sustained an exposure while delivering the baby of a crack-addicted mother; during the delivery, the mother's water broke in his face. (He is uncertain of the year this occurred, and can't locate the patient's records, which might allow him to check the mother's HIV status.) He also recalls a patient spitting in his eye with bloody saliva.

One week after his AIDS diagnosis, Steve began a three-

Figure 1. Sharps Injuries to Paramedics in the Field (20 injuries)

Of 43 needlestick and sharp-object injuries to paramedics/EMTs reported in the EPINet database (1993-2000), 20 occurred in the field (at an accident scene, in a home, or in an ambulance or helicopter during patient transport). The others occurred in a hospital or healthcare facility. The data below is for the 20 field-based injuries; for comparative purposes, we also provide EPINet data for injuries to healthcare workers (HCWs) in hospitals and other healthcare facilities (23,692 injuries). For paramedics in the field, 55% of injuries were caused by blood-filled needles, compared to 24% for facility-based HCWs. (Injuries to paramedics in hospital settings will generally have similar characteristics to those of ED personnel.)

Devices causing injury (top 3 devices):

Paramedics		Hospital HCWs	
IV catheters	50%	Disposable syringes	30%
Lancets	15%	Suture needles	12%
Phleb. needles	10%	Scalpel blades	7%

The fact that three devices account for 75% of injuries to paramedics may reflect the smaller variety of devices carried in medic bags and on ambulances.

Original purpose of devices causing injury (top 3):

Paramedics		Hospital HCWs	
Start IV/setup heparin lock	45%	Injection (IM/SQ)	17%
Fingerstick/heelstick	15%	Draw venous blood	15%
Draw venous blood sample	10%	Suturing	13%

When injury occurred:

	Paramedics	Hospital HCWs
Other after use, before disposal*	40%	22%
During use of item	20%	28%
While putting item in disposal container	15%	7%

**The high percentage of injuries in this category to paramedics may be due to the fact that it can be difficult to access a sharps container in an ambulance, especially during a code or severe trauma, or the sharps container may quickly become overfilled—again, especially during a code.*

drug treatment regimen (a protease inhibitor and two nucleoside reverse transcriptase inhibitors); he continues to take these drugs. His health is now fairly good, though he gets fatigued easily and has diarrhea from the drug cocktail, and arthritic joint pain from his infection.

In September 2000, Steve went back to full-time work as a firefighter, but his supervisor allowed him to move to a different station so he would only have to perform firefighting duties. He wanted to avoid working as a paramedic because of the risk to his compromised immune system.

Steve filed a workers' compensation claim shortly after his diagnosis. He could not prove through documentation (i.e., an exposure report and follow-up blood work) that he had been infected with HIV on the job; instead, his attorney tried to demonstrate to the workers' compensation board that paramedics are at high-risk for BBF exposures, and also that, because such exposures are a routine part of the job, most of the time they don't report them. Steve notes that many patients are treated by paramedics at the scene but never taken to a hospital—which makes it even harder to track them later if an occupational infection occurs.

The workers' compensation board ruled in his favor. But his employer filed several appeals and the case wasn't settled until July 2002, when the city decided to drop its court challenge.

Now Steve can rest assured that his medical expenses—including the drug therapy to suppress the virus, which costs around $1,500 a month—will be covered. Steve also received a small lump-sum payment from the workers' compensation provider for permanent partial disability resulting from his HIV infection. He was determined to be 40% impaired, based on factors such as side effects from the drugs he takes, arthritis and other conditions that are caused by his disease, and the fact that he can't have any more children.

In March 2003, Steve decided to permanently retire from his work as a firefighter. Even though he was no longer performing duties as a paramedic and had fewer exposures to patient blood and body fluids, he was still being exposed to carcinogens, and to airborne pathogens such as tuberculosis. He decided it wasn't worth the risk to his immune system, and gave up a job to which he had been devoted for 10 years. His most recent HIV tests brought good news: he had an undetectable virus load (under 50) and his t-cell count was 600, the highest it had ever been. He has felt better since leaving the job—less stressed and better rested. He now works part-time in a family business, and spends as much time as he can with his children.

In the fall of 2002, Steve and Melissa agreed to be interviewed for an article in the *Akron Beacon Journal*; up until that point, Steve had revealed his HIV status to only a small circle of friends and family members. In the interview, Steve expressed his hope that his story would help "shatter the AIDS stigma" and encourage firefighters and emergency care providers to be tested for HIV, as well as hepatitis B and C, on a regular basis. He notes that if he had had annual tests for HIV and his infection was detected earlier, his illness might not have progressed to the critical stage it reached in March 2000. His undetected HIV put his wife and children at risk as well: "Without question, the scenario could have been unbelievably worse. We could have been a family of four—all with AIDS. Maybe the next guy in line won't be as lucky as me."

When asked how paramedics and EMTs can better protect themselves, Steve says it's important to get over the

Figure 2. Body and Blood Exposures to Paramedics in the Field (30 exposures)

Of 55 blood and body fluid (BBF) exposures to paramedics reported in the EPINet database (1993–2000), 30 occurred in the field (at an accident scene, in a home, or in an ambulance or helicopter during patient transport). The data below is for the 30 exposures in the field; for comparative purposes, we provide EPINet data for BBF exposures to healthcare workers (HCWs) in hospitals and other healthcare facilities (6,318 exposures).

According to paramedic protocols, goggles, faceshields and gowns are required for certain procedures. However, paramedics appear to use personal protective equipment (with the exception of gloves) infrequently: 96% of paramedic BBF exposures in the field were to unprotected skin, compared to 84% for hospital-based HCWs. In these data, no paramedics in the field had on goggles when the exposure occurred, and only one had a faceshield on. Note that the least protected body part (eyes) was the one that had the highest number of reported exposures.

BBF came in contact with*:

	Paramedics	Hospital HCWs
Eyes	40%	53%
Mouth	33%	11%
Non-intact skin	33%	19%
Intact skin	30%	43%

**More than one part or area can be selected.*

Body fluids involved:

	Paramedics	Hospital HCWs
Blood/blood products	93%	66%
Vomit	20%	4%
BF other	10%	24%

Barrier items worn at time of exposure:

	Paramedics	Hospital HCWs
Gloves (single/double pr.)	90%	67%
Faceshield	3%	3%
Goggles	0%	3%

Other comparisons:

- 23% of BBF exposures to paramedics in the field exposed the worker to BBF for 15 minutes to over an hour; this compares to 7% for facility-based HCWs.
- 70% of BBF exposures to paramedics in the field were the result of direct patient contact, compared to 52% for facility-based HCWs.

mindset that "it can't happen to me": "Personal protective equipment—gloves and goggles especially—and exposure reporting need to be taken much more seriously." He says that while PPE such as gowns and safety glasses were typically available in his truck or ambulance, they were rarely worn; he doesn't recall goggles being available at all. "Paramedics and firefighters need access to better eye protection, such as goggles, but they also need to get in the habit of putting on protective equipment. It's hard to anticipate what you'll encounter at a scene; you need to be prepared *before* you get out of the truck."

Steve advises paramedics to "take all the precautions that you have at your disposal. Don't take unnecessary risks; when you anticipate massive exposures, make sure you're well protected before rushing into the scene." But, he admits, it can be difficult to take that extra step when the situation is critical.

Steve also says it's important to take seriously open lesions on your skin. "In our department, we weren't supposed to be working if we had open sores. But in the summer, especially, they're very common, and our uniform shirts had short sleeves. And that can make you more vulnerable if you have an exposure."

The risk of exposure to bloodborne pathogens for paramedics is, he says, "just as high—if not higher—as for emergency department workers in hospitals; yet so often we are much more poorly protected." And the protocol for exposure reporting is often not as well established for paramedics and EMTs as it is for hospital-based workers.

The CDC's statistics on documented and possible cases of healthcare workers with occupationally acquired HIV/AIDS bears this out. There are no paramedics listed among the "documented cases" (those in which there was a reported exposure and a complete sequence of follow-up blood work demonstrating seroconversion). However, 12 paramedic cases are listed in the "possible" category; these are ones where healthcare workers report a history of occupational exposure to blood or body fluids and no other risk factors for HIV infection, but where infection could not be attributed to a specific exposure.

As is clear from Steve's case, undocumented exposures can put paramedics at a significant disadvantage when they try to obtain workers' compensation for an occupational infection. For this reason, many states have now passed "presumptive eligibility" legislation for healthcare and emergency workers. It means that workers in these fields who contract certain diseases, such as HIV and hepatitis C (and, in the case of firefighters, certain types of cancer), are presumed to have acquired the disease on-the-job; it makes getting workers' compensation much easier.

Steve has worked on efforts to pass a presumptive eligibility law in Ohio; legislation has been introduced, but so far has not moved forward. "When you come on the job, you are told that you will be taken care of if you become injured or sick from the job," Steve commented to the *Akron Beacon Journal.* "When you are in your most vulnerable state, you shouldn't have to fight."

William P. Fiser, Jr.

First, Do No Harm:

An HCV-Infected Surgeon's Difficult Choice

By Jane Perry, M.A.

Vol. 6, no. 5, 2003

Dr. William Fiser is one of the few surgeons in the U.S. who has acknowledged publicly that he is infected with hepatitis C virus (HCV). Last year, he published a letter in *Infection Control and Hospital Epidemiology* that discussed surgeon-to-patient transmission of bloodborne pathogens.[1] He was also featured in an article in *Newsday*, a Long Island (NY) daily; the title was telling: "Deciding to Step Away."[2]

After becoming ill with hepatitis C, Dr. Fiser resigned his private practice and accepted a faculty appointment in the surgery department at the University of Arkansas for Medical Sciences, where he serves as research director for the pediatric cardiothoracic surgery division at Arkansas Children's Hospital and medical director for the Arkansas Regional Organ Recovery Agency in Little Rock.

What makes Dr. Fiser's experience unique is that, shortly after he was diagnosed with HCV, he learned that a patient he had operated on less than a year before—his office receptionist—had also been diagnosed with the disease.

These events raised compelling moral issues he couldn't ignore. They also sound an alarm for every ambulatory and hospital-based surgery center.

Multiple needlesticks

Dr. Fiser believes he was infected from an occupational sharps injury, since he had no other risk factors for HCV and sustained multiple needlesticks during his career. But he can't pinpoint a specific injury that caused his infection.

"Cardiac surgeons typically get many, many needlesticks over the course of their careers," Dr. Fiser observes. "We routinely use 80 to 100 suture needles per operative case." And, also typically, surgeons tend not to report their sharps injuries, which means there is no documentation if a surgeon is infected.

Dr. Fiser was diagnosed with HCV in April 1999. "I'd been having symptoms of fatigue and arthralgia and had seen a couple of internists. Ironically, though, my diagnosis was made when I had a medical exam for a life insurance application; it included a routine screening for hepatitis C, and I tested positive."

Although Dr. Fiser had been performing surgery for 20 years, and cardiovascular surgery for 15, up until that point he had never been tested for HCV. "I had my head buried in the sand as far as my risk of getting infected—although I was very compulsive about handwashing and universal precautions."

He now believes cardiac surgeons are at higher risk than most other surgical specialists for acquiring and transmitting HCV and other bloodborne pathogens. Cardiac surgery is lengthy, bloody, and requires working in a deep cavity, often with poor visibility.

"Dripping blood"

When Dr. Fiser learned that his receptionist was also infected with HCV, he thought back to her surgery 10 months earlier, an aortic arch replacement: "I was sewing a Dacron graft to the proximal common carotid using a 4-0 Prolene suture needle. In the middle of the procedure, the suture needle stuck my left forefinger. It was a pretty deep stick. I remember it well because my hand was dripping blood—some dripped into her mediastinum. I had to stop and change gloves."

Dr. Fiser was devastated when he realized he may have infected one of his patients, and he worried that others may have been infected as well. He contacted the Centers for Disease Control and Prevention (CDC) to request an investigation.

"They told me I had to go through the state health department. So I called the Arkansas Department of Health and reported my case and the possible transmission to one of my patients. The chief epidemiologist was disturbed that neither of our cases had been reported by the labs that did our blood work. But HCV reporting guidelines, which are based on CDC recommendations, only require reporting of acute hepatitis C cases."

Acute hepatitis C is defined as alanine aminotransferase (ALT) levels at least six times above normal limits. As Dr. Fiser wrote in his letter, this policy means that "most health departments record only a tiny fraction of all newly diagnosed cases of HCV." Further, "only a small minority of surgical patients infected with HCV become symptomatic and even fewer do so early enough to recognize any possible relationship to their prior surgery."[1]

The state epidemiologist submitted a request to the CDC for help in investigating Dr. Fiser's case, but the CDC refused because there wasn't a "cluster" of infections (i.e., more than one known patient potentially infected by the surgeon). This raises the obvious question: How is it possible to

New York cardiac surgeon infects 3, probably 7, and possibly more patients with HCV

In 2000, 67-year-old Joseph Carco of Hicksville, New York, had heart valve replacement surgery performed by Dr. Michael Hall, one of the top cardiac surgeons in New York state. The surgery seemed to be successful. But within weeks of the operation, Carco developed a virulent case of liver disease, and was diagnosed with hepatitis C virus (HCV).

By coincidence, at an HCV support group he met another patient of the same surgeon—an 83-year-old grandmother—who had also developed HCV after heart surgery. The surgeon was tested and found positive for HCV. It emerged that he had had HCV for at least 10 years—but had never been tested. The state epidemiologist's investigation revealed that he had almost certainly infected 3 patients, possibly 7, and perhaps more.

Dr. Hall is still performing surgery at the same hospital, although he has modified his technique to include double-gloving, blunt suture needles, and no-hands passing of surgical instruments. However, according to a Newsday article (3/30/03), he continues to open and close patients' chest cavities, despite being advised by experts to avoid these steps. A 1988 Lancet study found that 40% of cardiac surgeons puncture their gloves during sternal closure, which involves threading wires through holes in the breastbone and then tying them.

determine if a cluster of infections exists unless an investigation is conducted?

Taking matters in his own hands

At that point, Dr. Fiser decided to conduct his own investigation, at his own expense. He had already discussed with his receptionist the possibility that she acquired her infection from him, but he wanted a more definitive answer. He obtained her permission to have a lab company, National Genetics, compare a stored sample of her serum with a blood sample he submitted.

Earlier testing had indicated they had the same HCV genotype, and it seemed probable that Dr. Fiser was the source of her infection. He lived with that belief for more than two years.

But in 2002, after a long wait, he received results from RNA mapping that showed one difference in the RNA between their hepatitis C viruses. Based on this result, National Genetics concluded the two samples weren't a match.

"I still have my doubts," Dr. Fiser says, "because the circumstantial evidence is so strong. Also, the blood sample from my receptionist was drawn before her HCV treatment started, and mine was drawn after treatment—over one year apart. The virus is constantly mutating, which is why it can be so hard to clear, even with treatment. But the experts say it's probably not the same, and I have to accept their opinion."

Unsuccessful treatment

Shortly after Dr. Fiser was diagnosed with HCV, he suspended his surgical practice and started combination therapy with ribavirin and interferon, which continued for eight months. But he experienced severe side effects, including anemia and leukopenia. "My treatment had to be adjusted, stopped, then restarted. At one point I was receiving Epogen and Neupogen just so I could continue with therapy."

Ultimately, the treatment was unsuccessful; today his viral load remains very high (greater than 2 million viral particles per 1 ml of blood). Because of the serious complications he experienced, Dr. Fiser says further treatment, even at a low dose, is impossible. A liver biopsy in 1999 indicated he had grade three fibrosis—one stage short of cirrhosis. Since then, he's had regular liver surveillance (using ultrasound and other non-invasive methods) to check for hepatomas. He has up to a 5% chance per year of developing a liver tumor, which is an indication for immediate transplant.

Dr. Fiser's receptionist underwent HCV therapy at the same time he did. Her treatment was successful: her HCV level is now undetectable (fewer than 1,000 viral particles per ml of blood).

Difficult decision

After Dr. Fiser completed HCV therapy in December 1999, he returned to his surgical practice. But he worried about the risk he posed to his patients. He also continued to suffer from fatigue and other symptoms of hepatitis C, and found it difficult to keep up with his caseload.

Although the CDC doesn't recommend any restrictions for HCV-infected surgeons, he tried to avoid procedures such as ruptured abdominal aneurysms that typically are less controlled and more exposure prone. "I racked my brain trying to decide what to do; the CDC position was wrong for me. It kept coming back to the Golden Rule. I sure didn't want anyone operating on my family who had HCV, HIV, or hepatitis B."

By August 2000, he'd made up his mind: he decided to

Dr. Mark Davis, surgeon and OR safety consultant

"Infection of a healthcare worker—or a patient—with a bloodborne pathogen is as much a medical error as incorrect delivery of a medication which injures or kills a patient. Although safety devices that can prevent the types of sharps injuries most commonly incurred by surgeons and anesthesiologists have been commercially available for years, such injuries are still frustratingly common in operating rooms. High-yield, evidence-based prevention strategies include blunt-tipped suture needles, safety I.V. catheters, safety syringes, and needleless I.V. systems.

"Much has been written about the need for our healthcare delivery system to follow the example of the aviation industry—to implement safer systems that prevent adverse events. While portions of a surgical procedure may require more variation than landing an aircraft, some standardization of safety devices and safety protocols during surgery is clearly needed. These changes can be implemented without compromising patient care.

"I receive frequent phone calls and e-mail messages from OR managers and infection control professionals, telling me that safer systems for the OR aren't as widely implemented as they could be. Widespread education for all surgical care providers, supported by informed healthcare leadership, is the best hope for standardization of safety in our operating rooms.

"As healthcare institutions face increasing federal and consumer scrutiny and economic pressure, leaders of healthcare organizations understand the need to chart the course to patient and worker safety. The benefits of a safer workplace are undeniable—an improved bottom line, reduced liability, regulatory compliance and an improved ability to recruit, retain and protect our most precious resource, the healthcare worker."

—Dr. Davis, who practiced as a surgeon for 25 years, provides consulting and educational services in OR safety to healthcare institutions and industry. He is author of *Advanced Precautions for Today's OR: The Operating Room Professional's Handbook for the Prevention of Sharps Injuries and Bloodborne Exposures,* and founder of the website **ORprecautions.com.**

permanently close his surgical practice. He was only 47 years old, so it was a painful choice, made with little guidance or support. But he knew it was the right one for him.

Mandatory testing

Before learning he was infected with HCV, Dr. Fiser opposed mandatory testing of physicians for bloodborne pathogens. But he now thinks surgeons should be tested when they join a hospital's medical staff.

"Likewise, I think scrub nurses should be tested before they're hired," he wrote in his letter. "After that, both groups should be tested whenever there is a percutaneous injury or other significant blood exposure."

But, he says, "Most intraoperative, percutaneous injuries aren't reported or recorded. A more ethical approach would be for serology to be drawn and reported to the injured [worker] any time there is a percutaneous injury during a procedure."

Currently many, if not most, physicians avoid getting tested. Dr. Fiser attributes this to the lack of a "logical, well-thought-out plan" for how to manage infected physicians. He points to a Minnesota law, passed in 2000, as a good model: Physicians infected with HCV, hepatitis B, or HIV are assigned a monitor by the state health department and sign a contract in which they agree to eliminate exposure-prone procedures and make other practice modifications as recommended by the monitor.

Needed: Incentive to change

Dr. Fiser believes the most important safety measures surgeons can adopt are use of blunt suture needles and double-gloving. "Careful hand washing and universal precautions aren't enough. When there can be greater than two million viral particles of HCV per ml of blood, it doesn't take a large exposure to get infected."

Dr. Fiser commented in his letter that many cardiovascular surgeons and their assistants have the preconceived notion that they can't operate wearing double gloves. "I used to be one of those surgeons, but I am now certain that cardiac surgeons can operate effectively wearing double gloves."[1]

He thinks more effective barriers, such as puncture-proof gloves, need to be developed. "These kinds of innovations would be more expensive, and right now everyone is trying to contain cost. Blunt suture needles also need to be more widely used. But where is the incentive to change?"

Dr. Fiser believes that if there were more data regarding surgeon-to-patient transmission of pathogens, surgeons might be more inclined to adopt safety measures.

"If the CDC did turn over this rock and tried to define the extent of the problem, the data would be compelling. It's easy not to change when you have little or no data."

References

1. Fiser WP. Should surgeons be tested for blood-borne pathogens? *Infect Control Hosp Epidemiol.* 2002;23(6):296–297.
2. Rabin R. Deciding to step away. *Newsday.* 12/27/02, p. A03.

PART III

Other AEP Articles

Blood Salvage Machines Cause Blood Exposures to Operating Room Personnel

by Janine Jagger, M.P.H., Ph.D., and William P. Arnold III, M.D.
Vol. 1, no. 2, 1995

Blood salvage machines, also known as autologous blood recovery or autotransfusion machines, are used to recover, process, and reinfuse blood lost during surgery, thus minimizing the need for transfusion of donor blood in surgical cases in which significant blood loss is anticipated. Two models of these machines—Cell Saver 3 Plus and Cell Saver 4, both manufactured by Haemonetics Corporation of Braintree, Massachusetts—have recently been linked to numerous massive blood exposures of operating room personnel. Other blood salvage machines which are currently in widespread use, but haven't been associated with repeated blood exposures, include Haemonetics' Cell Saver 5, Medtronic-Electromedics' Elmd-500, COBE Laboratories' BRAT, and the Sorin Compact manufactured by Dideco. Blood salvage machines are usually operated by perfusionists, operating room technicians, or nurses; industry sources estimate that there are about 9,000 of these machines in use worldwide, with possibly one-half being Cell Saver models 3 Plus and 4. The machines range in price from $18,000 to $35,000.

Blood salvage machines operate in the following way: first, lost blood is recovered from the operative field by suction, and delivered to a processing bowl where it is centrifuged to separate the red cells. The red cells are then resuspended in a physiologic solution and pumped under pressure into a reinfusion bag; from there they are returned to the patient by intravenous infusion. Important design considerations for this and other equipment in which blood is pumped through tubing under pressure include secure junctions between tubing components and safeguards against excessive buildup of pressure that could result in the rupture of a blood circuit.

In June 1994 an anesthesiologist forwarded an incident report to the chair of the American Society of Anesthesiologists' (ASA) Committee on Occupational Health of Operating Room Personnel describing a massive blood exposure caused by a Cell Saver 3 Plus. The incident occurred during the course of a liver transplant, with a highly trained cardiovascular technician operating the machine. The technician did not notice that the tubing leading to the reinfusion bag had been clamped, and continued to pump processed cells to the infusion bag. This resulted in a pressure buildup in the tubing and finally a rupture in the junction between the tubing and the reinfusion bag. The rupture, according to the anesthesiologist's report, caused a "significant blood exposure to my entire face and upper torso even though I was three to four feet away from the Cell Saver." Two other similar incidents in the same institution with the same equipment were also identified.

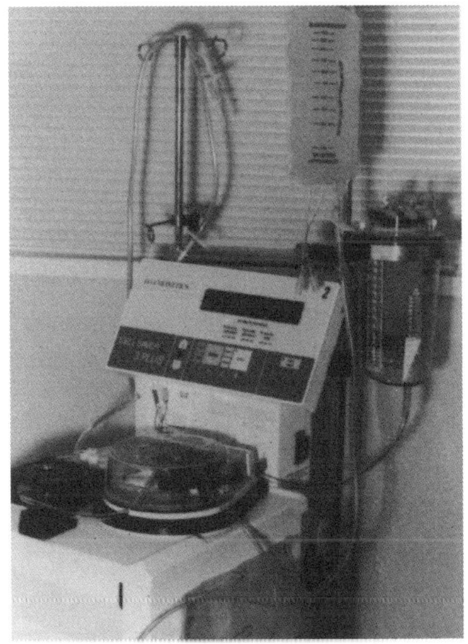

Haemonetics' Cell Saver 3 Plus

This report prompted an informal inquiry via the Internet of two list server groups dedicated to topics related to anesthesiology. The responses, reported back to ASA members[1], yielded descriptions of at least ten identical incidents all occurring with Cell Saver 3 Plus or Cell Saver 4. Because the incidents were suggestive of a widespread problem, a follow-up survey was conducted of hospitals participating in the data-sharing group of the Exposure Prevention Information Network (EPINet). In November 1994 a memo was sent to EPINet coordinators of 30 hospitals in diverse geographic locations asking what model of blood salvage machine was in use in their hospital, whether operating room personnel reported or recalled blood exposures associated with any make or model of these machines, and requesting descriptions of any such incidents. The results were as follows:

- 18 hospitals provided a complete response to the survey;
- 15 hospitals reported using Haemonetics' Cell Saver 3 Plus or Cell Saver 4; 11 of the 15 hospitals reported one or more blood line ruptures during surgical procedures, most occurring during the past three years;
- a total of 37 blood line ruptures were described by the 11 hospitals, with most hospitals reporting multiple incidents;

The following incidents illustrate that splashes of blood from high-pressure systems represent a potential source of bloodborne pathogen transmission to healthcare workers. Inadvertent interruptions of blood circuits caused by high-pressure ruptures or disconnections of tubing components are the most likely events leading to such exposures. Mucocutaneous blood exposures occurring as a result of the malfunction of such equipment have been associated with the transmission of bloodborne pathogens.

(1) A medical technologist in the United States was manipulating an apheresis machine that developed problems during the course of an outpatient procedure. Blood spilled from the machine on to her ungloved hands and forearms. She did not recall having open wounds on her hands, although she had dermatitis on her ear and may have touched it with a bloody hand. The patient was found to be HIV-positive. The healthcare worker's HIV seroconversion was documented three months after the incident. *(Centers for Disease Control. Update: Human immunodeficiency virus infections in health-care workers exposed to blood of infected patients.* MMWR. *1987; 36:285-289.)*

(2) An intensive care nurse in Italy sustained a massive exposure to the mucous membranes of her eyes, nose, and mouth and the intact skin of her scalp and face from the blood of an HIV-positive patient. The exposure occurred when the dome of the pressure-monitoring system broke as the nurse tried to clear a blockage in the patient's arterial catheter. The nurse's HIV seroconversion was documented 43 days after the incident. *(Gioannini P, Sinicco A, Cariti G, et al. HIV infection acquired by a nurse.* Eur J Epidemiol. *1988; 4:119-20.)*

(3) A dialysis nurse in Italy sustained a blood splash to her face and conjunctiva during a hemodialysis procedure; the exposure occurred following the rupture of a venous dripper. The patient was anti-HCV-positive; the nurse was found to be anti-HCV-positive four months after the exposure. *(Sartori M, La Terra G, Agliette M et al. Transmission of hepatitis C via blood splash into conjunctiva.* Scand J Infect Dis. *1993; 25:270-271.)*

- 16 of the 37 blood line ruptures resulted in direct blood exposures of operating room personnel, including the Cell Saver operators and those in close proximity, such as anesthesiologists;
- 21 of the blood line ruptures did not result in blood exposures to personnel, but caused widespread blood contamination of the operating room, including walls, ceilings, floors, and equipment;
- no similar incidents were reported from institutions using other makes or models of blood salvage devices.

Design Problems

The findings indicate design flaws both with the Cell Saver 3 Plus and Cell Saver 4 machines and with the disposable equipment needed for their clinical use. A critical problem is associated with three short tubes which are integral with the reinfusion bag and serve as ports [*see photograph*]. The large port carries blood from the machine to the reinfusion bag under pressure created by a roller pump; two smaller ports connect with tubing for intravenous reinfusion of processed blood. The three ports have similar manual clamps attached. The port carrying blood from the machine to the bag *must not be clamped* while the pump is in operation; when the bag is not being filled the roller pump prevents blood from returning to the machine. The clamp on the large port is provided *only* to prevent blood from leaking out of the bag once it is disconnected from the machine at the completion of surgery. Because the three ports on the reinfusion bag have similar clamps, however, it is possible to clamp the wrong line by mistake, especially during difficult cases when the operator is faced with multiple tasks. If the large port is clamped while the roller pump is running, blood in the tubing will be pressurized, thus placing the connection between the tubing and the port at risk of rupture. The owner's operating manual for Cell Saver 3 Plus and Cell Saver 4 provides this warning: "Avoid blocking any tubing carrying blood from the pump. A buildup of pressure in this tubing can result in the widespread dispersal of blood." But there is no feature on Cell Saver 3 Plus or Cell Saver 4 or on the reinfusion bag to prevent the operator from clamping the blood line at the wrong time, nor any kind of warning against doing so.

Another serious problem that is related to the design of the disposables is the type of junction used to connect the Cell Saver tubing to the reinfusion bag. The junction is a friction fit—that is, the two connecting pieces are simply pushed together; they have no positive locking mechanism, such as a Luer lock, which is found on other blood salvage machines. With no positive locking mechanism, the critical junction between the tubing and the blood bag is not secure and may separate if pressure increases in the tubing below the clamp.

In addition to the design problems with the disposables, the machines themselves are flawed in that they lack an interrupt feature to halt the pump if the blood line is inadvertently occluded. The Cell Saver 5 model, which was introduced in 1994, does incorporate a pressure sensor that shuts off the pump if pressure in the blood line increases beyond a critical level.

Remedies

While other blood salvage machines on the market incorporate safeguards to prevent high-pressure ruptures of blood lines, the large numbers of Cell Saver 3 Plus and Cell Saver 4 machines still in use continue to pose unnecessary risk to operating room personnel.

The results of the Internet inquiry and the survey of EPINet hospitals were both forwarded to the Haemonetics Corporation with a request for information on solutions to the problems identified. A company spokesperson stated that it is the hospital's responsibility to ensure that operators observe the manufacturer's instructions and warnings, and that no Haemonetics machine will expose the operator to blood from ruptured tubing if operated as directed in the manual. In addition, the company does not acknowledge that their machines place operating room personnel at risk of blood exposures and does not intend to retrofit either Cell Saver 3 Plus or Cell Saver 4 with the safeguards incorporated in Cell Saver 5.

Monitoring the safety of medical devices, including evidence of blood exposures from equipment such as blood salvage machines, is the regulatory responsibility of the Food and Drug Administration (FDA). An initial search of the FDA's Medical Device Reporting database did not identify any incidents of blood tubing ruptures from Cell Saver 3 Plus or Cell Saver 4. The FDA has requested documentation of the responses obtained from the Internet inquiry and the survey of EPINet hospitals, and is currently following up on these reports. There are several possible courses of action which the FDA can pursue. It can impose a mandatory recall which would require the manufacturer to notify hospitals of the problem and to alter the equipment in a manner approved by the FDA. The FDA can also request a voluntary recall, giving the manufacturer the option of whether or not to respond. Finally, a safety alert could be issued notifying hospitals of the hazard but imposing no requirement for a remedy on the part of the manufacturer. The FDA has not yet reached a final decision on possible action.

While a remedy is being sought for the hazards posed by Cell Saver 3 Plus and Cell Saver 4, AEP will continue to alert groups at risk through their professional associations. Operating room personnel who are exposed to blood from a high-pressure rupture of tubing on blood processing equipment, in addition to filing a blood exposure report with their institution's employee health department, can file an incident report with the FDA in compliance with the Safe Medical Devices Act of 1990.

References

1. Arnold W. Occupational hazard: Cell Savers spray anesthesiologists with blood. *ASA Newsletter*. 1994;58:27.

William P. Arnold III, M.D., is Associate Professor of Anesthesiology at the University of Virginia's Health Sciences Center, and Chair of the American Society of Anesthesiology's Committee on Occupational Health of Operating Room Personnel.

Workers' Compensation Benefits for Occupationally Infected Healthcare Workers in the United States

By Marshall Fowler and Patricia Miller Tereskerz, M.S., J.D., Ph.D.

Vol. 1, no. 3, 1995

A recent issue of *Virginia Lawyers Weekly*[1] reported the case of an HIV-infected LPN who filed a workers' compensation claim that was subsequently denied. The nurse claimed she sustained an injury from an HIV-contaminated needle when taking a blood sample from a jail inmate, and her physician testified that her low t-cell count was consistent with the timing of her injury. Nevertheless, the workers' compensation commission said that she had failed to prove that the employment-related needlestick was the source of her infection, since she could not provide evidence that she was HIV-negative prior to the stick. She had been hospitalized and received blood and blood products numerous times during the previous 11 years. In addition, because she could not identify the source patient of her exposure, she was unable to prove that he was HIV-positive.

This case carries a message for all healthcare workers: it is crucial to document at-risk blood exposures and follow post-exposure protocol. For most healthcare workers who are infected following an occupational blood exposure, the workers' compensation system will provide their only recourse. A basic understanding of how the system works, therefore, is important.

Understanding Workers' Compensation

Workers' compensation was established to provide financial support for an employee injured on the job. Before it became available, an injured employee had little recourse other than to file a personal injury suit against his employer, if the employer did not voluntarily pay for medical expenses and lost work time. Such suits were often drawn out and, even if successful, the employee remained without financial support until all appeals were decided. Workers' compensation laws were enacted to address these problems.

Workers' compensation is a no-fault insurance designed to protect employees and employers, and covers an estimated 87% of the U.S. work force. If an employee is injured or contracts an illness in the course of employment, workers' compensation will usually cover the costs of medical treatment and provide temporary or total disability payments. Employers pay the full cost of workers' compensation, but by participating in the system they receive immunity from personal injury law suits, since in most states workers' compensation provides the exclusive remedy for employees injured on the job.

The workers' compensation system thus has rewards and costs for both employees and employers. Employees have the advantage of being rapidly compensated for occupational injuries or illnesses, but in most cases they may not file suit against their employers, and critics say that workers' compensation does not always provide adequate compensation. While employers shoulder the cost of workers' compensation, they gain immunity from personal injury law suits.

Workers' compensation is a highly regulated system; the amount of compensation an employee can receive varies from state to state and is usually defined by state statute. It should be emphasized here that because workers' compensation is a state, not a federally, regulated system, there will be variations from state to state in the laws governing it.

Occupational Diseases

If a healthcare worker reports an at-risk exposure and is infected by a pathogen such as HIV or HBV, his or her workers' compensation claim will usually be processed as an occupational injury, because the subsequent illness can be related to a specific needlestick or other type of exposure. If the exposure has been fully documented, the compensation process should be straightforward. But healthcare workers who do not report exposures that result in infections will have the burden of proving that their illness is an "occupational disease," which can be difficult.

Most states recognize occupational diseases as compensable under workers' compensation law; each state, however, has its own definition of "occupational disease." Some states specify which diseases are covered, while others do not. The following definition of an occupational disease, from South Carolina's workers' compensation law, is representative: "*The words 'occupational disease' mean a disease arising out of and in the course of employment which is due to hazards in excess of those ordinarily incident to employment and is peculiar to the occupation in which the employee is engaged. A disease shall be deemed an occupational disease only if caused by a hazard recognized as peculiar to a particular trade, process, occupation or employment, as a direct result of continuous exposure to the normal working conditions thereof.*"

The South Carolina law lists conditions which would preclude an illness from being considered an "occupational disease," including "ordinary diseases of life to which the general public is equally exposed, *unless* such disease follows as a complication and a natural incident of an occupational disease or *unless there is a constant exposure peculiar to the occupation itself* which makes such disease a hazard inherent in such occupation" (emphasis added).

Several states have specifically recognized hepatitis B as an occupational disease in healthcare settings. In a number of published cases involving occupational hepatitis B, courts have decided in favor of claimants who can show that there is a greater probability that healthcare workers will contract hepatitis than non-healthcare workers, and who also can show that they had no other risk factors outside of employment for the disease. Even if a state's laws do not specify hepatitis B, hepatitis C, or HIV/AIDS as occupational diseases, workers' compensation boards usually acknowledge them as such if the employee has followed the appropriate reporting protocol and seroconverts to the pathogen within a recognized time interval.

Reporting Requirements

Healthcare workers should be familiar with their employer's reporting requirements. The Occupational Safety and Health Administration (OSHA) requires that all healthcare employers establish a postexposure protocol for employees who sustain at-risk blood or body fluid exposures. If an exposure is not reported and the postexposure protocol is not followed, a long delay or denial of workers' compensation benefits can result. Statutes of limitation governing workers' compensation vary, but in most cases they run for two years. If an occupational disease claim is filed without a documented exposure, an investigation will be necessary to rule out other routes of transmission.

The Centers for Disease Control and Prevention (CDC) provides guidelines for follow-up of healthcare worker at-risk exposures to potentially infected blood; these call for immediate testing following an exposure, which is important in determining whether the employee was already positive for HIV, HBV or HCV at the time of exposure.

A few states, such as California and Massachusetts, have proposed or passed "presumptive compensability" legislation, which presumes that, for certain occupations such as firefighters, police officers, and healthcare workers, infections from bloodborne pathogens such as HIV are job-related. These laws place the burden on the employer to disprove that an infection was not occupationally acquired. In most states, however, the burden is on the employee to follow reporting and postexposure protocols in order to document that an infection was contracted on the job.

Defining "Employees": Students and Other Healthcare Workers

Workers' compensation laws include specific definitions of "employer" and "employee" which could potentially exclude a healthcare worker from compensation. Each healthcare worker needs to know who his or her employer is; there is a difference between being employed by a hospital and working at a hospital. This distinction is particularly important for nursing students, medical students, residents, contract workers, and physicians who treat patients at a hospital but are not employed by it.

Hospitals sometimes require physicians to provide proof of workers' compensation coverage as part of their credentialing process; this is a standard risk-management procedure for most non-employees providing contract services at a healthcare institution or facility. If the non-employee does not have workers' compensation coverage, state law may require the hospital to cover that person as a "contractual employee."

With regard to students, most states have provisions in their workers' compensation laws that extend coverage to apprentices. In some jurisdictions, however, workers' compensation has been denied when a student received no remuneration other than training. Courts in other jurisdictions have reached the opposite conclusion, holding that workers' compensation laws apply to students because the training received is sufficient remuneration to classify the student as an apprentice employee. Generally, when a student receives some form of remuneration from a healthcare institution other than training, this is sufficient to invoke workers' compensation coverage.

Coverage

In most cases workers' compensation covers a legally mandated percentage of an employee's average weekly wage to compensate an employee for lost time from work; all states have minimum and maximum salary benefits. Although an employee does not receive 100% of lost salary, the workers' compensation benefits are tax exempt. If an employee's salary does not exceed the prescribed limits for compensation, he or she will usually receive about the same amount as take-home (net) pay. For those in higher tax brackets, however, the result is usually a net loss of income, and the same may be true for those at the lowest income levels. If an employee is partially or permanently disabled, he or she is given a lump-sum compensation based on a disability rating assigned by the treating physician.

Death benefit claims are also regulated by state workers' compensation laws. Benefits are usually based on a maximum-number-of-weeks formula; for example, 500 weeks times the employee's average weekly wage, or about ten years of salary. This amount may be reduced if lost time or disability benefits are paid as a result of the same claim. A healthcare worker can check with the workers' compensation benefits manager at his or her institution regarding the state's "average salary" for the purposes of workers' compensation; the figure is updated periodically, in some cases annually. The 1995 average weekly wage for South Carolina was $430.

Whether workers' compensation provides the exclusive remedy for an occupationally infected healthcare worker can usually be determined by calling the state workers' compensation board. In some states an employee may sue an employer for "serious and willful" injury or "willful intent" to cause an injury or exposure. These cases are usually difficult to prove, however, and may take years to settle. Furthermore, the workers' compensation provider can in some instances place a lien for reimbursement on any recoveries from a third-party action.

Gaps in Coverage

Many states do not require employers to carry workers' compensation if they have fewer than a given number of employees; the minimum number in some states is five. In a few states, including Texas and South Carolina, employers may "opt out" of workers' compensation altogether. Most facilities in these states do not choose this alternative, however, and do provide coverage. Healthcare workers need to determine if their employer has workers' compensation; if they are a contract worker at a healthcare facility, they should determine whether they are covered in this circumstance, and, if so, what the facility's reporting procedures are.

For employers who are not required to provide workers' compensation because they have too few employees, which is often the case for small medical practices or clinics, an Uninsured Employers' Fund is available in most states. The fund is managed by the state's workers' compensation division, and is provided specifically for this purpose. The state of Texas, however, does not have such a fund. Thus Dr. Patricia Wetzel, a physician who was occupationally infected with HIV while working at a hospital in Texas, found herself without coverage. The professional association that employed her did not have workers' compensation, there was no back-up Uninsured Employers' Fund, and the county hospital at which she worked denied her claim because she did not meet the definition of an employee.

Some employers who are required to carry workers' compensation do not. If there is any doubt about an employer's status or compliance with state laws, an employee can contact the state workers' compensation office, which maintains a current registry of certificates of coverage and notices of self-insurance that employers must submit to the state.

For occupationally infected healthcare workers who are not covered by workers' compensation or for whom workers' compensation benefits are inadequate, recourse is limited. Some healthcare workers have filed third-party lawsuits, such as a negligence claim against another healthcare worker whose action was related to the exposure, or a product liability claim against the manufacturer of a medical device associated with an exposure. The definition of "third party" may be relevant; for instance, in cases where a hospital is managed by an HMO, the HMO may be considered a third party. There have been relatively few third-party suits in the United States, however, for several reasons. The healthcare worker must find a lawyer who is willing to take the case, must be able to provide convincing evidence that the infection was the result of a specific occupational exposure, and must establish legally defensible grounds upon which to file suit. The outcome of lawsuits varies greatly, they take a long time to resolve, and are highly stressful; all of these factors act as disincentives to initiating a third-party suit.

Conclusions

The establishment over time of workers' compensation in the U.S. represented a significant advance in the protection of workers from the consequences of occupational disease and injury. But while the majority of American workers are now covered under workers' compensation plans, the requirements for employer participation and the definitions of compensable cases vary from state to state, and some workers remain uncovered, although they may be unaware of it. Because universal coverage has yet to be achieved, healthcare workers need to verify whether their employers have workers' compensation.

Because the majority of occupational blood exposures are never officially reported, the greatest obstacle facing many healthcare workers who are occupationally infected by a bloodborne pathogen and who seek workers' compensation is the failure to document their exposure. Without such evidence, the possibility of recourse to workers' compensation or any other remedy becomes remote. Therefore, it is of the greatest importance to aggressively educate healthcare workers about the necessity of reporting exposure incidents, and to emphasize the potential consequences of not doing so.

Adequate workers' compensation benefits need to be provided for workers at all income levels, and for workers with long-term illnesses such as those caused by bloodborne pathogens. The ultimate goal, however, is to prevent occupational transmission of bloodborne pathogens, which benefits both healthcare workers and workers' compensation insurance carriers.

Reference

1. *Virginia Lawyers Weekly*, 12/12/94, p. 27.

Hepatitis B Infection and Sharp-Object Injuries in Hospital Laundry Workers

By Connie Steed, R.N., C.I.C., and Ludwig Lettau, M.D., M.P.H.

Vol. 1, no. 5, 1995

Photo from OSHA's Hospital e-Tool website (Laundry Module).

Introduction

Healthcare workers who most commonly sustain injuries due to contaminated sharp objects in healthcare facilities include nurses, physicians, surgical technicians, and phlebotomists. But employees in the "other" category are also affected by sharps injuries. These "others" include housekeepers, food service assistants, laundry workers and a variety of other workers, most of whom do not use needles or other sharp objects in the course of their work. In this report, we describe the occurence of sharps injuries and the prevalence of hepatitis B infection in one of those groups—laundry workers—in a large community teaching hospital in South Carolina. The report should serve as a reminder that those in the "other" category are at real risk of infection from bloodborne pathogens from needlesticks.

Background

The Greenville Hospital System (GHS) includes a 675-bed community teaching hospital and five affiliated healthcare institutions with a total of 500 additional beds. GHS is serviced by a single laundry department with 50 employees. In late 1989, two laundry workers who sustained occupational needlestick injuries were found to be HBsAg-positive after baseline testing. An environmental assessment was then conducted in the laundry department to look for potential bloodborne pathogen hazards. When needle boxes were observed in the dirty laundry sorting area, employees explained that such containers were needed in order to properly dispose of contaminated sharp objects commonly encountered during the sorting process. Laundry workers were then asked to set aside all sharps and unusual objects found in dirty linen for a period of five days. An alarming number of such objects was collected over that period.

This experience prompted investigation of the prevalence of hepatitis B virus (HBV) infection and the occurrence of sharps injury among GHS laundry workers. In 1990 we also conducted a mail survey on the experience of other hospital laundries. In late 1992, a follow-up mail survey of hospital laundries was done to assess whether significant changes in the rates of sharp-object injury and utilization of hepatitis B vaccination had occurred as a result of the OSHA bloodborne pathogens standard, issued in 1991.

Investigation of GHS Laundry Workers

Methods

A confidential questionnaire was developed for collecting information from laundry workers and interviews were

Figure 1. Hepatitis B Status and History of Sharps Injuries by Years of Laundry Work

Greenville Hospital System Laundry Workers
June 1990, 46 cases

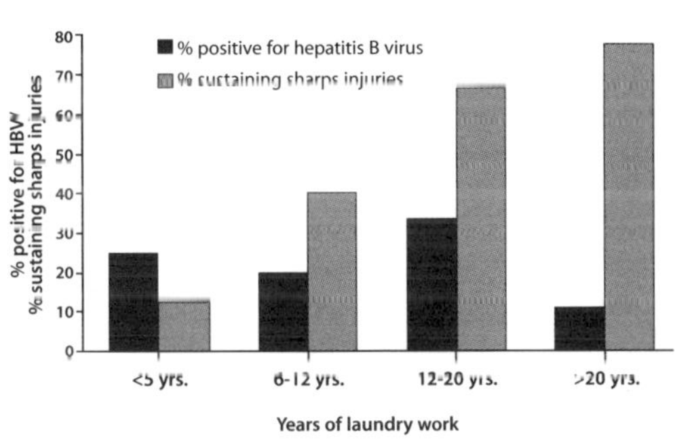

Figure 2. Unusual Items Found in Dirty Linens in Hospital Laundries

*Based on 1990 and 1992 Surveys of 52 Hospital Laundries** in 14 States*

Personal Items: Dentures/teeth*(11), shoes, watch*(3), eyeglasses*(7), billfold/money, rings/jewelry*(2), teddy bear/doll*(2), pens*, book, hearing aid*(2), razor blade*, cigarettes*, lighter*, play gun*, "Walkman" radio*, video games*

Patient Care: Bed pan*(5), needles/syringes*(5), whole needle disposal box*, thermometers*(3), box of hazardous waste*, suction container*, pillows, I.V. bags/bottle/tubing*(9), I.V. poles*(2), egg crate, blood pressure cuff*(2), stethoscope*(2), trays, slings

Critical Care: Heart monitor*(3), life support system, fetal monitor*, respirators, telemetry unit, airway*, trauma pants*, pacemaker*, oxygen regulator*(2), oxygen cylinder*, oxygen tank stand*

Surgical Supplies: Complete surgical instrument tray*(4), scalpels*(4), endoscope(2), endoscope with camera attached*(2), miscellaneous surgical instruments*(9), dental instruments*(2), bone drill*, orthopedic traction device*

Body Parts: Placenta(3), finger, finger with ring*

Miscellaneous: Motors, bottles, whole sink, metal bed elevators, wheelchair parts(2), x-ray vests(2), mop wringer, butcher knife*(2), beer can, soda can, box cutters, paper towels

Notes:
( ) indicates number of laundries reporting the item
* indicates item was reported at least once in 1992 survey; items without an asterisk appeared in 1990 survey only.
** 1990 survey: 21 hospital laundries; 1992 survey: 31 hospital laundries

conducted by one of the authors. Information requested included demographic data, laundry and other prior healthcare employment history, potential non-occupational exposure to HBV, past medical/social history (liver disease, number of lifetime sex partners, history of sexually transmitted diseases), and HBV vaccination status. An HBV seroprevalence survey was done by screening for hepatitis B core antibody, with subsequent testing for HBsAg in those who were core antibody positive.

Results

Of 49 participating laundry workers, 85% were black and 65% were female. Of the 46 who were interviewed, none had major non-occupational risk factors for HBV. Overall, 39% (18/46) had a history of at least one prior sharps injury. This percentage increased with years of laundry work (see **Figure 1**). None had received hepatitis B vaccination. Eleven of 46 (24%) tested positive for hepatitis B core antibody, and five were HBsAg positive. No correlation was found between HBV infection and years of laundry work (**Figure 1**).

Follow-up/Changes in Practice

As a result of these investigations, a number of measures were taken to minimize exposure of laundry workers to contaminated sharp objects and to reduce the risk of HBV transmission. Laundry workers without evidence of HBV infection (n=35) were offered hepatitis B vaccination, with an acceptance rate of 100%. Educational programs were conducted for healthcare workers who used sharps, emphasizing proper disposal techniques and the potential consequences of improper disposal. Utility gloves were added to the personal protective equipment worn by laundry workers when in the contaminated laundry sorting area. A needleless I.V. system was implemented in 1991 throughout the Greenville Hospital System, and a safety I.V. catheter was introduced in 1992, in order to reduce the total number of sharp objects that could potentially injure employees, both at the point of use and downstream in the disposal process. Ongoing monitoring for inappropriate objects mixed with dirty linen is conducted by the laundry department manager, and reports are forwarded to the infection control department when results of monitoring indicate the need for further action.

Questionnaire Surveys of Other Hospital Laundries

Methods

Mail surveys were conducted in June 1990 and December 1992 (before and after the promulgation of the OSHA bloodborne pathogens standard). The pre-OSHA-standard survey was distributed to laundry managers attending two regional meetings, and the post-standard survey was mailed to laundry managers for all hospitals in South Carolina, North Carolina, Virginia and Georgia. Both surveys asked the same questions regarding type and size of hospital(s) serviced, number of laundry employees, HBV vaccination program and vaccine acceptance rate, incidence of sharps found in laundry, other inappropriate or unusual objects found in laundry, and incidence of reported employee sharps injuries. The 1992 survey requested laundries to report data for the period July-December 1992.

Results

The 1990 survey of 21 hospital laundries in 14 states found a monthly average of 66 sharps in dirty laundry (range 0.5-400). The range of sharps injuries reported annually was 0-15, with an average of three per year. The 1992 survey had 31 respondents (42% response rate), who reported a monthly average of 35 sharps in dirty laundry (range 0-350). The range of sharps injuries per year in 1992 was 0-24, with an average of 2.7 (yearly injury rate obtained by doubling injuries reported for the six-month period). A large number of inappropriate or unusual items were found in dirty hospital laundry (see **Figure 2**). Items mentioned frequently included: dentures, needles/syringes, whole needle containers, bedpans, I.V. bags/bottles/tubing, scalpels, other surgical instruments, and body parts. The 1990 survey found that 66% of hospitals (14/21) offered hepatitis B vaccination, with an estimated 54% of employees vaccinated. The 1992 post-OSHA-standard survey found that 97% (30/31) of hospitals offered HBV vaccination, with an estimated 70% of employees vaccinated.

Conclusions

While the high prevalence of HBV was more likely related to demographic factors than to contaminated sharp-object injuries among GHS laundry workers[1], the fact that contaminated sharps and other objects continue to be found in significant quantities in dirty hospital laundry supports the use of HBV vaccine in laundry workers. It also indicates the necessity of identifying methods—such as the use of engineering controls and personal protective equipment—that will reduce the number and hazard of downstream sharps injuries to laundry workers. Continued education of all healthcare workers on proper disposal of equipment is crucial, since sharps injuries among laundry workers cannot be reduced without the cooperation of the clinical workers who use the sharp devices.

Concern for the safety and health of laundry workers was the driving force behind this study, but there is another reason—an economic one—for educating hospital employees about inappropriate objects in dirty laundry. Such carelessness can result in a significant loss of valuable equipment and supplies for hospitals.

References

1. McQuillan GM, Townsend TR, Fields HA, et al. Seroepidemiology of hepatitis B virus infection in the United States, 1976 to 1980. *Am J Med.* 1989;87(3A):5S–10S.

See Also:

- Centers for Disease Control. Summary of the agency for toxic substances and disease registry report to Congress: the public health implications of medical waste. *MMWR.* 1990;39(45):822–824.
- Penney JP, Gordon ME. Laundry hazards [letter]. *Lancet.* 1982;1(8284):1309–1310.

Hepatitis C Virus in Healthcare Workers

by Patricia Miller Tereskerz, M.S., J.D., Ph.D., Nicola Petrosillo, M.D., Vincenzo Puro, M.D.*, and Janine Jagger, M.P.H., Ph.D.*

Vol. 2, no. 1, 1995

**Dr. Petrosillo and Dr. Puro are from the Centro di Riferimento AIDS, Ospedale Lazzaro Spallanzani, Rome, Italy.*

Overview and Epidemiology

Hepatitis C virus (HCV) is a single-strand RNA virus[1] which was previously called non-A non-B (NANB) hepatitis until the specific identification of HCV was reported in 1989.[2] HCV is characterized by the potential for reinfection with subtypes of the virus, and the absence of clearly defined immunity following infection.[1] It is now recognized that HCV accounted for 60%–90% of what was formerly known as NANB hepatitis.[3]

Currently, about 150,000 new hepatitis C virus (HCV) infections occur in the United States annually.[4] It is estimated that 1.5%, or nearly 2,200, of these cases are occupationally acquired.[5] The population prevalence of HCV is unknown since most studies have examined select populations rather than a sample of the general population. However, HCV prevalence appears low, since a rate of .6% was found among blood donors in the U.S.[1]

Individuals at increased risk for HCV include hemophiliacs, intravenous drug users, recipients of transfusions and organ transplants, and individuals undergoing hemodialysis.[6-9] Sexual transmission among heterosexuals and homosexuals may occur[10]; evidence for sexual transmission weakens, however, when other risk factors are controlled.[1] The rate of perinatal transmission in infants born to anti-HCV-positive mothers varies from 0-13%. High titers of HCV RNA in mothers appear to favor transmission.[11] Other household contact studies have shown HCV infection rates of 0-20% among children living with chronic HCV patients.[11] HCV has also been transmitted by human bites.[12,13]

Clinical Presentation and Diagnosis

Following an average incubation period of 50 days (range 15-150 days), a patient presenting with HCV may experience general malaise and weakness followed by anorexia, nausea, vomiting, and dull right upper quadrant pain.[14] However, there is considerable variation in the clinical manifestations of HCV, and there are no specific symptoms distinguishing one type of hepatitis from another.[5]

The mortality rate for all HCV infections is 5-7%.[1] Of the 150,000 individuals infected each year in the U.S., it is estimated that only 37,500 become symptomatic; however, 93,000 may ultimately develop chronic liver disease and 30,700 more may develop cirrhosis.[1] It is estimated that from 50%-90% of HCV infections will lead to chronic infections and intermittent viremia, which in turn can lead to cirrhosis or hepatocellular carcinoma after a clinical latency period of two decades or more.[1]

Recombinant alpha-interferon has been shown, at least in some cases, to be effective in treating chronic HCV, by reducing aminotransferase abnormalities, and by decreasing both inflammation and the level of HCV viremia. More than half of patients responding to this treatment, however, have a relapse after completion of therapy.[1] In addition, adverse reactions and the lack of data on whether interferon will reduce the risk of developing cirrhosis or liver cancer complicate treatment.[1]

Diagnostic Tests

Following the identification of HCV, tests quickly became available to diagnose HCV infection. In 1989 the development of the first-generation enzyme-linked immunosorbent assay (ELISA-1) for HCV was announced.[15] The ELISA-1 tests serum for the presence of IgG antibody against the c100-3 antigen, and allows for detection of antibodies within about 15 weeks of infection. There is a seronegative window phase, during which infected patients have negative test results. The first-generation HCV assay was found to be 80%–90% sensitive (detected 80%–90% of those who were positive), but was associated with false-positive results.[1,16] Consequently, second-generation tests were sought to overcome these shortcomings.

Second-generation assays (ELISA-2), with additional recombinant antigens from the viral genome, became available in 1992.[17] Whereas the first-generation anti-HCV assays detect antibody to the c100-3 antigen, the second-generation assays combine c100-3 with another nonstructural protein, c33c, to form a new antigen known as c200. The second-generation assays also detect antibody to c22-3 antigen in the viral core.[17]

The second-generation tests have higher sensitivity and allow for earlier diagnosis because the antigens that are tested for are usually detectable before antibodies to c100-3 appear.[1] The second-generation tests are up to 98% sensitive, improving sensitivity over the first-generation test by about 20%.[17] To further improve sensitivity and specificity, third-generation assays have been developed, but they have not yet been approved for commercial use in the U.S.[18,19] Recombinant immunoblot assay (RIBA) is also available and is often used as a supplementary confirmatory test to limit false-positive results. Reverse transcription/polymerase chain reaction (RT-PCR) is used to detect HCV RNA, and the results of this assay may have a bearing on management of patients. Because of its low reliability, results of HCV PCR should be interpreted with caution.[20]

Prevalence of HCV in Healthcare Workers

In the United States, the overall prevalence of HCV in healthcare workers is 1.4%-1.6%, a rate similar to that seen in volunteer blood donors.[1,21-23] By comparison, 3%-35% of healthcare workers studied in the 1970s had antibodies to HBV; and subsequent studies of the seroprevalence of HIV demonstrated that .4% of healthcare workers at risk for blood exposures were HIV positive.[1] It is hypothesized that the difference in occupational risk between HBV and HCV is due to the fact that HBV circulates in higher titers, rendering the risk of HBV transmission following a needlestick between 6% and 30%.[24,25]

In a U.S. study[4] in which retrospective testing of serum samples drawn in 1983 from 1,677 hospital employees was conducted, 1.4% of the employees had antibodies to HCV. In this study, factors significantly associated with the presence of HCV antibody included antibody to HBV core antigen, a history of blood transfusion, and needlestick injuries. Maintenance staff and food service employees had significantly higher rates of antibody to HCV. No relationship was found between antibody to HCV and ethnic group. In addition, there was no association between antibody to HCV and years in the same occupation or in the same hospital work area.

A British study reported a lower rate (.28%) than the U.S. study of HCV-positive healthcare workers among those at risk for exposures to bloodborne pathogens.[26] In Italy, on the other hand, an HCV prevalence rate of .85% was found among 937 healthcare workers, leading the authors to conclude that hospital personnel appear to be at no greater risk for HCV infection than the general population.[27] These researchers observed no significant difference between the HCV prevalence rate in HBV-positive versus HBV-negative healthcare workers.

Another Italian study compared the rate of HCV in healthcare workers to factory workers and found no significant difference in the prevalence of infection, suggesting the risk factors for HCV are primarily socioeconomic.[28] More recently, two Italian studies which did not account for sexual or other behavioral risk factors such as I.V. drug use found 2% and 2.2% prevalence rates of HCV among healthcare workers tested in 1985 and 1992, respectively.[29,30] In these studies, history of blood transfusion was the only risk factor significantly associated with HCV infection. In general, European studies have concluded that the prevalence of HCV in healthcare workers is between 1%-4%.[23-26,31-34]

In an Indian study of 50 pathology department personnel, 17 anesthetists, 11 cardiologists, nine gastroenterologists and three internists, no HCV antibodies were reported.[35]

A Japanese report found the prevalence rate of HCV infection to be 4.3% (5/115) in medical staff, 2.2% (15/670) in nurses, and 5.5% (10/183) in acupuncturists. The prevalence of HCV infection among those with direct patient contact was slightly, but not significantly, higher.[36]

When subsets of healthcare workers with increased risk of exposure to blood are studied, the prevalence of HCV-positive healthcare workers is higher. A Belgian study reported an HCV prevalence rate of 4.1% in hemodialysis nurses.[37] The authors ruled out non-occupational variables as risk factors for the high prevalence of HCV and suggested that the high rate may have been associated with the long duration of employment.

There are conflicting reports about the risk of HCV infection among dentists and oral surgeons. In one study, dentists and oral surgeons were found to be at increased risk, with HCV antibody found in 1.7% (8/456) of dentists compared to .13% (1/723) for controls. Oral surgeons were at even greater risk than other dentists, with HCV antibodies found in 9.3% (4/43) of oral surgeons compared with .97% (4/413) for dentists.[38] In contrast, a study conducted in Taiwan found that there was no increased risk of HCV infection in dentists, with a positivity rate of .6% (3/461) using a second-generation assay.[39] Likewise, a study of dental surgeons in south Wales did not show any HCV-positive samples among the 94 dental surgeons tested.[40]

Risk of HCV Infection to Healthcare Workers Following Needlestick Injury

There have been several cases reported in the medical literature of HCV transmissions following needlestick injuries to healthcare workers. Prior to the development of diagnostic tests for HCV, there was a report of transmission of NANB hepatitis via a needlestick in a medical student who suffered a deep puncture wound of a finger with a contaminated needle. The source patient had a history of kidney transplant and multiple transfusions, and was undergoing immunosuppressive therapy.[41]

By 1990, the first case of HCV transmission following a needlestick injury was reported in a surgeon who was stuck by a needle from a patient known to be an HIV-positive I.V. drug user.[42] Another 1990 report described HCV transmission in a nurse who sustained a deep puncture wound to her finger from a contaminated needle. The source patient was on hemodialysis and the injury occurred two weeks after the patient experienced an increase in serum alanine aminotransferase levels.[43]

Another report involved an incident in 1972 of HCV transmission via a needlestick injury to a nurse's finger while she was removing a hypodermic needle from a patient's arm. Sera were stored from the date of the needlestick and from six weeks following the needlestick. Serum drawn immediately after the injury was negative, but the serum drawn six weeks after the injury was positive for anti-HCV. The nurse remains positive today.[44]

A report from a Spanish institution found evidence of HCV seroconversion in a physician who punctured his finger with biopsy forceps that were contaminated with extracellular fluid and blood.[45]

Another reported case demonstrates the need for a test which can directly detect infectious HCV particles, both for screening blood donors and for the early diagnosis of HCV infection.[46] In this case, a nurse was injured by a contaminated needle used on a hemodialysis patient who had received a unit of packed red cells from a volunteer blood donor. First- and second-generation assays were performed on the patient's and the nurse's blood; both were negative for HCV antibodies at the time of exposure. However, 17 weeks after the transfusion, the patient's ALT concentration rose, and HCV RNA was detectable at 23 weeks. The findings remained negative for anti-HCV using first- and second-generation assays until more than 32 weeks after transfusion. The nurse became symptomatic 13 weeks after the needlestick, and serum HCV RNA was detectable. But she remained negative for anti-HCV until she seroconverted three weeks later. Retrospective study of the blood donor showed no indication of HCV infection by commercially available ELISA tests, but HCV sequences were detectable by polymerase chain reaction.

In addition to anecdotal reports of HCV seroconversions following needlesticks, there have been several studies of the risk of HCV transmission following such injuries. An Italian study, which included nine dialysis centers, found a 39.4% prevalence of HCV among patients. The investigators followed 61 healthcare workers who sustained needlestick injuries, 29 who experienced mucous membrane contaminations, and 40 who had non-intact skin contacts with blood from HCV-positive patients. None of the HCV-exposed healthcare workers seroconverted to HCV during a ten-month follow-up period.[47] A subsequent Italian study looked at 646 HCV-exposed healthcare workers who were seronegative for HCV at the date of exposure and who were followed for six months; this study documented four (1.2%) HCV seroconversions, including two surgeons and one nurse who sustained needlestick injuries from blood-drawing needles, and a nurse injured by a needle used for intramuscular injection.[48]

Another Italian study evaluated the risk of HCV infection following 225 exposures to HCV-positive blood in hospital personnel. Three cases of HCV seroconversion (1.3%) occurred following needlestick injuries.[49] The first, in a surgeon, involved a needle from a patient with a history of transfusions who was anti-HCV positive. The second, also in a surgeon, was caused by a needle from an HIV- and HBV-positive intravenous drug user. The third case involved a sanitation worker who was stuck by a needle inappropriately discarded in a trash bag.

Early Japanese studies using first-generation tests for HCV found a 4% transmission rate for HCV following a needlestick injury where the source patient was anti-HCV positive.[50] Subsequent studies conducted in Japan and the U.S. using the more sensitive second-generation tests found the risk of HCV transmission following a needlestick injury to be 6-10%.[4,51] In the U.S. study, 1,387 source patients (12.7%) were anti-HCV positive[5], and a 6% HCV transmission rate was found. However, these data were not collected prospectively on a single cohort, but were based on sera samples that had been stored for up to 10 years. This compares to a median rate of 1.6% across studies of HCV needlestick transmission.[48]

The incidence of clinical NANB hepatitis among healthcare workers in the U.S. study[5] was three times higher than for the general population. However, direct comparison of the U.S. data with population data should be undertaken with caution, because NANB hepatitis is the most underreported of all types of hepatitis, and the incidence of NANB hepatitis varies widely among and within geographical areas.[48] On the other hand, Italian researchers have followed up serologically 2,622 healthcare workers for one year and have observed three seroconversions (0.1%) in subjects who did not acknowledge occupational or community risk factors.[30] They found that source patients with renal or liver disease, HIV infection, or a history of I.V. drug use were more likely to have HCV infection than other patients, and that the highest rate of reported exposure to HCV occurred in the emergency department.

The difference in rates reported in these studies likely reflects differences in study designs, diagnostic methods, exposure mechanism (hollow-bore vs. solid needles), number of cases followed, and infectivity of different HCV strains.[52]

Consequently, the risk of HCV transmission from a single needlestick injury appears to be lower than the 7-30% risk of contracting HBV.[53] Using a deterministic model that takes into account the anti-HCV seroprevalence among dialysis patients, the rate of occupational exposures among dialysis healthcare workers, and the probability of HCV transmission, the risk of infection with HCV among dialysis healthcare workers has been calculated as 0.06 and 0.0087 per 10,000 dialysis procedures performed for percutaneous injuries and mucous membrane contaminations, respectively.[54]

Consequences and Prevention of HCV Infection

Currently there is no vaccine available for HCV. Because HCV is characterized by genetic and serological heterogeneity, by the absence of a clearly defined protective immune response after infection, and by the possibility of reinfection, the development of a vaccine is difficult.[1,55] There has been some success reported in a study in which chimpanzees were protected following immunization with recombinant DNA derived from HCV.[1] However, it will likely be several years at least before a vaccine is available.

Furthermore, it has not been determined whether the administration of immune serum globulin provides protection against parenteral exposure to HCV.[1]

The studies cited above demonstrate the need for prospective, comprehensive surveillance in which follow-up serologies are reported for HIV, HCV and HBV so that valid comparisons may be made regarding the incidence, prevalence and mechanisms of transmission for each virus. In addition, research indicates that parenteral exposure to HCV via needlestick injuries represents a significant transmission risk for healthcare workers. This underscores the need for protective devices which reduce needlestick risk; in addition, good disposal systems which provide puncture-resistant containers close to the point of use should be routinely used.

References

1. Lemon S, Brown E. Hepatitis C Virus. In *Principles and Practice of Infectious Disease*. Ed. Mandell GM, Bennett JE, Dolin R. 1995;1474–1486.
2. Choo Q-L, Kuo G, Weiner AJ, et al. Isolation of a cDNA clone derived from a blood-borne non-A, non-B viral hepatitis genome. *Science*. 1989;244:359–62.
3. Rubin RA, Falestiny M, Malet PF. Chronic hepatitis C (Review). *Arch Intern Med*. 1994;154:387–392.
4. Polish LB, Tong MJ, Co RL, et al. Risk factors for hepatitis C virus infection among health care personnel in a community hospital. *Am J Infect Control*. 1993;21:196–200.
5. Lanphear BO, Linneman CC, Cannon CG. Hepatitis C virus infection in healthcare workers: risk of exposure and infection. *Infect Control Hosp Epidemiol*. 1994;15(12):745–750.
6. Esteban JI, Esteban R, Viladomiu L, et al. Hepatitis C virus antibodies among risk groups in Spain. *Lancet*. 1989;ii:294–96.
7. Mosley JW. Hepatitis B and non-B. Epidemiologic background. *JAMA*. 1975;233:967–969.
8. Francis DP, Hadler SC, Prendergast TJ. Occurrence of hepatitis A, B, and non-A/non-B in the United States. CDC sentinel county hepatitis study I. *Am J Med*. 1984;76:69-74.
9. Alter MJ, Gerety RJ, Smallwood LA, et al. Sporadic non-A, non-B hepatitis: frequency and epidemiology in an urban US population. *J Infect Dis*. 1982;145:886 93.
10. Alter MJ, Hadler SC, Judson FN, et al. Risk factors for acute non-A, non-B hepatitis in the United States and association with hepatitis C virus infection. *JAMA*. 1990;264:2231-2235.
11. Alter MJ. Occupational exposure to hepatitis C virus. A dilemma. *Infect Control Hosp Epidemiol*. 1994;15:742-744.
12. Dusheiko GM, Smith M, Scheuer PJ. Hepatitis C virus transmitted by human bite. *Lancet*. 1990;336:503–504.
13. Figueiredo JFC, Borges AS, Martinez R. Transmission of hepatitis C virus but not human immunodeficiency virus type 1 by a human bite. *Clin Infect Dis*. 1994;19:546–547.
14. Hus HH, Feinstone SM, Hoofnagle JH. Acute viral hepatitis. In *Principles and Practice of Infectious Disease*. Ed. Mandell GM, Bennett JE, Dolin R. 1995; 1137.
15. Kuo G, Choo Q-L, Alter HJ, et al. An assay for circulating antibodies to a major etiologic virus of human non-A, non-b hepatitis. *Science*. 1989;244:362–364.
16. Chien DY, Choo Q-L, Tabrizi A. Diagnosis of hepatitis C virus (HCV) infection using an immunodominant chimeric polyprotein to capture circulating antibodies: Reevaluation of the role of HCV in liver disease. *Proc Natl Acad Sci USA*. 1992;89:10011–5.
17. Alter HJ. New kit on the block:evaluation of second-generation assays for detection of antibody to the hepatitis C virus. *Hepatology*. 1992;15(4):350–353.
18. Uyttendale S, Clayes H , Mertens W, et al. Evaluation of third-generation screening and confirmatory assays for HCV antibodies. *Vox Sang*. 1994;66:122–129.
19. Tedeschi V, Seeff LB. Diagnostic tests for hepatitis C: where are we now. *Ann Intern Med*. 1995;123:383–385.
20. Zaaijer HL, Cuypers HTM, Reesink HW, et al. Reliability of polymerase chain reaction for detection of hepatitis C virus. *Lancet*. 1993;341:722–724.
21. Gerberding J. Incidence and prevalence of human immunodeficiency virus, hepatitis B virus, hepatitis C virus, and cytomegalovirus among health care personnel at risk for blood exposure: final report from a longitudinal study. *J Infect Dis*. 1994;170:1410–1417.
22. Cooper B, Krusell A, Tilton RC. Seroprevalence of antibodies to hepatitis C virus in high-risk hospital personnel. *Infect Control Hosp Epidemiol*. 1992;13:82–85.
23. Stevens CE, Taylor PE, Pindyck J, et al. Epidemiology of hepatitis C virus, a preliminary study in volunteer blood donors. *JAMA*. 1990;263:49–53.
24. Grady GV, Lee VA, Prince AM, et al. Hepatitis B immune globulin for accidental exposures among medical personnel: final report of a multicenter controlled trial. *J Infect Dis*. 1978;138:625–638.
25. Seeff LB, Wright EC, Zimmerman HJ, et al. Type B hepatitis after needlestick exposure: prevention with hepatitis B immune globulin: final report of the Veterans Administration cooperative study. *Ann Intern Med*. 1978;88:285–293.
26. Zuckerman J, Clewley G, Griffiths P, et al. Prevalence of hepatitis C antibodies in clinical health-care workers. *Lancet*. 343:1618–1620.
27. Di Nardo V, Bonaventura ME, Chiaretti B, Petrosillo N, Puro V, Ippolito G. Low risk of HCV infection in healthcare workers. (Letter). *Infection*. 1994;22:115.
28. De Luca M, Ascione A, Vacca C. et al. Are health-care workers really at risk of HCV infection? (Letter). *Lancet*. 1992;339:1364–1365.

29. Petrosillo N, Puro V, Ippolito G et al. Hepatitis B virus, hepatitis C virus and human immunodeficiency virus infection among healthcare workers: a multiple regression analysis of risk factors. *J Hosp Infect*. 1995;30: 273–285.
30. Puro V, Petrosillo N, Ippolito G, et al. Occupational hepatitis C virus infection in Italian healthcare workers. *Am J Publ Health*. 1995;85:1272–1275.
31. Struve J, Aronsson B, Frenning B. Prevalence of antibodies against hepatitis C virus infection among healthcare workers in Stockholm. *Scand J Gastroenterol.* 1994;29:360–362.
32. Hofmann H, Kunz C. Low risk of healthcare workers for infection with hepatitis C virus. *Infection*. 1990;18:286-288.
33. Campello C, Majori S, Poli A. Prevalence of HCV antibodies in health-care workers from northern Italy. *Infection*. 1992;20(4):224–226.
34. Libanore M, Bicocchi R, Ghinelli F, et al. Prevalence of antibodies to hepatitis C virus in Italian healthcare workers. *Infection*. 1992;20:50.
35. Amarapurkar DN. Prevalence of hepatitis C antibodies in health-care workers (letter). *Lancet*. 1994;344:339-340.
36. Nakashima K, Kashiwagi S, Hyashi J. Low prevalence of hepatitis C virus infection among hospital staff and acupuncturists in Kyusha, Japan. *J Infect*. 1993;26:17-25.
37. Jadoul M, Akrout M EL, Cornu, C. Prevalence of hepatitis C antibodies in health-care workers (Letter). *Lancet*. 1994;344:339.
38. Klein RS, Freeman K, Taylor PE. Occupational risk for hepatitis C virus infection among New York City dentists. *Lancet*. 1991;338:1539-1542.
39. Kuo MY, Hahn LJ, Hong CY. Low prevalence of hepatitis C virus infection among dentists in Taiwan. *J Med Virol*. 1993;40:10-13.
40. Herbert A-M, Walker DM, Davies KJ, et al. Occupationally acquired hepatitis C virus infection (Letter). *Lancet*. 1992;339:305.
41. Mayo-Smith M. Type non-A, non-B and type B hepatitis transmitted by a single needlestick. *Am J Infect Control*. 1987;15(6):266–267.
42. Vaglia A, Nicolin R, Puro V, Ippolito G, Bettini C, de Lalla F. Needlestick hepatitis C virus seroconversion in a surgeon. *Lancet*. 1990;336:1315-1316.
43. Schlipkoter U, Roggendorf M, Cholmakow K, et al. Transmission of hepatitis C virus (HCV) from a haemodialysis patient to a medical staff member. *Scand J Infect Dis*. 1990; 22:757-758.
44. Seeff LB. Hepatitis C from a needlestick injury. *Annals Intern Med* (Letter). 1991; 115(5):411.
45. Perez-Trallero E, Cilla G, Saenz JR. Occupational transmission of HCV. *Lancet*. 1994; 344:548.
46. Cariana E, Zonaro A, Primi D. Detection of HCV RNA and antibodies to HCV after needlestick injury. *Lancet*. 1991;337:850.
47. Petrosillo N, Puro V, Ippolito G, et al. Prevalence of hepatitis C antibodies in healthcare workers. (Letter). *Lancet*. 1994;344:340.
48. Puro V, Petrosillo N, Ippolito G. The risk of hepatitis C seroconversion after occupational exposures in healthcare workers. *Am J Infect Control*. 1995;23(5):273–277.
49. Marranconi F, Mecenero V, Pellizzer GP. HCV infection after accidental needlestick injury in healthcare workers. *Infection*. 1992;20(2):111.
50. Kiyosawa K, Takeshi S, Tanaka E. Hepatitis C in hospital employees with needlestick injuries. *Ann Intern Med*. 1991;115(5)367–369.
51. Mitsui T, Iwano K, Masuko K. Hepatitis C virus infection in medical personnel after needlestick accident. *Hepatology*. 1992;16(5):1109–1113.
52. Puro V, Petrosillo N, Ippolito, G, Jagger J. Hepatitis C virus infection in healthcare workers. *Infect Control Hosp Epidemiol*. 1995;16:324-325.
53. Occupational Safety and Health Administration. Occupational exposure to bloodborne pathogens: proposed rule and notice of hearing. *Fed Regist*. 1989;54:23042–23139.
54. Petrosillo N. Puro V, Jagger J, et al. The risks of occupational exposure and infection by human immunodeficiency virus, hepatitis B virus, and hepatitis C virus in the dialysis setting. *Am J Infect Control*. IN PRESS.
55. Zuckerman AJ, Zuckerman JN. Prospects for hepatitis C vaccine. *J Hepatology*. 1995;22 (Suppl):97–100.

Reducing Pain During Arterial Blood Drawing—Minimizing Patient Movement

by Richard M. Levitan, M.D.
Department of Emergency Medicine, New York Medical College, Lincoln Hospital
Vol. 2, no. 1, 1995

Editor's note: The use of local anesthetics prior to placement of intravascular devices is common practice among anesthesia personnel; other healthcare workers, however, may not be aware of this practice and its potential to reduce patient movement.

Traditional strategies for reducing needlestick injury have focused on engineering controls and healthcare worker behavior modification. Examples of engineering controls are self-sheathing needles and needle-free intravenous connection mechanisms; the principle change in work practice controls has been avoidance of needle recapping. A third strategy that warrants further investigation is the appropriate use of local anesthetics when performing skin punctures with large-gauge hollow-bore needles.

Growing evidence from studies on HIV seroconversion suggest that injuries from large-gauge hollow-bore needles are associated with a higher risk of seroconversion than solid needles or small-gauge hollow-bore needles. Procedures that require the use of large-gauge hollow-bore needles include phlebotomy, arterial blood gas (ABG) sampling, and placement of intravenous lines. Despite advances in engineering controls, needles are still required for skin penetration during these procedures; patient movement secondary to pain occurs frequently as a result, and contributes to an increased risk of healthcare worker injury. The use of local anesthetics prior to such procedures can significantly decrease the pain of skin puncture and should decrease patient movement as well, thereby reducing the risk of needlestick injury.

Local anesthetics can be topically applied or injected. The most commonly used topical anesthetic is EMLA, which is an emulsion of lidocaine and prilocaine. This agent has been extensively studied for pediatric venipuncture. It must be placed 60 minutes prior to the procedure and works equally well as a cream or a patch.[1,2,3]

More commonly, a small volume (0.25 ml) of local anesthetic is injected using a 25-, 27- or 30-gauge needle. Lidocaine, either 1% or 2%, is the usual agent. The addition of epinephrine causes vasoconstriction and increases the duration of action. This is not necessary or desirable for phlebotomy, arterial sampling, or I.V. catheter placement. The pain of administration is related to the speed and depth of injection.[2] Injections into the superficial dermis, resulting in a weal, are more painful than subdermal injections. Injections faster than 2 seconds are more painful than those given over 10 seconds. Subdermal injections, however, require 5 to 6 minutes for full anesthesia to pinprick to take effect, whereas intradermal injection produces immediate anesthesia. The pain of infiltration can be further attenuated by warming or buffering the lidocaine solution prior to injection.[2,4,5] Lidocaine warmed to 98.6 degrees Fahrenheit is less painful on injection than the same solution at 68 degrees. The addition of sodium bicarbonate in a 1:10 ratio by volume to lidocaine results in less pain on injection than lidocaine alone.

Some clinicians argue that administration of lidocaine prior to I.V. placement results in more pain than the single stick of the procedure itself. Several studies have proven that intravenous cannulas as small as 22 gauge are significantly more painful than a subcutaneous injection of lidocaine, and that the pain of cannulation is significantly lessened by prior lidocaine injection.[6,7,8] These studies included not only patients' verbal assessment of pain, but also observed responses such as facial grimacing and arm and leg jerking. Moreover, the placement of I.V. catheters without anesthesia has been shown in the operating room to result in significant cardiovascular responses (i.e., increases in rate-pressure product and mean arterial pressure) that can be eliminated by prior administration of lidocaine.[9] In each of these studies the prior administration of lidocaine had no negative impact on the success rate of venous cannulation.

Carefully controlled studies on the utility of lidocaine prior to radial blood gas sampling have yet to be performed. The radial artery site used for ABG sampling is located on the palmar aspect of the hand, just proximal to the flexor crease at the wrist. As opposed to venous cannulation sites, which are further up the arm or on the dorsum of the hand, this area of the body has significantly more sensitivity to pinprick. It is logical to assume that if lidocaine has been shown to attenuate pain responses in the less sensitive areas, it would work as well or better at the more sensitive site for radial ABGs. From anecdotal experience with patients who have had the procedure many times, it is clear that patients prefer prior infiltration with lidocaine. As with venous cannulation, the small amount of lidocaine used does not seem to adversely impact the success of the procedure.

Although the use of locally injected lidocaine increases the number of needles used, the small-gauge needles used for local infiltration are much less hazardous in terms of seroconversion risk than the relatively larger needles used for venous cannulation and blood gas sampling. While topical anesthesia avoids a second needle, it is more expensive and requires time for absorption.

Before the use of local anesthetics can be recommended as a method for reducing healthcare worker risk, further study is needed on pain response and patient movement as an independent risk factor for needlestick injury.

References

1. Robieux I, Eliopoulos C, Hwang P, et al. Pain perception and effectiveness of the eutective mixture of local anesthetics in children undergoing venipuncture. *Ped Res.* 1992;32(5):520-523.
2. Edlich R, Rodeheaver GT, Thacker JG. Local and regional anesthesia for wound repair. In Tintinalli J, Ed. *Emergency Medicine: A Comprehensive Study Guide*, 4th ed. New York: McGraw Hill, 1996, 270-272.
3. Joyce TH. Topical anesthesia and pain management before venipuncture. *J Ped.* 1993;122:S24-29.
4. Brogan GX, Giarrusso E, Hollander JE, et al. Comparison of plain, warmed, and buffered lidocaine for anesthesia of traumatic wounds. *Ann Emerg Med.* 1995;26:121-125.
5. Bainbridge LC. Comparison of room temperature and body temperature local anaesthetic solutions. *Br J Plas Surg.* 1991;44(2):147-148.
6. Harrison N, Langham BT, Bogod DG. Appropriate use of local anesthetic for venous cannulation. *Anaesthesia.* 1992;47:210-212.
7. Van den Berg AA, Abeysekera RMMS. Rationalising venous cannulation; patient factors and lignocaine efficacy. *Anaesthesia.* 1993;48:84.
8. Van den Berg AA, Prabhu NV. Rationalising venipuncture pain: comparison of lignocaine injection, butterfly (21 gauge and 23 gauge) and Venflon (20 gauge). *Anaesth Intensive Care.* 1995;23:165-167.
9. Langham BT, Harrison DA. The pressor response to venous cannulation: attenuation by prior infiltration with local anaesthetic. *Brit J Anaesth.* 1993;70(5):519-521.

Needlestick Hazards for Anesthesia Personnel

by Arnold J. Berry, M.D., Professor of Anesthesiology, Emory University, and Elliott S. Greene, M.D., Associate Professor of Anesthesiology, Albany Medical College

Vol. 2, no. 2, 1996

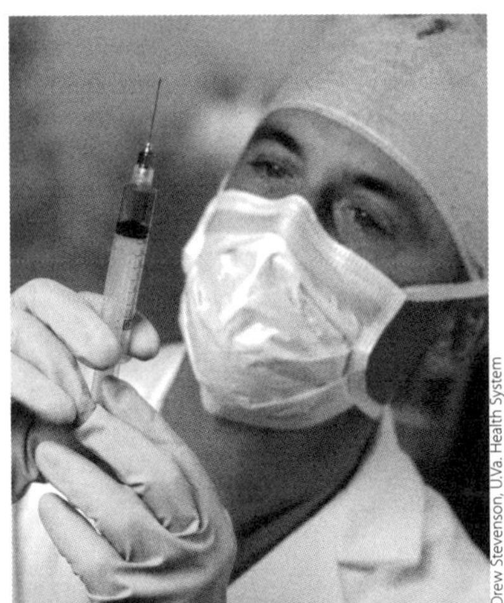

Drew Stevenson, U.Va. Health System

Background

In the United States, the anesthetic care of patients undergoing surgical, diagnostic, or other therapeutic procedures is provided by physician anesthesiologists (MDA), Certified Registered Nurse Anesthetists (CRNA), Anesthesiology Assistants (AA), and, in teaching institutions, by residents and students. The procedures performed by anesthesia personnel put them at risk for percutaneous or mucocutaneous exposures to blood and other body fluids.[1] This is supported by serologic surveys which have confirmed that anesthesia personnel are at high risk for occupationally acquired hepatitis B. In a series of studies performed prior to the general availability of the hepatitis B vaccine, MDA, CRNA, and AA personnel at several university-affiliated centers had a 19% seropositivity rate for hepatitis B markers[2], while physicians in anesthesia residency programs had a rate of 13%.[3] Since prevalence of hepatitis B may be considered as a surrogate indicator of the risk of significant exposures to blood and body fluids, anesthesia personnel also appear to be at increased risk for occupational transmission of other bloodborne viruses.

Anesthesia Procedures

To develop better strategies for prevention of blood exposures in anesthesia personnel, it is necessary to understand the procedures they perform and the associated risks. Many anesthetic procedures require the use of large-gauge, hollow-bore needles.[1,4] There are now several lines of evidence from studies of human immunodeficiency virus (HIV) transmission in healthcare workers (HCWs), and from laboratory investigations of simulated needlestick injuries, suggesting that transmission of bloodborne viruses is more likely after percutaneous injuries with blood-filled hollow-bore needles than with solid needles or mucocutaneous exposures.[5-10] This is consistent with the finding that hollow-bore needles carry a larger volume of blood and can transfer a greater number of viral particles than solid needles. Therefore, percutaneous injury with these devices would likely result in a greater risk of transmission of bloodborne pathogens. Cardo et al. have demonstrated that for HCWs who seroconverted after a percutaneous exposure to HIV-infected blood, risk factors included use of a large-gauge (greater than 18) hollow needle, deep percutaneous injury, and visible blood on the needle.[8] When these data were combined with comparable data from France and the United Kingdom, the risk factors for HIV seroconversion included a deep injury, visible blood on the device, a procedure involving a needle placed directly in a vein or an artery, and terminal illness in the source patient.[10]

Anesthesia personnel routinely use large-gauge, hollow-bore needles. The majority of patients having an anesthetic require an intravenous catheter for administration of anesthetic agents, intravenous fluids, blood or blood products. When large blood losses or intraoperative fluid shifts are anticipated, anesthesia personnel will insert one or more large-bore (14g to 9 Fr) intravenous catheters for rapid fluid administration. A local anesthetic is usually administered via a 25g or 26g needle prior to these invasive procedures, since the patient is often awake during vascular cannulation. When frequent blood sampling or continuous systemic arterial pressure monitoring is required, intra-arterial catheterization using a 20g or 22g catheter is performed.[11] The technique of placing a percutaneous intra-arterial catheter requires that the high-pressure blood return be controlled while the catheter is connected to flush solution. During this maneuver, the catheter stylet is usually placed on an adjacent work surface and must be retrieved later for disposal. This practice can result in accidental injury; if the catheter is inadvertently left on a patient's bed, it could also injure a worker at a later time.

For patients having major surgical procedures or for those with significant cardio-pulmonary pathology, a central venous or pulmonary artery catheter may be indicated for assessment of myocardial function. Often these devices are inserted prior to the induction of anesthesia to assess and optimize hemodynamics. Prepackaged kits for percutaneous placement of these central venous catheters contain multiple syringes and needles. At the conclusion of the procedures, the needles, including hollow-bore blood-filled needles, must be removed from the tray for disposal in appropriate containers. Again, anesthesia personnel may be injured while performing this task; in addition, needles left on the procedure tray may inadvertently be placed in a regular trash container, creating the possibility of a downstream injury to housekeeping staff.

For some surgical procedures, regional anesthesia (spinal or epidural) or nerve blocks may be indicated, or these techniques may be combined with general anesthesia for postoperative pain control. Procedural kits for administering regional or nerve block anesthesia contain several hollow-bore needles for administration of local anesthetic or placement of indwelling catheters. The sharps from these kits pose risks of percutaneous injury during use and disposal.

During the course of an anesthetic, multiple incremental intravenous injections of anesthetic agents are required. This necessitates recapping needles used to access intravenous administration sets. Alternatively, "piggy-back" infusions may be used requiring connections into primary intravenous lines. Many practitioners are now using needleless or protected needle systems for infusions and intravenous administration of anesthetic agents.

Types of Devices

In a study to quantify the types of needles used by anesthesia personnel, puncture-resistant needle disposal containers located in operating rooms on anesthesia carts or in the preoperative holding area were opened and the contents categorized by needle type.[4] Ninety-five percent of the sharps were hollow-bore needles, with the single largest categories being 18g to 23g hollow-bore needles and intravenous catheter stylet needles. When compared to the needles purchased for use throughout the hospital, anesthesia personnel used a larger proportion of 25g-26g hollow-bore needles, intravenous catheters, and spinal and epidural needles.[4] This pattern is compatible with the frequent use of small-gauge needles for intradermal local anesthesia prior to insertion of intravenous catheters or administration of spinal or epidural anesthesia. Comparing annual purchasing records for the hospital and the department of anesthesiology, anesthesia personnel used 19% of all intravenous catheters, 32% of central venous pressure insertion kits, and 29% of all pulmonary artery catheter introducer kits.[11]

Shortcomings of Data

There have been several observational studies attempting to characterize blood exposures to operating room personnel.[12,13] Although some of these exposures have included anesthesia personnel, most of the recorded exposures occurred in the operating room to non-anesthesia personnel. Many of the invasive procedures performed by anesthesia personnel take place in locations other than the operating room, such as the preoperative holding area, post-anesthesia care unit, critical care unit, radiology suite, labor and delivery suite, and pain clinics. Therefore, exposures occurring during actual surgery represent only a fraction of total exposures that might occur in relation to a surgical case. For example, to improve efficiency of turnover between operative cases, most centers now have a preoperative holding area where anesthesia personnel insert intravenous, intra-arterial, central venous, and pulmonary artery catheters for hemodynamic monitoring, and where they also perform nerve blocks and regional anesthetics prior to transporting the patient to the operating room. Observational studies limited to the operating room would not capture data on percutaneous injuries or other exposures to anesthesia personnel occurring in these other settings.

Prospective Study

Information on the types of devices and circumstances of contaminated percutaneous injuries (CPI) to anesthesia personnel was obtained during a study conducted at 11 university-affiliated hospitals.[14] Anesthesia personnel reported 105 CPI that were associated with 250,708 anesthetics and 12 CPI associated with non-anesthetic procedures. Ninety-four percent of the injuries involved a needle (79% hollow-bore). The needle devices responsible for the injuries were: needles on disposable syringes (42%), intravenous catheter stylets (21%), suture needles (20%), and others (17%). Based on the procedures performed, 30% of CPI involved blood-filled, hollow-bore needles; of that fraction, 46% were used to insert intravenous catheters, 29% to insert central venous or pulmonary artery catheters, 14% to obtain venous blood samples, and 11% to insert intra-arterial catheters.

Better Protection Needed

There are still many needs for safety improvements in the specialized equipment used by anesthesia personnel. Although many needleless or protected needle devices have been developed to reduce needlestick injuries in healthcare workers, only some of these are applicable to the practice of anesthesiology.[1] For example, safety intravenous catheters cannot be used for intra-arterial insertion or used with the Seldinger technique for central venous cannulation. Protective features needed for intravenous catheters include a syringe attachment that would allow for aspiration through the stylet needle during insertion, and a mechanism to control splattering of blood from the catheter, especially when used for intra-arterial catheterization. There are no regional anesthesia needles (spinal or epidural) with safety features. Since multiple injections are often made when administering local anesthesia, the design of an acceptable safety product would require maintenance of sterility and protection against injury between steps of the procedure.

Because of the procedures that they perform, anesthesia personnel are at higher risk of percutaneous injuries from blood-filled, hollow-bore needles than other operating room personnel. Injuries from these devices are associated with an increased likelihood of transmission of bloodborne pathogens compared to injuries from other contaminated needles or from mucocutaneous exposures. By understanding devices used by anesthesia personnel and the mechanism of needlestick injuries, new engineering controls or work practices can be developed to limit exposures in this specialized group of healthcare workers.

References

1. Berry AJ, Greene ES. The risk of needlestick injuries and needlestick-transmitted diseases in the practice of anesthesiology. *Anesthesiology*. 1992;77:1007.
2. Berry AJ, Isaacson IJ, Kane MA, et al. A multicenter study of the prevalence of hepatitis B viral serologic markers in anesthesia personnel. *Anesth Analg*. 1984;63:738.
3. Berry AJ, Isaacson IJ, Kane MA, et al. A multicenter study of the epidemiology of hepatitis B in anesthesia residents. *Anesth Analg*. 1985;64:672.
4. Berry AJ. The use of needles in the practice of anesthesiology and the effect of a needleless intravenous administration system. *Anesth Analg*. 1993;76:1114.
5. Centers for Disease Control and Prevention. Table 11, Healthcare workers with documented and possible occupationally acquired AIDS/HIV infection, by occupation. *HIV/AIDS Surv Rep*. 1995; 7(1):15.
6. Mast ST, Woolwine JD, Gerberding JL. Efficacy of gloves in reducing blood volumes transferred during simulated needlestick injury. *J Infect Dis*. 1993; 168:1589.
7. Shirazian D, Herzlich BC, Mokhtarian F, et al. Needlestick injury: blood mononuclear cells, and acquired immunodeficiency syndrome. *Am J Infect Control*. 1992;20:133.
8. Cardo DM, Srivastava PU, Ciesielski C, et al. Case-control study of HIV seroconversion in healthcare workers after percutaneous exposures to HIV-infected blood.(abstract) *Infect Control Hosp Epidemiol*. 1995;16:536.
9. Bennett NT, Howard RJ. Quantity of blood inoculated in a needlestick injury from suture needles. *J Am Coll Surg*. 1994; 178:107.
10. Centers for Disease Control and Prevention. A case-control study of HIV seroconversion in health-care workers after percutaneous exposure to HIV-infected blood-France, United Kingdom, and United States, January 1988-August 1994. *MMWR*. 1995;44:92.
11. Berry AJ. Injury prevention in anesthesiology. *Surg Clin North Am*. 1995;75:1123.
12. Panlilio AL, Foy DR, Edwards JR, et al. Blood contacts during surgical procedures. *JAMA*. 1991;265:1553.
13. Gerberding JL, Littel C, Tarkington A, et al. Risk of exposure of surgical personnel to patients' blood during surgery at San Francisco General Hospital. *N Engl J Med*. 1990; 322:1788.
14. Greene ES, Berry AJ, Jagger J, et al. Prospective multicenter study of needlestick and other percutaneous injuries in anesthesia personnel.(abstract) *Infect Control Hosp Epidemiol*. 1995;16:537.

Nursing Students at Risk

Survey of Nursing School Graduates Reveals Injury Risk and Shortcomings in Post-Exposure Protocols for Student Nurses

by Susan D. Schaffer, Ph.D., F.N.P., C.S.

Vol. 2, no. 3, 1996

Introduction

Nurses account for the largest number of documented occupational HIV seroconversions in the United States[1], in part because they are a large occupational group, but also because they perform a large proportion of high-risk needle procedures.[2] Although nursing students perform many of the same procedures as RNs, little is known about their at-risk blood exposures during clinical training. Because of the seriousness of the infection risks, however, there is increased interest in determining the causes and outcomes of exposures, and in improving strategies for prevention.

Nursing students have not been a primary focus of published studies, although a British study compared student nurses with registered nurses and found that they were twice as likely to experience a sharps-related injury (17.0 compared to 8.1 incidents per 1,000 employees per year).[3] A U.S. study found that, during a 24-month period, 21 of 180 nursing students (11.7%) at a large medical center sustained percutaneous injuries.[4]

Of additional concern is the adequacy of post-exposure follow-up provided to nursing students. Because students are not employees, they are not subject to the 1991 OSHA bloodborne pathogens standard. In many institutions nursing students do not report exposures to employee health services, but rather to student health services or private healthcare facilities where their numbers and outcomes are difficult to track.[5]

Study Description

In order to better document at-risk exposure incidents among nursing students, a questionnaire using modified EPINet methodology[6] was mailed to all registered nurses (1,580) who passed the 1993 nurse licensure examination in Virginia. The questionnaire asked about percutaneous and mucocutaneous exposure incidents that occurred while the nurses were students, and about environmental factors related to these incidents. "Exposure incidents" were defined as percutaneous contact with blood or with body fluid contaminated with blood, or splashes of blood or blood-contaminated body fluid to mucosa or non-intact skin. Respondents were also asked to describe post-exposure follow-up procedures in place for student nurses, and any morbidity that might have resulted from their exposures.

Exposure Incidents

Five hundred and eighty completed questionnaires were returned, representing a 36.7% response rate. Forty-two (7.2%) of the respondents described a total of 56 exposure incidents during training, a rate of .1 total incidents per 4.5 semesters, which is the mean duration of student clinical training. Thirty-four of the exposures were percutaneous and 22 were mucocutaneous. The percutaneous exposure rate was .06 exposures per student per 4.5 semesters.

Causes of Exposure Incidents

Forty-four respondents provided detailed descriptions of their exposures (31 percutaneous and 13 mucocutaneous). The majority of these incidents occurred during medical-surgical rotations in patients' rooms. Ten of the percutaneous injuries occurred during use of a sharp device: five during insertion of a needle through the skin, and five between steps of a multi-step procedure (**see Figure 1**). Seventeen percutaneous incidents occurred post-use: eight during disposal, four during disassembly, and five during recapping. Three incidents occurred due to improper disposal (for example, a needle left on a table), and one involved a used razor. Eleven of the 31 percutaneous injuries (35.7%) involved needles used to access intravenous lines.

Of the 13 mucocutaneous incidents described, seven reported exposures to non-intact skin, five experienced eye splashes, and two reported mouth/nose exposures. Four incidents occurred during routine procedures in which personal protective equipment should have been used in accordance with CDC guidelines[7], but was not. The remaining incidents would have been difficult to anticipate, such as a dropped blood tube.

Reporting

Of the 44 described incidents, 24 percutaneous and 9 mucocutaneous exposures were reported to the clinical facility or school of nursing; 11 (25%) were not reported. All but one of the 11 respondents who did not report their exposure incidents (and one who reported to the facility, but not to the instructor) said it was because of concern about negative faculty reaction. Seven respondents noted that they did not know how to report, or believed the incident to be low-risk.

Follow-up

The post-exposure follow-up of nursing students was compared to that required by the OSHA bloodborne pathogens standard. OSHA regulations require testing of the source patient and healthcare worker for HIV and hepatitis B (HBV), repeat testing of the healthcare worker when the source patient is HIV- or HBV-positive, post-exposure prophylaxis of the healthcare worker according to U.S. Public Health Service recommendations, counselling related to exposure risks, and evaluation of reported illnesses.[8]

Sixteen incidents were eliminated because they may have occurred prior to the March 1992 implementation of the OSHA standard. Of the 17 remaining incidents, 12 were not followed according to OSHA guidelines. None of the nine mucocutaneous incidents was followed according to OSHA guidelines. Most respondents reported that they were not counselled and that HIV/HBV testing was not performed on themselves or the source patient. Of particular concern were two respondents exposed to blood or bloody sputum from known HIV-infected patients; neither received baseline or follow-up HIV or HBV testing.

Morbidity

Although there were no occupational HIV or HBV infections reported among respondents, two respondents claimed that eye exposures resulted in other types of infections within two weeks of the incident. One respondent reported respiratory symptoms and pneumonia following a blood splash to her eyes; the other reported fever, abdominal pain, elevated liver function tests and positive blood culture for an unspecified organism after an HIV-positive patient with pneumonia coughed sputum into her face, including her eyes. She stated that although she received baseline testing for HBV, she received no counselling or testing for HIV. Because respondents were anonymous, it was not possible to obtain further documentation or verification of reported infections.

Figure 1. Mechanism of Percutaneous Injuries in Student Nurses

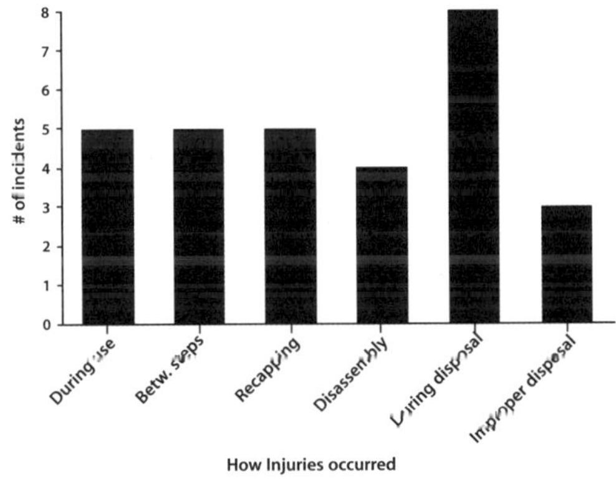

International Healthcare Worker Safety Center, University of Virginia

Discussion

Although percutaneous injury rates among nursing students in this study were lower than those reported in previous studies, the rates reported in different studies may not be directly comparable. Different denominators have been used for rate calculation (for instance, clinical contact hours, or number of needle procedures performed). Differences in underreporting rates in various studies is another factor that can affect comparability.

Although the respondents in this study were self-selected and may not be fully representative of all nursing students, two important conclusions can be drawn. First, while exposure to bloodborne pathogens is one of the most serious workplace hazards faced by nursing students, their exposure risks are not well documented by the appropriate institutional authorities. This study highlights the need for nursing schools to develop post-exposure protocols and to educate student nurses about reporting procedures. Policies must be developed that afford the same protections for nursing students as those already mandated for practicing nurses, in compliance with the OSHA bloodborne pathogens standard.

Second, two respondents attributed occupational infections to eye exposures. Although splash incidents usually generate less concern than sharp-object injuries, the potential for pathogen transmission from mucous membrane, and specifically conjunctival, exposures clearly exists. One occupational HIV infection and one HCV infection reported in the literature have been attributed to eye exposures.[9,10] The importance of providing protective eyewear and other personal protective garments and enforcing their use during student clinical experiences should be stressed.

Acknowledgement

This study was supported in part by grants from Becton Dickinson and Company and Sigma Theta Tau, Epsilon Chi Chapter.

References

1. Centers for Disease Control and Prevention. Table 16, Healthcare workers with documented and possible occupationally acquired AIDS/HIV infection, by occupation. *HIV/AIDS Surv Rep.* 1995;7(2):20.
2. DeCarteret JC. Needlestick injuries: An occupational health hazard for nurses. *AAOHN J.* 1987;35(3):119–123.
3. Anonymous. Sharps injuries in hospitals. *Occup. Health.* 1982; 34(11):502–508.
4. Yassi A, McGill M. Determinants of blood and body fluid exposure in a large teaching hospital: Hazards of the intermittent intravenous procedure. *Am J Infect Control.* 1991;19(3):129–135.

5. Tereskerz PM, Crane C. Percutaneous injuries among medical students. *Adv Exposure Prev.* 1995;1(5):10–11.
6. Becton-Dickinson Safety Compliance Initiative. Exposure prevention information network. Stone Mountain, GA. 1992.
7. Centers for Disease Control. Guidelines for prevention of transmission of human immunodeficiency virus and hepatitis B virus to health-care and public-safety workers. *MMWR.* 1989;38(S-6):2–33.
8. Department of Labor/OSHA. Occupational exposure to bloodborne pathogens: Final rule. *Fed Regist.* 1991;56(235):64004–64182.
9. Gioannini P, Sinicco A, Cariti G, et al. HIV infection acquired by a nurse. *Eur J Epidemiol.* 1988;4:119–120.
10. Sartori M, La Terra G, Agliette M, et al. Transmission of hepatitis C via blood splash into conjunctiva. *Scand J Infect Dis.* 1993;25:270–271.

Trends in U.S. Patents for Needlestick Prevention Technology

by Donald Kelly (formerly Patent Examining Group Director, United States Patent and Trademark Office) Vol. 2, no. 4, 1996

The hypodermic syringe, which celebrated its 150th anniversary in 1995, is a remarkably simple piece of technology that revolutionized the practice of medicine. Decade after decade, design enhancements have been introduced to meet new therapeutic and diagnostic needs, improve patient comfort and safety, and increase manufacturing efficiency. For all the benefits that syringes and medical needles have brought to patients, however, they have also been a vehicle for the transmission of infectious diseases to healthcare workers. The past decade has witnessed a proliferation of innovations in needle technology as independent, corporate, and academic inventors around the world, motivated by the AIDS pandemic, have risen to the challenge of preventing needlestick injuries. The following article documents historic trends in U.S. patent activity relating to needlestick prevention and provides insight into the process by which ideas are transformed into better products.

One of the most satisfying motives for pursuing innovation in technology is an inventor's desire to solve a life-threatening problem. This has been a driving force for hundreds of inventors who have directed their passionate energies towards solving one of the thorniest problems facing healthcare professionals: the hazard of needlestick injuries. The tragedy of occupational transmission of potentially deadly pathogens has drawn the attention of corporate and independent inventors, whose efforts are reflected in the rapid rise in U.S. patents in the field of needle safety and needle disposal technology.

Few problems have prompted a response of similar magnitude. Matched only by the accelerated pace of innovation during the energy crisis of the 1970s, and more recently by advances in genetic research, the wave of needle safety patent applications crossing the desks of examiners at the Patent and Trademark Office (PTO) has been truly remarkable. More than 1,000 U.S. patents have been issued in the field of needlestick prevention technology in the past decade, over 800 of which were issued in the past five years (**Figure 1**). This compares to more than 6,000 patents issued for syringe designs in general since 1970, the year the PTO began keeping computerized records. The Patent Office now has one examiner devoted full-time to needlestick prevention technology, and several others who specialize in this area part-time.

The structure of the U.S. patent system motivates inventors to disclose new inventions to the public. The reward for disclosure is the grant of exclusive rights for a period called the "patent term." Recent legislation extended this term from 17 to 20 years from the original filing date of the patent application.

An independent inventor's cost for applying for a patent runs from $3,200 to $4,200, including a filing fee, an issue fee, and attorney's fees. Fees for corporations are higher. After a patent is issued, an inventor must pay maintenance fees at specified intervals; those fees total about $3,000 over a 12-year period. Failure to pay maintenance fees terminates all patent rights. Foreign nationals as well as American citizens can apply for U.S. patents, but U.S. patents are enforceable only in this country. U.S. patents give inventors the right to exclude others from making, using or selling the covered invention in the U.S. They provide the patent holder with a measure of market control and a means for attracting investment in the new technology.

An early syringe patent issued to Charles T. Sage in 1843, designed to deliver "an irritant when used in producing inflammation in the cure of hernia." Courtesy of the U.S. Patent and Trademark Office.

For many years the PTO, located in Arlington, Virginia, has maintained a rapidly growing database known as the "patent search file." The database is organized by classes

Figure 1. U.S. Patents Issued on Needlestick Safety Devices 1984 through 1995

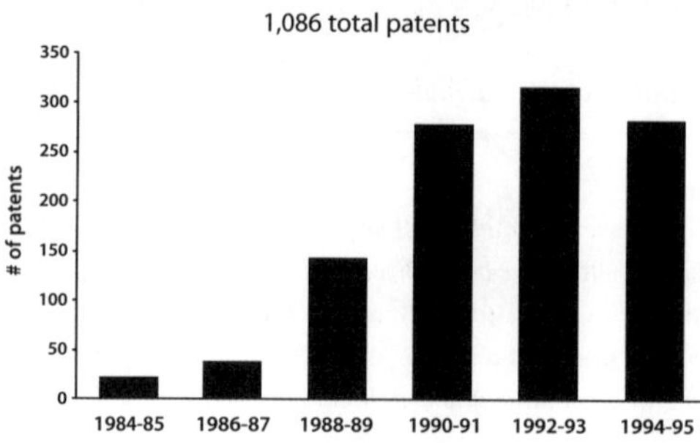

21% of patents were issued to inventors in foreign countries. Top 10 foreign countries issued patents, in order: Canada, France, Italy, U.K., Taiwan, Japan, Australia, Germany, the Netherlands, Spain.

Source: D. Kelly, U.S. Patent and Trademark Office

and subclasses of technology-specific categories. Taken as a whole, this collection defines the technological world as we know it. Within each class, or combination of classes and subclasses, we can determine the characteristics of discrete technological advances, and analyze technology trends. Class 128, subclass 919, and Class 604, subclasses 110, 192, and 263 of the patent search file include most patents related to the prevention of needlestick injuries. These classes include inventions such as non-reusable hypodermic needles, end-caps, sheaths and safety shields, retractable needles, barrel traps, flashback ventilation caps, sleeve adapters, needle grinders and melters, passive sliding sleeves, retracting cannulas and locks, protective catheters, and a multitude of related devices and techniques.

This growing body of technology shares some interesting characteristics. First, nearly 56% of these devices and techniques come from independent, non-corporate inventors, as compared to 20% for other kinds of technology. Many of these independent inventors are close to the frontlines of healthcare and are driven by concern for the life-threatening consequences of bloodborne pathogen transmission. Another common characteristic, based on anecdotal reports of patent examiners, is that inventors in this field are more interested in making their solutions known to the public than in making money from their ideas. This is unusual in our profit-driven economy, and speaks volumes about the culture of frontline caregivers.

A third distinguishing characteristic of patents in the field of needlestick prevention has to do with country of origin. Nearly 80% of needlestick-related patents are issued to American inventors, as compared to approximately 52% for patented technology overall. While the needlestick problem is certainly global in scope, the solutions appear to be U.S.-driven, as is the case with most technology in the medical field. The top five foreign countries with patent holders in this field are Canada, France, Italy, the United Kingdom, and Taiwan.

Although more than 1,000 U.S. patents have been issued in the field of needle safety technology in the past decade, far fewer safety devices have made it into the hands of healthcare workers. As in most other fields of technology, only a fraction of patented needle safety technology translates into marketable products. Holding a patent and bringing a new product to market are two different things. Some patents describe well-intentioned but unrealistic technology that would be difficult or impractical to produce. Other patents are not intended for production but are filed in order to gain broader patent protection for devices already in development. The willingness of a company to develop and market a new product depends, in part, on how secure its patent protection is. Does the patent provide broad coverage of the technology or is it possible to circumvent the patent with a minor design modification? How many years of the patent term remain before it expires? Is the patent vulnerable to legal challenge? Some good devices never make it past these hurdles. And those that do must still survive the manufacturing, marketing, and distribution demands before staking a claim in today's competitive healthcare market place.

Our analysis of patenting trends related to needlestick prevention technology shows that the surge of patent activity has leveled off in recent months, a trend most apparent in the individual inventor category. The reason for this recent trend is not known. But we do know that the creative efforts already directed toward solving this problem have been immense. The records of the U.S. Patent and Trademark Office provide historic evidence of the existence of the technology and the will to make the healthcare workplace safe from the risk of occupational transmission of bloodborne pathogens.

Simultaneous Transmission of HIV and HCV from Single Needlestick Injury

by Jane Perry, M.A.

Vol. 2, no. 7, 1996

The New England Journal of Medicine recently reported a case of simultaneous transmission of HIV and HCV from a single needlestick injury.[1] It is the first such case documented in the medical literature, although there has been one other case of simultaneous HIV and HCV transmission reported in Italy (G. Ippolito, personal communication), and anecdotal reports of an occupational exposure in France resulting in HIV, HBV and HCV infection.

The case reported in the *Journal* involved a 48-year-old healthcare worker who sustained a deep injury in July 1990 while drawing blood from a patient with AIDS. The report states that "blood also spilled from the collection tube into the spaces between the cuffs of the healthcare worker's gloves and her wrists and onto her hands, which were chapped with open cracks." The source patient had a history of intravenous drug use and, at the time of the needlestick, was receiving zidovudine therapy and was not known to have HCV infection. The healthcare worker, who reported no behavioral or transfusion-related risk factors for HIV, declined postexposure prophylaxis with zidovudine. Because the source patient was not initially known to be HCV-infected, no baseline HCV testing was done on the healthcare worker.

In addition to the simultaneous transmission of two bloodborne pathogens, the case was remarkable in that seroconversion times for both HIV and HCV were unusually long (in the case of HIV, one of the longest reported to the CDC), and also for the rapid progression of disease in the healthcare worker. HIV seroconversion occurred between 8 and 9-1/2 months after exposure, and seroconversion to HCV between 9-1/2 and 13-1/2 months after exposure. Eighteen months after HIV seroconversion and 28 months after the needlestick, the healthcare worker developed hepatic coma and progressive renal failure, and died. The postmortem exam showed cirrhosis of the liver with no evidence of opportunistic infection or cancer. This is in contrast to a retrospective study cited in the article that found the mean time to the development of cirrhosis after HCV infection to be 20 years.[2] The article noted that the unusually rapid course of disease in this case may have been related to acquiring two infections at once, since there is evidence of "pathogenic interaction" between the two viruses.

The strains of HIV and HCV infecting the source patient and the healthcare worker were compared by polymerase chain reaction and genetic sequencing, and found to be closely related.

A Public Health Service task force on the management of occupational exposure to HIV considered this case when reviewing data on length of HIV seroconversion window, according to a personal communication from a CDC official cited in the article. The guidelines developed by the group, published in the *MMWR*[3], did not recommend routine HIV serologic follow-up beyond six months postexposure, because of the unlikelihood of detecting a new infection. However, in the case of a simultaneous exposure to HIV and HCV, evaluation for delayed seroconversion may be advisable.

References

1. Ridzon R, Gallagher K, Ciesielski C, Mast, EE, Ginsberg MB, Robertson BJ, Luo C-C, DeMaria A. Brief report: simultaneous transmission of human immunodeficiency virus and hepatitis C virus from a needle-stick injury. *N Engl J Med.* 1997;336:919–922.
2. Kiyosawa K, Sodeyama T, Tanaka E, et al. Interrelationship of blood transfusion, non-A, non-B hepatitis and hepatocellular carcinoma: analysis by detection of antibody to hepatitis C virus. *Hepatology.* 1990;12:671–675.
3. CDC. Update: provisional Public Health Service recommendations for chemoprophylaxis after occupational exposure to HIV. *MMWR.* 1996;45:568–472.

Occupationally Acquired HIV: The Vulnerability of Healthcare Workers Under Workers' Compensation Laws

by Patti Miller Tereskerz, M.S., J.D., Ph.D., and Janine Jagger, M.P.H., Ph.D.

Vol. 3, no. 3, 1998

"One of the issues for someone like me, getting workers' comp, is that I cannot sue my employer .... [T]rying to discuss compensation intellectually for something like this is a challenge, because there is none. There is no compensation."[1]

—Jane Doe (pseudonym of a nurse at San Francisco General Hospital who is HIV positive as a result of an occupational needlestick injury)

Introduction

There are over four million healthcare workers in the United States at risk of occupational infection from bloodborne pathogens, including human immunodeficiency virus (HIV), hepatitis B virus (HBV) and hepatitis C virus (HCV).[2] Approximately 800,000 needlesticks and other sharp injuries from contaminated medical devices occur in healthcare settings each year, of which an estimated 16,000 are HIV-contaminated.[3] Injuries occurring outside of hospitals in clinics and private office settings remain undocumented, and there is serious under-reporting of accidental exposures to blood and body fluids, with as many as 34% of exposed workers not reporting percutaneous injuries and up to 75% not reporting mucocutaneous exposures.[4] Such underreporting may result from healthcare workers' fear of reprisal and job discrimination or from institutions' failure to provide effective reporting systems. Most healthcare workers will sustain, on average, several needlesticks or other blood exposures during their careers. Many of those exposed will file claims for workers' compensation benefits.

Healthcare workers are required to work with inherently dangerous medical devices that put them at risk of infection from bloodborne pathogens. Yet many are unaware of the limitations of the workers' compensation system until they apply for benefits.

Workers' compensation is governed and defined by state law. Disabled employees receive benefits for qualifying occupational illnesses and injuries, regardless of any fault of the employee. The employee also does not have to prove employer negligence to receive benefits. Employers, in turn, are usually insulated from liability for an occupationally acquired illness or injury.

Workers' compensation benefits are potentially available to healthcare workers occupationally infected by bloodborne pathogens. This article focuses on healthcare workers who are already occupationally infected with HIV, and uses HIV as a vehicle to examine limitations of workers' compensation laws. However, many of the shortcomings identified apply to other occupational diseases as well.

Limitations of Workers' Compensation Laws

The Exclusive-Remedy Provision

Workers' compensation laws have an exclusive-remedy provision that precludes employees from bringing private suit against employers to recover damages for an occupational disease or injury.[5] There are few published cases in which workers' compensation laws have been applied to HIV exposures or infections. Of the few decisions handed down, however, workers' compensation has been upheld as the exclusive remedy for employees whose occupational exposure to HIV resulted in compensable workers' compensation claims.[6,7] The practical result of the exclusive-remedy provision of workers' compensation laws has been to insulate healthcare institutions from liability for failing to provide available medical technology incorporating injury- or exposure-prevention features or for continuing to provide medical products known to be unsafe for employee use.

This problem is best illustrated by a Montana case[7] in which the claimant brought suit against the medical center where he worked on the grounds that it required him to use a defective medical device, which was responsible for an injury he sustained. The claimant was injured by a needle included with an arterial blood gas kit that he used to draw blood from a patient infected with HIV. The claimant's supervisor agreed to accept from a Radiometer America, Inc., sales representative, at reduced or no cost, residual stock of arterial blood gas kits known to have a manufacturing defect that presented a risk of needlestick injury to users. The defective kits were the only ones available for use in the unit where the claimant worked.

The claimant alleged that the sales representative and supervisor explained to some, but not all, of the workers using the kits that the devices were defective and had to be handled in a particular way to avoid injury. The inventoried kits were stored in boxes displaying the handwritten comments, "Free rejects?," "Yes," and the warning, "When using needle may screw past it's [*sic*] mark into hub of serenge [*sic*]—can still use, but be aware of this."[7] Despite evidence that the claimant's injury was caused by the manufacturing defect, the court held that the workers' compensation exclusive-remedy provision barred the claim.

This case illustrates that under the exclusive-remedy

provision of workers' compensation laws, there is little legal incentive for healthcare institutions to purchase safety equipment that would reduce employees' injury risks. Most workers' compensation laws shield healthcare institutions from liability even when, to save money, defective equipment has been knowingly purchased and used and has resulted in injury to an employee. As one legal analyst states:

> Where some employers can avoid more costly protections for their employees without incurring additional liability, they usually will do so. Employers generally will act only if given the monetary incentive to do so.[8]

While empirical data suggest that employers who fail to protect their employees can incur increased premiums for workers' compensation insurance, the myriad of complicated factors that determine an employer's premium rate do not always accurately reflect the employer's safety record.[9] Both aggregate safety data and empirical studies support the conclusion that the high costs of workers' compensation have not motivated large numbers of employers to take injury prevention seriously, and occupational injury rates have not declined.[9]

Although negligence claims are precluded, there are two recognized exceptions to the exclusive-remedy provision.[10] First, if an employer intentionally injures an employee, the employee may elect to bring a private suit against the employer.[11] However, an employer's failure to institute safety measures, or even the intentional removal of existing safety measures, has not been sufficient to demonstrate an intent to injure the employee, even if such actions result in injury. Therefore, the intent to injure must be distinguished from the intent to remove or not to institute a safety measure.[10]

The second exception, recognized by some states, applies when the employer fails to comply with the terms of the Workers' Compensation Act or is not a subscriber to a workers' compensation program.[10] An employer who does not participate in a workers' compensation program is not protected against suits for negligence and other causes of action.

Meeting the Definition of an Occupational Disease

Whether an employee who has contracted an occupational infection qualifies for workers' compensation benefits depends on the relevant statutory definition of compensable occupational disease or injury. Hepatitis B has been used as a model for HIV on issues of legal liability.[12] Therefore, it is likely that any precedents set for occupationally acquired HBV will serve as the basis for court decisions for occupationally acquired HIV.

Conflicting decisions have been handed down in cases in which the court was asked to decide whether HBV met the definition of an occupational disease under workers' compensation laws. Decisions have varied as a result of different statutory language in the jurisdictions where the cases are filed.

Some states have strict and specific definitions of "occupational disease." For example, in a Georgia case, *Fulton-DeKalb Hosp Authority v Bishop*, benefits were denied on the grounds that the claimant was unable to show that he or she did not have substantial exposure to the disease outside of employment or was unable to prove that HBV is not an "ordinary disease of life" to which the public is exposed.[13]

Other states have less stringent statutory requirements. In a Florida case, *Wuesthoff Memorial Hospital v Hurlbert*,[14] the following criteria were laid out by the court:

1. the disease must be actually caused by employment conditions that are characteristic of and peculiar to a particular occupation;
2. the disease must be actually contracted during employment in the particular occupation;
3. the occupation must present a particular hazard of the disease occurring so as to distinguish that occupation from usual occupations, or the incidence of the disease must be substantially higher in the occupation than in usual occupations; and
4. if the disease is an ordinary disease of life, the incidence of such a disease must be substantially higher in the particular occupation than in the general public.[14]

In this case, the court found that there was substantial evidence that the claimant's HBV was occupationally acquired and held that it was not necessary to demonstrate that a specific exposure caused the infection. The court held that the following evidence was sufficient to satisfy the four requirements outlined above: (1) conditions characteristic of the claimant's occupation as a laboratory technologist exposed him to HBV; (2) the claimant contracted the disease while employed in the laboratory; (3) the incidence of HBV is substantially higher in the claimant's occupation than in other occupations; and (4) the incidence of HBV is higher in the claimant's occupation than in the general public.[14]

The Florida statute in this case (like the Georgia statute recited in *Fulton-DeKalb*) defined an occupational disease as excluding all ordinary diseases of life to which the public is exposed. However, the Florida statute included a qualifying provision, which reads, "unless the incidence of the disease is substantially higher in the particular trade, occupation, process, or employment than for the general public."[11] This qualifier allowed the Florida appellate court to rule in favor of the claimant.

A North Carolina appellate court reached a similar conclusion in awarding workers' compensation benefits to the widow and dependents of a deceased laboratory technician who often handled blood specimens that spilled on his fingers and who contracted serum hepatitis.[15] The court rejected the notion that an illness cannot qualify as an occupational disease because it is not unique to the injured employee's profession, noting that this has been the practice in several other jurisdictions.[16] Furthermore, this court did not require proof of a causal connection between a specific exposure and the resulting infection, noting that the assumption of causation must frequently be based on circumstantial evidence, including:

1. the extent of exposure to the disease or disease-causing agents during employment;
2. the extent of exposure outside employment;
3. absence of the disease prior to the work-related exposure as shown by the employee's medical history.[15]

Meeting the definition of an occupational disease is a common problem for those seeking worker's compensation. A Department of Labor study on occupational diseases found that "only five percent of those severely disabled from an occupational disease receive workers' compensation benefits," in part because of the difficulty of establishing the work-relatedness of the illness.[17] Only 5% of income support for those severely disabled by an occupational disease is provided by workers' compensation, with the major sources of support coming from Social Security (53%), pensions (21%), veterans benefits (17%), welfare (16%), and private insurance (1%).[17] (Some people receive support from more than one source.) Twenty-five percent of severely disabled workers receive no income support payments at all.[17] For those disabled workers who receive workers' compensation payments, the average disabled worker must wait one year before receiving the first benefits payment.[17]

Amount of Workers' Compensation for Occupationally Acquired Diseases

Recovery under workers' compensation laws usually does not provide full restitution. In contrast, while the claimant in a civil suit has no guarantee of success, damages, if they are recovered, may compensate the victim not only for lost income and medical expenses but also for pain and suffering. The goal of providing recovery in civil suits is to make victims "whole," that is, to compensate them to the fullest extent possible for their loss. In comparison, workers' compensation generally only pays for medical expenses and a portion of the victim's wages, and provides no compensation for pain and suffering. In addition, a workers' compensation claimant is precluded from recovering punitive damages in the face of gross negligence.

Courts have rendered different decisions with regard to allowing claims for emotional distress following potential exposure to HIV.[18-20] Scholarly review of this topic recommends against allowing such claims.[21]

The Virginia Workers' Compensation statute[22] is typical and will be used to illustrate the limited level of compensation provided for a worker who is incapacitated as a result of occupational exposure to HIV.

In Virginia, an employee's compensation is fixed at 66.66% of his or her average weekly wages for the year prior to the date of injury, with a minimum of not less than 25% and a maximum of not more than 100% of the average weekly wage in the state.[23] This means that healthcare professionals with higher incomes receive only a small percentage of their current income.

In part, the rationale for providing only 66.66% of the worker's salary is that the compensation benefits, unlike wages, are not taxed. Unfortunately, this provision has an adverse impact on those who earn the lowest wages. Healthcare workers in the lowest tax brackets may pay little or no federal and state income tax. Consequently, at the lowest end of the salary scale, workers may be compensated at a level that represents as much as a one-third reduction in net income.

Second, in Virginia there are no provisions for lifetime benefits for those who contract HIV as a result of occupational exposure, except in the unusual case where the HIV infection manifests itself in ways for which there is permanent coverage as, for example, when the individual has permanent loss of both eyes or limbs as a result of the infection.[24] In Virginia, the maximum period of coverage is 500 weeks for the typical claimant with occupationally acquired HIV.[25]

Third, the right to compensation is lost in Virginia unless a claim for compensation is filed within two years after a positive test for HIV infection.[26] This statutory provision is particularly unreasonable when applied to employees who become HIV-positive but remain asymptomatic for years. The determining factor as to when a claim must be filed should be when benefits are needed, not when seroconversion occurs. Time limitations for filing a claim are also a problem for other occupational diseases where there is a considerable period of time between occupational exposure and manifestation of the disease.[17]

The problem of inadequate compensation is not limited to occupationally acquired HIV. In 1980, the Department of Labor reported that, even though workers' compensation benefits are supposed to replace two-thirds of an employee's lost income, a worker who became totally disabled for life and was able to prove that the disability was occupationally induced received an average of approximately $9,700 in total compensation compared with expected future earnings of $77,000.[17]

An HIV-positive healthcare worker runs the risk of revealing his or her HIV status when filing a workers' compensation claim and may jeopardize job security and confidentiality. Confidentiality is of extreme importance in the United States, where a worker's right to employment is not guaranteed by law and where there remain serious problems with discriminatory practice in employment and in obtaining and keeping health insurance.[27-30] Because disclosure of HIV status is required in order to qualify for benefits, healthcare workers face a difficult choice: either risk forgoing benefits by failing to file a claim within the designated time period, or risk the potentially adverse consequences of revealing HIV status before becoming symptomatic.

The confidentiality of workers' compensation records is

an issue of current concern. In the past, workers' compensation records have been placed in the public domain for the purpose of supporting workplace-safety improvements.[31] In Virginia, a statute was adopted barring public access to these records. However, the records may still be open to the Employment Commission and any party that convinces the Workers' Compensation Commission of a legitimate interest in the records.[32] Hence, a healthcare worker who may need future benefits has real disincentives to filing a claim owing to concerns about the incomplete protection of confidentiality.

The extensive struggle of Jane Doe, who is quoted at the beginning of this article, to maintain the confidentiality of her workers' compensation records is illustrative. Initially, no assurance was provided that the number of individuals with access to her records would be limited. For two years, while this issue was being resolved, she received no benefits. Finally, as a result of media attention, the mayor of San Francisco intervened to restrict to four the number of people at the workers' compensation office who would know her identity.[1]

Conclusions

Workers who are occupationally exposed to or infected by HIV and other bloodborne pathogens are vulnerable to being left without adequate workers' compensation benefits, or even without any benefits at all. A major hurdle for healthcare workers is meeting the definition of an occupational disease, which varies among jurisdictions and in some cases is so restrictive that infected healthcare workers are unlikely to qualify for benefits. The issue for those meeting the statutory requirements is one of obtaining adequate compensation.

Healthcare workers have limited recourse for redressing these deficiencies. Workers' compensation laws shield healthcare institutions from liability from civil suits for occupationally acquired HIV. The exclusive remedy provision minimizes employers' legal and economic incentives to purchase equipment or institute engineering systems that reduce exposure risk if the safer products or systems cost more than conventional products.

More equitable remedies are needed for healthcare workers who are occupationally exposed to or infected by HIV. We offer several recommendations for redressing these deficiencies.

Education of healthcare workers regarding the importance of reporting an occupational exposure to blood or body fluids is crucial, since ineligibility for workers' compensation benefits can be the direct consequence of failing to submit an incident report of an adverse blood exposure. All health professionals need to know the provisions contained in their state's workers' compensation laws, and should consider when collective efforts to amend restrictive qualifying criteria might be warranted on the basis of epidemiologic data. Professional organizations can play an important role in educating their members and in undertaking initiatives to amend existing legislation where necessary.

Revision of workers' compensation laws is needed to broaden options for redress for all workers, not just those with occupationally acquired HIV. Potential claimants should be allowed the option of either filing civil suit or applying for workers' compensation, when it can be demonstrated for any occupational disease or injury that the employer: (1) knowingly or recklessly removed safety equipment from the workplace; (2) failed to provide available and effective safety equipment in the workplace; or (3) knowingly or recklessly introduced unsafe equipment into the workplace. Legal scholars have criticized the workers' compensation system for its exclusive-remedy provision, because it allows unwarranted immunity for employers' willful and wanton misconduct for more than 10 years.[8,33] Changes in the workers' compensation system could productively begin with remedies for occupationally acquired HIV. The proposed amendments would provide legal incentives that currently do not exist for employers to provide protective equipment for employees.

We recommend the following changes in workers' compensation laws to provide more equitable remedies for workers with occupationally acquired HIV. We believe that these changes will also provide needed incentives for employers to supply protective equipment to employees.

First, the definition of "occupational disease" should be broadened where necessary so that every jurisdiction will allow compensation for employees who can reasonably demonstrate that they did not acquire the disease outside of work and that their occupation put them at increased risk of the disease. Provisions that compel the claimant to prove that a specific occupational incident resulted in acquiring the disease and provisions that require the claimant to prove that the occupationally acquired disease is not an ordinary disease of life should be eliminated. This recommendation is not limited to HIV but should extend to all occupationally acquired diseases, given the problematic nature of qualifying for benefits for many occupationally acquired diseases.[17]

Second, the time limit for filing a claim should be increased for asymptomatic workers who become HIV positive as a result of an occupational exposure. The time limit in such cases should be extended until the employee is disabled. This recommendation applies to states that do not have what is known as a discovery statute (which allows claims to be filed when the occupational disease is discovered or manifests itself), but instead impose an arbitrary time limit after exposure for filing claims. Further study is necessary to determine where time limits need to be extended for other occupational diseases in which substantial time can elapse between exposure and manifestation of the disease.

Third, benefits should be provided for a worker who has

occupationally acquired HIV and is totally disabled, for as long as the worker is unable to work.

Fourth, confidentiality of claimants receiving benefits for occupationally acquired HIV must be guaranteed by withholding their identities from public disclosure. Disclosure should be limited to those with an absolute need to know the claimant's identity in order to administer benefits or medical care.

The need for special statutory requirements for HIV with regard to the third and fourth proposed amendments is justified on the basis of the extraordinary characteristics of HIV. Few diseases carry the stigma that HIV does, even now. Therefore, special statutory provisions are warranted to assure that HIV claimants' privacy rights are guaranteed. Otherwise, few valid claims will be made because of the claimant's fear of potential job discrimination and stigmatization by colleagues. Likewise, benefits should be provided for occupationally acquired HIV for the entire period of disability, because few diseases are so completely devastating with certain mortality. Precedent exists for awarding lifetime benefits in instances when occupational injuries have been particularly devastating.[34]

Increased attention to the problem of preventing occupational blood exposures is an important way to reduce the need of healthcare workers for workers' compensation benefits. It has been estimated that over 80% of needlesticks are potentially preventable by existing means that have yet to be implemented.[3] Measures to prevent occupational blood exposures and infections should be aimed at enforcing existing Occupational Safety and Health Administration regulations requiring employers to provide "engineering controls"[35] when such technology is commercially available. In addition, healthcare facilities must provide reporting systems that motivate and encourage workers to report injuries.

While HIV has been the focus of this article, at least 20 other bloodborne pathogens have been documented in the medical literature as being occupationally transmitted to healthcare workers—hepatitis B and hepatitis C being the most serious and frequently transmitted among them.[36] As was discussed, many of the issues raised here are also relevant to occupational infections by these other pathogens. The broad scope of this occupational hazard only serves to emphasize the need for change in preventing occupational blood exposures and in providing adequate and accessible compensation for those bearing the burden of risk at the front lines of healthcare.

References

1. AEP interview: Jane Doe, R.N. *Adv Exposure Prev.* 1995;1(2):5,10–11.
2. Occupational Exposure to Bloodborne Pathogens; Final Rule, 56 *Fed Regist* 64055 (1991) (codified at 29 CFR Part 1910.1030).
3. Jagger J, Pearson D. Universal precautions: still missing the point on needlesticks. *Infect Control Hosp Epidemiol.* 1991;12:211–213.
4. Roy E, Robillard P. Underreporting of accidental exposures to blood and other body fluids in health care settings=an alarming situation. In: *Proceedings of the Conference on "Bloodborne Infections: Occupational Risks and Prevention.* Paris, France:1995. Abstract.
5. Larson A. Exclusiveness of compensation remedy. In: *The Law of Workmen's Compensation.* Vol. 2A(suppl). New York, NY:Times Mirror Books;1995.
6. *Vallery v S. Baptist Hosp.,* 630 So 2d 861 (La. App. 1994).
7. *Blythe v Radiometer America Inc et al,* 262 Mont 464, 866 P2d 218 (1993).
8. Schroeder TD. Workers' compensation: expanding the intentional tort exception to include willful, wanton, and reckless employer misconduct. *Notre Dame Law Rev.* 1983;58:890–910.
9. Spieler EA. Symposium: new challenges in occupational health. Perpetuating risk? Workers' compensation and the persistence of occupational injuries. *Houston Law Rev.* 1994;31:119–264.
10. Abbott D. Workplace exposure to AIDS. *Maryland Law Rev.* 1989;48:212–245.
11. Larson A. Misconduct of employer. In: *The Law of Workmen's Compensation.* Vol. 2a(suppl) § 68. New York, NY:Times Mirror Books;1995.
12. Miller P, O'Connell J, Leipold A, et al. Potential liability for transfusion-associated AIDS. *JAMA.* 1985;253:3419–3420.
13. *Fulton-Dekalb Hosp Authority v Bishop,* 185 Ga App 771, 365 SE2d 549 (1988).
14. *Wuesthoff Memorial Hospital v Hurlbert,* 548 So2d 771 (1st Dist Ct Fla 1989), adopting test set out in Broward Indus Plating Inc v Weiby, 394 So2d 1117, 1119 (1st DCA Fla 1981).
15. *Booker v Duke Med Center,* 297 NC 458, 256 SE2d 189 (1979). [p475NC; p200SE]
16. *Young v City of Huntsville,* 342 So2d 918 (Ala Civ App 1976) cert denied, 342 So2d 924 (Ala 1977); *Aleutian Homes v Fischer,* 418 P2d 769 (Alas 1966); *State ex rel Ohio Bell Telephone Co v Krise,* 71 OO2d 226, 327 NE2d 756 (1975); *Underwood v Nat'l Motor Castings Division,* 329 Mich 273, 45 NW2d 286 (1951); *County of Cook v Ind Comm'n,* 54 Ill 2d 79, 295 NE2d 465 (1973); *Evans v Indian Univ Medical Center,* 121 Ind App 679, 100 NE2d 828 (1951); *Peterson v State,* 234 Minn 81, 47 NW2d 760 (1951); *Vanore v Mary Immaculate Hosp,* 260 App Div 820, 22 NY2d 350 (1940) *aff'd* 285 NY 631, 33 NE2d 446 (1941): Note, Occupational Disease and the Hospital Employee: A Survey. *Memphis State Univ Law Rev.* 1975;5:368.
17. Assistant Secretary for Policy, Evaluation and Research. *An Interim Report to Congress on Occupational Diseases.* Washington, DC: US Department of Labor; 1980.
18. Matsumoto AV. Reforming the reform: mental stress

claims under California's workers' compensation system. *Loyola at Los Angeles Law Rev.* 1994;27:1327–1365.

19. Zuber J. Comment: the employment-related emotional distress morass: confusing signals from California's courts and legislature. *Pacific Law J.* 1990:21:1035–1067.
20. Carpenter ML. Note: Petersen v Sioux Valley Hospital: reckless infliction of emotional distress. *SD Law Rev.* 1993;38:359–378.
21. Mariner WK. AIDS Phobia, public health warnings, and lawsuits: deterring harm or rewarding ignorance? *Am J Pub Health* 1995; 85:1562–1568.
22. Va Code Ann. §65.2–100 et seq. (Michie 1991).
23. Va Code §65.2–500 (Michie 1991).
24. Va Code Ann. §65.2–503(C) (Michie 1991).
25. Va Code Ann. §65.2–500(D)(Michie 1991).
26. Va Code Ann. §65.2–406(A)(4) (Michie 1994S).
27. Ostolaza Y. Severino v. North Fort Myers Fire Control District: AIDS discrimination in the workplace—will disclosure leave HIV-infected workers jobless? *Univ Miami Law Rev.* 1 1992;47:241–266.
28. McMartin JV. AIDS (HIV) and insurance: discrimination against HIV-infected individuals. *WL* 1990;357778.
29. Schatz B. The AIDS insurance crisis: underwriting or overreaching? *Harvard Law Rev.* 1987;100:1782–1805.
30. Glantz LH, Mariner WK, Annas GJ. Risky business: setting public health policy for HIV-infected health care professionals. *Milbank Q.* 1992;70(1)43–79.
31. Baig EF. Arising out of and in the course of employment: AIDS and worker's compensation law. *J Fla Bar.* 1994;68:75–79.
32. Va Code Ann. §65.2–903(1994S).
33. Marlow M. Exclusive remedy provision in the workers' compensation system: unwarranted immunity for employers' willful and wanton misconduct. *SD Law Rev.* 1985;31:161–170.
34. Va Code Ann.§65.2-503(C)(1–3).
35. 29 CFR 1910.1030(1994).
36. Jagger J, Hunt E, Brand-Elnaggar J, et al. Rates of needlestick injury caused by various devices in a university hospital. *N Engl J Med.* 1988;319:284–288.

CDC Statistics on Occupationally Acquired HIV

Vol. 3, no. 3, 1998

Location of Percutaneous Exposure in 46* Healthcare Workers with Documented Occupationally Acquired HIV
United States though June 1997

Location	No.
Intensive care unit	11
Hospital room/ward	10
Outpatient clinic/office	5
Clinical lab	3
Hospice/home care	3
Operating room	2
Emergency room	2
Long-term care facility	2
Pre-op room	1
Morgue/autopsy room	1
Dialysis center	1
Nonclinical lab	1
Outside by dumpster	1
Unknown	3
TOTAL	**46**

Procedure/Device Involved in Percutaneous Exposure in 46* Healthcare Workers with Documented Occupationally Acquired HIV
United States though June 1997

Procedure/device	
Phlebotomy	14
Intravenous insertion (IV catheter)	8
Arterial blood sampling	4
Withdrawing blood from central line	2
Broken glass of blood collection tube	2
IM or SC injection	1
Manipulation of specimen	1
Administration of meds or blood products	1
Manipulation of an I.V. line	1
Transvenous pacemaker insertion	1
Lesion aspiration	1
Dialysis fistula needle	1
Biopsy needle	1
Elutriator needle	1
Scalpel used in autopsy	1
Other/unknown *(3 were needles protruding from needle disposal units)*	6
TOTAL	**46**

Source: Russ Metler, CDC, personal communication, October 1997.

**Note: Out of 52 documented cases of HCWs with occupationally acquired AIDS/HIV, 6 involved mucocutaneous exposures.*

Legal Implications of Needlestick Injuries

by Patti M. Tereskerz, J.D., Ph.D.

Vol. 3, no. 5, 1998

The following article is based on a presentation made at the 1997 meeting of the Intravenous Nurses Society, and subsequently published in the Journal of Intravenous Nursing *(vol. 20, no. 6S, p.S25). It is reprinted, in abridged form with permission.*

Introduction

I will lay out some of the important legal issues associated with needlestick injuries. But first, I want to recall the words of two nurses upon learning they were occupationally infected with HIV. We must never lose sight of the human dimension of the problem—the names and faces behind the statistics.

Here is the reaction of Lynda Arnold, who was infected as a result of a needlestick she sustained while starting an I.V. line:

> Huber told her to come in and sit down. She shut the door behind the young nurse. "I'm not," Arnold said. "Lynda," Huber said, "I don't want to tell you, but you are." "I'm not, I can't be, I'm not!" Arnold cried. "Lynda, sit down," Huber said. Huber moved her chair close to her, took her hands and said, "Lynda, you are HIV positive." It was the darkest moment of Arnold's life. It was as if the world ended. She wanted to get up and run; she wanted the nightmare to end. Lynda, you are HIV-positive.[1]

The next words are those of Marie Jasmin, a French nurse who contracted HIV as a result of a needlestick injury she sustained while drawing blood.

> I go to the lab to have my blood drawn. My HIV results are negative. Six weeks later, during a bout of fever, extreme fatigue and swollen lymph glands, I am retested. They send me to a professor for my results, who announces, "You are infected. You are seropositive." I can't breathe, I ache, I scream. They send me to a psychologist. He tells me, they killed you. I think of my children who are four and nine.[2]

We now know that the two categories of devices most likely to transmit bloodborne pathogens are vascular access catheters and blood-drawing devices. These devices were used by Lynda and Marie, respectively, when their injuries occurred.

Safer Devices: Widely Available, Under-Utilized

The encouraging news is that most needlestick injuries can be prevented. Ten years ago, it was first reported that the design of needles and other sharp devices could be modified to prevent approximately 90% of sharp injuries.[3] Industry has been very responsive, with more than 1,000 U.S. patents issued for devices with injury-prevention features.[4] All major U.S. medical device companies have safety devices on the market.

The issue that remains is whether safety devices work. Recently published data provide evidence that safety I.V. catheter needles and blood-drawing devices reduce needlestick injuries. One study showed that the ProtectIV™ I.V. Catheter Safety System reduced needlestick injuries by 84%[5]; a CDC study concluded that the Safety Lok™ Wing Needle reduced needlestick injuries by 23%, the Punctur-Guard™ Phlebotomy Needle reduced injuries by 76%, and the Venipuncture Needle-Pro™ reduced injuries by 66%.[6] All of these were significant reductions in needlestick injury rates. It is important to emphasize, however, that there is a critical need for further studies to evaluate safety devices on the market. Just because a device has "safety" on the label does not necessarily mean it is effective.

This leads us to the most challenging and disturbing question related to needlesticks: If the technology is available to prevent most of these injuries, why hasn't it been implemented? Market penetration of safety syringes and I.V. catheters is only about 10% and 20%, respectively (see table following this article). It is alarming that thousands of healthcare workers sustain potentially life-threatening injuries each year which could have been prevented.

The answer as to why safety technology hasn't been universally implemented is complicated, but it rests in large part with the law. I will begin by summarizing one area that has an important impact upon the implementation of safety technology: workers' compensation laws. I will go on to discuss court cases and regulations that have a significant bearing upon the issue of needlestick injuries and the implementation of safety devices.

Limitations of Workers' Compensation Laws

Workers' compensation, which is governed by state law, has many limitations, but one of the most important in terms of the implementation of safety devices is the exclusive remedy provision, a common feature in most states. The exclusive remedy provision means that, if covered healthcare workers contract a bloodborne pathogen as a result of an occupational exposure, they cannot file suit against their employer under most circumstances in order to recover damages for occupational disease or injury. The tradeoff for employees is that they, in turn, do not have to prove their employer was negligent in order to receive workers' compensation benefits.

The exclusivity provision is probably the most important reason why healthcare institutions have not purchased safety devices in larger numbers. The provision insulates institutions covered under workers' compensation from liability so that there is no economic incentive for them to buy safety devices, which are more expensive than conventional ones.

For the most part, then, healthcare workers are precluded from bringing suit against their employer. Relying on workers' compensation, they may find that they do not even qualify for those benefits if they cannot successfully meet the statutory definition of an occupational disease. Even if they do receive benefits, the compensation is usually inadequate and does not provide for lifetime benefits (particularly problematic in the case of HIV), and their confidentiality may be breached in the process of obtaining those benefits.

Products Liability Suits

For many occupationally infected healthcare workers facing a life-threatening illness with inadequate benefits, the only choice left has been to bring a products liability suit against the manufacturer of the device which they were using when they were injured, on the grounds that the device was unreasonably unsafe and should not have been on the market. In most instances these suits have been settled out of court and, because of confidentiality agreements, the terms of the settlements cannot be known.

There has, however, been one published products liability case. In the *Riley* case,[7] the court found that the traditional I.V. catheter at issue was not unreasonably dangerous. I would emphasize, however, that the outcome in *Riley* is not controlling precedent in another jurisdiction. A completely different result could be reached by a court considering a case involving any conventional sharp device which has an effective safety counterpart on the market.

Two important determinants of outcome in a products liability suit are how frequently the injury occurs and the severity of injury. The *Riley* court's analysis focused primarily on the rate of HIV transmission per device used, to reach the conclusion that the rate of injury was so small that it did not justify holding the manufacturer responsible. However, the picture would have been very different had the court instead: (1) emphasized the rate of injury per healthcare worker over the course of a career; (2) taken into account the problem of underreporting; (3) included HBV and HCV in its analysis; and (4) considered that not all needlestick injuries are created equal in terms of the risk of transmitting a bloodborne pathogen. The Riley court also placed little weight on the fact that an "injury" from a needlestick is not limited to the possible transmission of a bloodborne pathogen, but also includes the months of testing and possibly toxic prophylactic treatment the injured worker must endure.

It is ironic that the entire burden of liability has been placed upon the medical device industry, while healthcare institutions are allowed to use non-safety devices which injure healthcare workers with impunity. Industry has been the one entity that has invested heavily in safety technology. The problem is not a lack of safety devices on the market, but a failure to buy the devices that are available. Workers' compensation laws should be amended to allow, under some circumstances, injured workers to bring a civil action against their employers, so that the liability burden is appropriately shared by those who are responsible for not providing healthcare workers with safety devices.

OSHA's General Duty Clause

The bloodborne pathogens standard[8] (BPS) issued by the Occupational Safety and Health Administration (OSHA) in 1992 states: "Engineering and work practice controls shall be used to eliminate or minimize employee exposure [to bloodborne pathogens]. Where occupational exposure remains after institution of these controls, personal protective equipment shall also be used." "Engineering controls" are defined within the BPS, as sharps, disposal containers, and self-sheathing needles that isolate or remove the bloodborne pathogen hazard from the workplace.

While the BPS has been an important enhancement to healthcare worker safety, it has been criticized for not specifically delineating what safety devices are required. Consequently, healthcare institutions may believe they are unlikely to be cited for not using safety devices, since specific requirements have not been set forth. However, such reasoning fails to consider that there is also a "general duty" clause under OSHA, which holds employers responsible for providing a place of employment that is free from recognized hazards that are causing or likely to cause death or serious physical harm to employees.[9]

In order to fully understand the meaning of an OSHA regulation, it is necessary to consider not just the regulatory language itself but also cases interpreting that language. In this context, the case of *Continental Oil Co. v. OSHA*, in which the level of safety required in the workplace under the general duty clause was considered, is important.[10] The court held that a workplace cannot just be "reasonably" free of a hazard, or merely as free as the average workplace in the industry; rather, it found that there is a general and common duty to bring *no* adverse effects to the life and health of employees throughout the course of employment. The same court concluded that it is not necessary for safety equipment to be commonly used in the industry before failure to use that equipment will establish a violation of the general duty clause. It found that safety equipment is not infeasible merely because it is expensive. The court went on to uphold a citation in this case, even though the safety equipment at issue was expensive and was not used by the entire industry.

In ruling as it did in the *Continental* case, the court reasoned that if industry custom or practice were the sole criterion for establishing a workplace safety standard, an entire industry could avoid liability by maintaining inadequate safety devices.

Can a healthcare institution be cited under the general duty clause for failing to use devices designed to prevent needlestick injuries? Three requirements must be met to demonstrate a general duty violation: (1) the hazard is likely to cause death or serious bodily harm; (2) the hazard must be recognized by either the employer or the industry; and (3) a feasible method must be available to attenuate the hazard.[11] Does occupational transmission of HIV, HBV, or HCV as a result of needlestick injuries from conventional devices meet these requirements? I think the answer is yes, because: (1) infection with HIV, HBV or HCV is life-threatening; (2) the risk of such injuries is widely recognized, as evidenced by the large volume of medical literature on the subject and the passage of the BPS itself; and (3) there is obviously a feasible means to attenuate the hazard, given the number of safety products on the market. What about the higher cost of these products? Does that make them infeasible? The standard will not be enforced if a court finds that it would be economically devastating to the healthcare industry to enforce its use. However, the court in *American Dental Assoc. v. Martin* has already held that enforcement of the BPS is a feasible means for institutions to attenuate the hazard. The court, however, did not extend this holding to the home healthcare setting.[12]

Limitations of the Bloodborne Pathogens Standard

The BPS is an important step in the right direction, but it does have limitations. First, some needlestick and sharp injuries are not required to be reported under OSHA rules; consequently, OSHA does not know all of the injuries that are occurring and cannot appreciate the magnitude of the problem. Under OSHA requirements, an incident must be reported if it is: work related; involves loss of consciousness, transfer to another job or restriction of motion; results in the recommendation of medical treatment beyond first aid; or results in the diagnosis of seroconversion.[13] A study at the University of Virginia concluded that only a small percentage of the percutaneous and mucocutaneous exposures that occurred among a network of study hospitals were actually OSHA reportable.[14] Further, most of the cases meeting the OSHA reporting criteria were the ones least likely to result in the transmission of bloodborne pathogens.

Another problem, reported in a law review article discussing the BPS, is that OSHA has a limited number of compliance officers.[15] The officers have not aggressively enforced the standard through unannounced inspections, allowing most small employers a grace period to achieve compliance. Because of the limited number of officers available, most enforcement efforts come in response to complaints about a specific incident.

Safety Violations and Criminal Law

Given the above discussion, the question arises: Are there *any* legal reasons to invest in safety? While healthcare institutions in general have not been held legally accountable under civil law, the corporate officers and agents who run those institutions, and the institutions themselves, may be held culpable under criminal law for safety violations that result in death.

Death resulting from occupational transmission of bloodborne pathogens from needlestick injuries is a relatively new employment hazard compared to traumatic deaths that occur in other industries. I am not aware of any criminal cases to date involving occupational deaths associated with needlestick injuries. However, there have been criminal sanctions against corporate officers and agents for deaths resulting from safety violations in other industries. This will be an area to pay particular attention to in the future.

Criminal prosecution of officers and agents can be pursued in two ways. The first is under state laws, where the trend is to prosecute corporate officers and managers who run the workplace or institution for crimes such as murder or manslaughter. The second way is under federal law, for willful violation of the Occupational Safety and Health Act resulting in the death of an employee.

With regard to state prosecutions, the case of *People v. O'Neil* is illustrative.[16] In February 1983, a worker at a film extraction plant died from acute cyanide poisoning resulting from the inhalation of fumes. Evidence at trial established that plant workers were not informed that they were working with cyanide, were not warned of the danger of breathing cyanide gas, and were not given safety instructions or adequate protective clothing. Corporate agents responsible for operating the plant under these conditions were convicted of murder and reckless conduct. The corporation was convicted of reckless conduct and voluntary manslaughter, and three of the corporate agents received 25 years in prison each for murder and 14 concurrent 364-day imprisonment terms for reckless conduct. This case was appealed due to a legal error and remanded for retrial. To avoid re-trial, a plea bargain was reached in which shortened prison sentences were granted.

In a similar case, *Sabine Consol., Inc., et al. v. State*,[17] a corporate employee and the corporations's president were convicted of criminally negligent homicide resulting from the death of two employees when a trench in which they were working collapsed. Safety regulations regarding proper support of the trench had not been followed. In *People v. Deitsch*,[18] a corporate president, his brother who ran the corporation, and a manager were indicted and charged with the crimes of manslaughter in the second degree, criminally negligent homicide, and reckless endangerment in the first degree arising from a fire that erupted in a textile warehouse where safety measures were not observed. The indictments were upheld by the appellate court. These cases have led one legal commentator to write: "Local prosecutors in Los Angeles County, Milwaukee County, and Travis County, Texas (Austin) have instituted investigatory teams that are routinely sent to the scene of every traumatic occupational death... In essence they are treating every occupational death as a potential homicide."[19]

With regard to criminal penalties under OSHA, any employer who "willfully violates any standard, rule, order, or regulation which causes the death of an employee shall, upon conviction, be punished by a fine of not more than $10,000 or by imprisonment for not more than six months, or by both."[20] Two issues immediately come to mind. How is "willful" defined, and who or what constitutes an "employer"? The court in *U.S. v. Dye Construction* Co.[21] considered the first issue and determined that neither OSHA nor related regulations required moral turpitude for conviction—that is, they did not require proving that the employer intended to harm the employee. The court in *U.S. v. Cusack*[22] addressed the second issue, holding that an officer's or director's role in a corporation may be so pervasive that the officer or director is the "employer" subject to criminal penalties of OSHA violations which resulted in the death of an employee.

Conclusion

In summary, the legal issues associated with needlestick injuries that have been reviewed here point to a need for changes in laws and regulations and for tort reform that will promote investment in healthcare worker safety. The law will likely be the engine which drives the economic decision of institutions to purchase safety devices. Hopefully, through legal and policy reforms, the use of safety devices will one day become the legal standard; this, in turn, will be the catalyst to make safety devices the industry standard.

References

1. Hawkes J. Accidental stick of a needle leaves young nurse's life forever changed. *Intelligencer Journal.* February 19, 1996: 1–7.
2. Jasmin M. Moi, Marie Jasmin. *Adv Exposure Prev.* 1995;1(6):9.
3. Jagger J, Hunt EH, Brand-Elnaggar J, Pearson RD. Rates of needle-stick injury caused by various devices in a university hospital. *N Engl J Med.* 1988;319:284–288.
4. Kelly D. Trends in U.S. patents for needlestick prevention technology. *Adv Exposure Prev.* 1996;2(4):7–8.
5. Jagger J. Reducing occupational exposure to bloodborne pathogens: where do we stand a decade later? *Infect Control Hosp Epidemiol.* 1996;17:573–575.
6. Centers for Disease Control. Evaluation of safety devices for preventing percutaneous injuries among health-care workers during phlebotomy procedures–Minneapolis-St.Paul, New York City, and San Francisco, 1993-1995. *MMWR.* 1997;46:21–25.
7. *Riley v. Becton Dickinson Vascular Access, Inc.*, 913 F.Supp. 879 (EDPa.1995).
8. 29 CFR 1910.1030(1994).
9. 29 USC §654(a)(1)(1982).
10. *Continental Oil Co.v.OSHA*, 630 F.2d.446(6th Cir.1980).
11. 630 F.2d 446 (6th Cir. 1980).
12. *American Dental Assoc.v. Martin, et al.*, 984 F.2d 823 (7th Cir. 1993).
13. OSHA Instruction CPL 2-2.44C, March 1992.
14. Jagger J, Balon M, Tereskerz PM. Recordkeeping and the OSHA bloodborne pathogens standard: What hospitals record vs. what is required. *Adv Exposure Prev.* 1995;1(4):1,6–7.
15. Schinagl H. The Occupational Safety and Health Administration's Bloodborne Pathogen Standard: An important first step toward protecting employees from the risks of occupational exposure. *Seton Hall Legis J.* 1993;17:541.
16. *People v. O'Neil*, 194 I11 App. 3d 79, 550 N.E. 2D 1090 (1990).
17. *Sabine Consol., Inc. et al. v. State*, 756 S.W. 2d 865 (Texas 1988).
18. *People v. Deitsch*, 97 A.D.2d 327, 470 N.Y.S.2d 158 (1983).
19. Bros CL. A fresh assault on the hazardous workplace: corporate homicide liability for workplace fatalities in Minnesota. *Wm. Mitchell L. Rev*; 1989; 15:287–326.
20. 29 USC §666(d-i)(1982).
21. United States v. Dye Construction Co., 510 F.2d 78 (10th Cir.1975).
22. *United States v. Cusack*, 806 F. Supp. 47 (D.N.J. 1992).

Estimated Incremental Hospital Cost of Safety Features by Device Type for Needles Most Often Associated with Healthcare Worker HIV Seroconversions

Vol. 3, no.5, 1998

(a device manufacturer's analysis)

Purpose of Needle	Primary Device Used	Non-Safety Price (ea.)	Safety Price (ea.)	Annual Cost for "Average" Hospital			% of Market Using Safety
				Non-Safety	Safety	Incremental	
Draw venous blood	Vaccuum tube blood collection (phlebotomy) needle	0.10	0.33	$6,500	$22,000	$15,500	<10%
	Winged steel (butterfly) needle	0.65	0.90	$11,000	$15,000	$4,000	30+%
Start I.V. line	IV Catheter	0.75	1.75	$25,000	$58,500	$33,500	20+%
Inject meds (IM/subQ/ draw blood/ transfer fluids)	Hypodermic needle/syringe*	0.05	0.25	$16,500	$83,500	$67,000	<10%

- This analysis includes the most common needle devices used to penetrate tissue or enter patient veins/arteries. It does not include specialty needles or needle substitutes used for I.V. line connections.
- According to CDC figures, 57% of healthcare worker HIV infections are from needles used for drawing blood or starting I.V. lines (blood-filled needles). (Ref.: CDC. Procedure/device involved in percutaneous exposure in 46 healthcare workers with documented occupationally acquired HIV. AEP. 1998;3[3]:33.)
- An average hospital (250-300 beds) can protect against 30% of the risk from devices associated with HIV infection through a $19,500 annual incremental cost for safety phlebotomy devices (vacuum tube blood collection needles and winged-steel [butterfly] needles).
- Approximately 48% of the risk can be addressed by adding safety I.V. catheters.
- Today, over 50% of hospitals purchase needleless connector systems for their I.V. line connections. 1997 sales have been estimated at $400 million, with a growth rate over 10% per year. An average hospital may typically pay an incremental $60,000 to $100,000 or more for these systems per year. Needles used for I.V. line connections are associated with less than 10% of HIV seroconversions, per CDC data cited above.
- Prices used in this analysis are estimates; actual prices may vary by device and individual purchaser.

* *Less than 10% of hypodermic needles become blood-filled through accessing veins during blood draws.*

Source: BioPlexus, Inc. based on market share data and internal industry knowledge.

Institutional Liability for Needlestick Injury

by Patricia M. Tereskerz, J.D., Ph.D., Richard D. Pearson, M.D., and Janine Jagger, M.P.H., Ph.D.. *Vol. 4, no. 2, 1999*

On December 18, 1997, a Connecticut jury found Yale University School of Medicine negligent in its training and supervision of a first-year resident who was infected with HIV after sustaining an injury from a needle used on an AIDS patient in 1988. The jury awarded the infected physician $12.2 million.[1,2] This verdict illustrates the catastrophic consequences a needlestick injury can have on both the injured party and the affiliated training institution.

This is a landmark case not only because of the verdict but also because of the legal means by which the verdict was obtained. Generally, workers' compensation is the exclusive remedy available to employees who contract a disabling occupational disease and who are covered under workers' compensation programs.[3-5] In return for guaranteed benefits for qualifying diseases, regardless of fault or an omission on the employee's part, state laws preclude employees from bringing civil suit against their employers.[3,4] In this case, the physician was able to circumvent workers' compensation exclusivity provisions by bringing suit against Yale University School of Medicine, rather than the hospital where she was a contractual employee covered under workers' compensation.

Because the residency program was advertised under the university's name and correspondence related to the residency program was sent under Yale University School of Medicine's name, the plaintiff successfully argued that the university also had a duty to assure that she was trained and supervised adequately and that it was negligent in meeting that obligation.

In the past, university-affiliated medical schools considered that they assume limited liability for employees who contracted bloodborne pathogens as a result of an occupational exposure.[3] Following this verdict, the potential for liability is increased significantly if residents in training can circumvent the exclusivity provisions of workers' compensation laws, which apply to the hospital where they are employed, by bringing suit against the affiliated university.

This verdict underscores (1) the importance of adequate training and supervision in the appropriate use of needle devices that have the potential for transmitting bloodborne pathogens; (2) the need for institutions to reduce the risk of injury by using safer needle devices that have become available recently; and (3) the immediate need for universities to seek legal counsel in carefully structuring promotion of their residency programs.

References

1. Greenberg B. Intern who got HIV on job sues Yale. *Rocky Mountain News*. 12/9/97:A27.
2. *Doe v Yale University. Superior Court of New Haven*, CT. Docket no. CV 90-0305365 S (appeal pending).
3. Tereskerz PM, Pearson RD, Jagger J. Occupational exposure to blood among medical students. *N Engl J Med*. 1996;335:1150-1153.
4. Tereskerz PM, Jagger J. Occupationally acquired HIV. The vulnerability of healthcare workers under workers' compensation laws. *Am J Public Health*. 1997;87(9):1558-1562.
5. Tereskerz PM. Legal implications of needlestick injuries. *J Intraven Nurs*. 1997; 20(6S):S25-S31.

International Researchers Publish Study on

Worldwide Cases of Occupational HIV Infection

by Jane Perry, M.A.

Vol. 4, no. 4, 1999

In a study recently published in *Clinical Infectious Diseases*, researchers from Italy, Great Britain and the United States described characteristics of reported cases worldwide of occupationally acquired HIV infection in healthcare workers. It is the most comprehensive effort thus far to collect and review case descriptions of occupational HIV infection from around the world. The study looked at all cases of occupationally acquired HIV infection, with or without documented seroconversion, reported worldwide through September 1997, and compared the findings with a Centers for Disease Control and Prevention (CDC) international case control study (*New England Journal of Medicine* 1997; 337:1485-1490).

To identify cases, the authors conducted extensive literature reviews and searches of databases such as MEDLINE and the AIDSline databases of MEDLARS (National Library of Medicine). They also reviewed epidemiological bulletins from surveillance centers in the U.S., Canada, Australia, the Caribbean, the United Kingdom, France, Spain, the Netherlands, and Switzerland. For research purposes, healthcare workers were defined as any persons, including students and trainees, who worked in a healthcare, clinical or HIV laboratory setting.

Table 1. Cases of documented and possible occupational HIV infection among healthcare workers, by country, reported through September 1997.

	No. (%) of infections		
Country	**Documented**	**Possible**	**Total**
United States	52 (55.4)	114 (67.1)	166 (62.9)
France	11 (11.7)	27 (15.9)	38 (14.3)
United Kingdom	4 (4.2)	9 (5.3)	13 (4.9)
Mexico	–	9 (5.3)	9 (3.4)
Italy	5 (5.3)	–	5 (1.9)
Australia	4 (4.2)	–	4 (1.5)
Spain	5 (5.3)	–	5 (1.9)
South Africa	3 (3.2)	1 (0.6)	4 (1.5)
Germany	3 (3.2)	3 (1.8)	6 (2.3)
Belgium	2 (2.1)	1 (0.6)	3 (1.1)
Canada	1 (1.1)	2 (1.1)	3 (1.1)
Holland	–	2 (1.1)	2 (0.8)
Switzerland	2 (2.1)	–	2 (0.8)
Denmark	–	1 (0.6)	1 (0.4)
Israel	–	1 (0.6)	1 (0.4)
Argentina	1 (1.1)	–	1 (0.4)
Zambia	1 (1.1)	–	1 (0.4)
TOTAL	**94 (100)**	**170 (100)**	**264 (100)**

Cases of occupational HIV infection were divided into two categories, documented and possible. A documented case was defined as one in which the healthcare worker was "reported to have percutaneous or mucocutaneous exposure to biological material from a patient with HIV infection or to cultured HIV, with (1) documented seroconversion (i.e., absence of HIV infection at the time of exposure, confirmed by negativity of a test for HIV antibody just after or before the exposure, and positivity of a test for HIV antibody performed at postexposure follow-up) or (2) a high degree of relatedness between the source viral strain and the strain isolated from the exposed healthcare worker, as demonstrated by biomolecular assay." A possible case was defined as one in which the healthcare worker was "reported to have (1) a history of exposure to biological material from an HIV-infected patient or a history of work in a setting of high HIV prevalence or involving frequent contact with biological material potentially contaminated with HIV and (2) no evidence of behavioral or transfusion risks for HIV, even if the exposure was not reported at the time of the incident, or there was no record of a negative result of a test for HIV antibody (i.e., no documented seroconversion)."

The authors identified 94 cases of documented and 170 cases of possible occupational HIV infection. As shown in **Table 1**, the U.S. accounted for 62.9% and Europe for 28.4% of documented and possible cases; of European countries, France was the highest with 14.3% of cases. Less than 5% of documented cases were reported from African countries. The authors noted that among the possible cases, six surgeons and two midwives had worked for lengthy periods in African countries during the early stages of the AIDS pandemic. Also, according to the authors, "there was a striking absence of reports of occupationally acquired infection from countries in the Indian subcontinent and Southeast Asia"; they concluded that the number of reported cases "must underestimate the number of healthcare workers worldwide who have acquired HIV occupationally."

Table 2 shows worldwide cases by occupation. Nurses represented 52.1% of documented cases, followed by clinical lab workers with 18% of cases. Based on case descriptions, the authors characterized the majority of documented infections as occurring "in nurses or clinical

Table 2. Cases of documented and possible occupational HIV infection among healthcare workers, by occupation, reported through September 1997.

	No. (%) of infections		
Occupation	**Documented**	**Possible**	**Total**
Nurse	49 (52.1)	45 (26.5)	94 (35.7)
Laboratory worker, clinical	17 (18.0)	19 (11.2)	36 (13.6)
Laboratory worker, nonclinical	3 (3.2)	4 (2.3)	7 (2.6)
Physician, nonsurgical	9 (9.6)	17 (10.0)	26 (9.8)
Physician, surgical	1 (1.1)	14 (8.2)	15 (5.7)
Health aide/attendant	1 (1.1)	15 (8.8)	16 (6.1)
Housekeeper/maintenance worker	3 (3.2)	8 (4.7)	11 (4.2)
Emergency medical technician/paramedic	–	10 (5.9)	10 (3.8)
Dental worker	–	9 (5.3)	9 (3.4)
Embalmer/morgue technician	–	2 (1.2)	2 (0.7)
Respiratory therapist	1 (1.1)	2 (1.2)	3 (1.1)
Dialysis technician	1 (1.1)	3 (1.8)	4 (1.5)
Technician, surgical	2 (2.1)	3 (1.8)	5 (1.9)
Other/unspecified	7 (7.4)	19 (11.2)	26 (9.8)
TOTAL	**94 (100)**	**170 (100)**	**264 (100)**

laboratory workers (70.2%) after contact with infected blood (89.4%) from patients with AIDS (76.5%), by percutaneous exposure (88.3%) during a procedure involving placement of a device in an artery or vein (68%)."

In contrast to the CDC's international case control study, which found four variables—a needle placed in the source patient's artery or vein, visible blood on the device, a deep injury, and terminal illness in the source patient—to be significantly associated with occupational transmission of HIV via needlesticks, this study was only able to document the significance of the first, a needle placed in the source patient's artery or vein. Lack of detail in case descriptions prevented the authors from evaluating the other factors.

The authors emphasized the need for "enhanced safety" during vascular access procedures: "Six of the documented HIV seroconversions occurred after injuries from intravenous catheter stylets, despite the fact that intravenous catheters represent only a small fraction of all needle devices used during the delivery of healthcare."

One finding of note was that 18 cases of HIV infection occurred despite a complete or partial course of postexposure prophylaxis (PEP) with zidovudine. Possible explanations include the emergence of zidovudine-resistant strains of HIV; delayed, low-dose or incomplete treatment with zidovudine; and massive viral inoculum. The authors observed that most public health agencies in developed countries now recommend the use of combination drug therapy for PEP. Another finding was that seroconversion occurred within six months of exposure in 57 (61%) of the 94 documented cases; of these, 42 were found to be seropositive at 3 months postexposure.

Based on data from these worldwide case reports, the authors drew the following conclusions about occupational exposure to HIV:

"(1) All healthcare workers, regardless of job category or the health care setting in which they work, face a low but real risk of occupational infection from HIV exposure.
(2) Infections most often occur following puncture injuries from blood-filled, hollow-bore needles, but have also been caused by cuts from solid objects and contamination of nonintact skin or mucous membranes by at-risk biological substances.
(3) At-risk biological substances include first and foremost blood, as well as other substances to which universal precautions apply.
(4) Adherence to universal precautions, modifications in procedural techniques, and improvements in the designs of sharp medical instruments are critical prevention measures for creating a safer workplace.
(5) All HIV-positive patients, regardless of stage of infection and including those in the window period, present a transmission risk.
(6) The exposed healthcare worker must be followed serologically and clinically for at least six months following exposure, as recommended in the postexposure follow-up protocol.
(7) The efficacy and long-term safety of postexposure antiretroviral prophylaxis must be further studied."

Reference

Ippolito G, Puro V, Heptonstall J, Jagger J, De Carli G, Petrosillo N. Occupational Human Immunodeficiency Virus infection in healthcare workers: Worldwide cases through September 1997. *Clinical Infectious Diseases* 1999;28:365-383. Tables reprinted by permission of University of Chicago Press; ©1999, Infectious Diseases Society of America.

Percutaneous Injuries in Anesthesia Personnel: Results of a Multicenter Prospective Study

by Jane Perry, M.A.

Vol. 4, no. 5, 1999

In the first prospective study of needlestick injuries in anesthesia personnel, recently published in the journal *Anesthesiology*, researchers analyzed contaminated percutaneous injuries (CPI) in healthcare workers performing anesthesia and anesthesia-related tasks over a two-year period at 11 university hospitals. They calculated CPI rates, with corrections to compensate for injury underreporting, and estimated the risk of infection with human immunodeficiency virus (HIV) and hepatitis C virus (HCV) for this group of healthcare workers. They also studied the devices and mechanisms associated with CPIs in anesthesia personnel, and determined the fraction of CPIs that were potentially preventable. Based on these data, the authors formulated some prevention strategies for the practice of anesthesiology.

Data were collected prospectively on injury report forms; this was followed by a retrospective survey to determine reporting rates. Injury report forms were included in the study only if the device causing injury was contaminated with a patient's blood or body fluid. An injury was classified as "high risk" if, based on the purpose of the device involved, the injury was likely to have resulted in a larger volume of blood inoculum compared with other types of injuries. Thus, the researchers defined a high-risk CPI as one involving a "blood-contaminated hollow-bore needle where it was likely that the needle lumen was filled with undiluted blood (i.e., intravascular catheter insertion or obtaining patient blood)."

Over the two-year period, 138 CPIs were reported; of those, 125 were associated with 361,943 anesthetics, and 13 with an unknown number of nonanesthetic procedures. Based on the retrospective reporting-rate surveys, the authors calculated injury reporting rates of 29% for anesthesia residents, 19% for attending anesthesiologists, 23% for certified registered nurse anesthetists (CRNAs), and 64% for student registered nurse anesthetists (SRNAs). Corrected CPI rates for all anesthesia personnel were 0.27 CPIs per year per person, 0.42 CPIs per year per FTE, and 1.35 CPIs per 1,000 anesthetics. By job category the rates for corrected CPIs per year per person were 0.33 for residents, 0.31 for attendings, 0.21 for CRNAs, and 0.077 for SRNAs.

Thirty percent of CPIs (42 of 138) fell into the "high risk" category—injuries from blood-filled, hollow-bore needles used for intravascular catheter insertion or obtaining blood. This compares to a 25% rate of high-risk injuries for hospital personnel in general found in another, nationwide study. The authors found that 79% of high-risk CPIs (33 of 42) were potentially preventable, had a safety device been used.

Most CPIs (73%) occurred in the operating room; 9% occurred in the preoperative holding area, 5% in the postanesthesia care unit or a procedure room, and 13% in other areas. Twenty-eight percent of injuries occurred during use of the device, 23% between steps of a multistep procedure, 14% during or after needle recapping, 25% after the device was used but before disposal, and 9% during disposal.

Intravascular catheter insertions accounted for 28% of injuries; injection of intradermal local anesthesia, 21%; suturing of intravascular catheters, 19%; cutting, 9%; intravenous tubing access, 7%; administration of spinal, epidural, or caudal anesthesia, 5%; peripheral nerve block, 4%; obtaining blood, 4%; intramuscular or subcutaneous injections, 2%; and others, 2%.

Table 1 shows the number of CPIs by personnel group. Residents had the highest number of CPIs, followed by attendings, CRNAs, and SRNAs.

Table 2 breaks down devices causing CPIs by needle vs. non-needle devices and, under needle devices, by hollow-bore, solid or unknown. It also shows the percentage of CPIs that were contaminated with blood in each device category. **Table 3** gives greater detail on needle devices causing CPIs, including total number of devices used during the study period.

The authors developed an infection risk model for HIV and HCV, and applied it to the 56,208 anesthesia personnel in the U.S. They estimated a rate of 0.56 HIV infections per year and 5.18 HCV infections per year for this group, with 17 HIV infections and 155 HCV infections expected

Table 1. CPIs in Anesthesia Personnel Groups

Anesthesia Personnel Group				
Group	Personnel* (n)	FTE (n)	All CPIs [n (%)]	High-risk CPIs [n (%)]
Residents	372	231.8	73 (53)	18 (43)
Attendings	311	189.1	37 (27)	12 (29)
CRNAs	198	129.6	19 (14)	8 (19)
SRNAs	91	50.5	9 (7)	4 (10)
TOTAL	972	601.0	138 (100)	42 (100)

CPIs = contaminated percutaneous injuries; FTE = full-time equivalents; CRNAs = certified registered nurse anesthetists; SRNAs = student registered nurse anesthetists.
** All full-time and part-time anesthesia personnel.*

Table 2. Devices Causing CPIs* and Type of Contaminant†

Type of Device	CPIs from Device [n (%)]	Blood-contaminated CPIs from Device [n (%)]
Needle devices	131 (95)	128 (97)
Hollow-bore needles	104 (75)	102 (77)
Solid needles	26 (19)	26 (20)
Unknown type of needle	1 (0.7)	—
Nonneedle devices‡	7 (5)	4 (3)
TOTAL	138	132

CPIs = contaminated percutaneous injuries
**No CPIs were caused by safety devices.*
† A total of 132/138 (96%) CPIs involved devices contaminated with blood. Of the other CPIs—6/138 (4%)—1 involved cerebrospinal fluid on a spinal needle; 1, saliva on scissors (potential for blood); 1, a needle for intravenous tubing injection (potential for blood); 3, unknown type of contaminant.
‡ A total of 4 scalpel blades (blood-contaminated); 3, others.

over a 30-year period. They also calculated the average risk of HIV infection for all anesthesia personnel in their study, and found it was 0.0016% per year per FTE worker. They noted that this is in the range reported for emergency room workers.

Another finding of note was related to recapping. The authors observed that, although injuries from recapping for nonanesthesia personnel have decreased from 33% in 1986 to 4% in 1994-95, in their study 14% of CPIs were related to recapping. (The incentive for recapping remains high in anesthesia practice because of the need to give incremental doses of anesthetics.) They speculated that many of these injuries were due to two-handed recapping, and emphasized the need for a one-handed recapping technique.

Based on the above data, a typical profile of an anesthesia-related CPI would be a resident or attending anesthesiologist injured in an operating room with a hollow-bore, blood-contaminated needle. A typical high-risk CPI was from a blood-contaminated hollow-bore needle used for intravascular catheter insertion (and presumed to be blood filled). Most CPIs were potentially preventable, and fewer than half were reported to hospital health services.

The authors' recommendations for prevention of CPIs in anesthesia personnel include the use of safety-engineered devices; eliminating the use of unnecessary sharp devices and reducing to a minimum the use of necessary ones wherever possible; developing safer methods for procedures involving sharp devices; and use of standard precautions. The authors observed that, "Although safety devices do not exist for use in all types of procedures performed during the practice of anesthesiology, practitioners should evaluate available safety devices as possible replacements for the sharp devices they use that lack safety features."

Reference

Greene ES, Berry AJ, Jagger J, Hanley E, Arnold WP, Bailey MK, Brown M, Gramling-Babb P, Passannante AN, Seltzer JL, Southorn P, Van Clief MA, Venezia RA. Multicenter study of contaminated percutaneous injuries in anesthesia personnel. *Anesthesiology.* 1998;89:1362-1372. ***[Tables reprinted with permission.]***

Table 3. CPIs from Needle Devices, Number of Devices Used, and Device-Specific Injury Rates

Type of Device	CPIs (n)	Corrected CPIs* (n)	Devices Used (total for entire study period) (n)	Corrected CPIs/100,000 Devices Used* [n (95% CI)]†
Needle on syringe				
Disposable syringe	55	211.54	1,512,411	13.99 (10.54–18.21)
Prefilled cartridge	1	3.85	112,016	3.43 (0.087–19.13)
Arterial blood gas	0	0	5,251	0 (0.00–219.43)
Other type	1	3.85	79,663	4.83 (0.12–26.90)
Intravascular device				
IV catheter	27	103.85	541,395	19.18 (12.64–27.91)
Winged needle (butterfly)	0	0	6,507	0 (0.00–177.07)
Suture needle	26	100.00	45,393	220.30 (143.91–322.79)
Unattached hollow-bore needle	13	50.00	2,391,430	2.09 (1.11–3.58)
Nerve block needle	4	15.38	8,206	187.48 (51.08–480.02)
Epidural needle	1	3.85	48,610	7.91 (0.20–44.08)
Spinal needle	1	3.85	38,129	10.09 (0.26–56.20)
IV infusion connection needle	1	3.85	13,608	28.26 (0.72–157.48)
Others	1	—	17,095	—
TOTAL	131	503.87	4,819,714‡	—

CPIs = contaminated percutaneous injuries; I.V. = intravenous.
** Corrected for CPI underreporting.*
† The 95% confidence interval for mean corrected CPIs/100,000 devices.
‡ A total of 2,380,130 devices/yr. The number of devices used does not include any safety devices.

The Push Towards Safety: Manufacturers Respond to Changes in Regulatory and Legislative Climate

by Jane Perry, M.A.

Vol. 5, no. 1, 2000

In the last year, important state and federal regulatory and legislative changes have helped to ensure that healthcare workers will have unprecedented access to safety-engineered sharps devices. For manufacturers of safety devices, these new initiatives have also signaled a major shift in the safety device market.

California's new state OSHA regulations, triggered by A.B. 1208, went into full effect in July 1999; they require that needles and other sharp devices provide "engineered sharps injury protection." Since California represents approximately 10% of the U.S. medical device market, this regulatory change resulted in a major surge in demand for safety devices. During 1999, Tennessee and Texas also passed needlestick prevention laws, although in the case of Texas, the legislation covered only public or state healthcare facilities. In November 1999—and perhaps most important for the safety device market nationwide—OSHA issued a revised compliance directive that requires OSHA inspectors to evaluate healthcare facilities' use of safety-engineered sharps devices in enforcing the federal bloodborne pathogens standard. Also in November 1999, NIOSH/CDC issued an alert on needlesticks urging healthcare employers to use safety-engineered sharps devices and to establish comprehensive programs to reduce needlestick injuries. Finally, on January 4, 2000, New Jersey Governor Christine Todd Whitman signed into law what appears to be the most comprehensive safety needle legislation thus far. New Jersey represents about 3-1/2% of the medical device market in the U.S.; together with California, that means that, at a minimum, approximately 14% of the medical device market should have converted to safety-engineered sharp devices by the end of this year.

These regulatory and legislative changes are unquestionably good news for healthcare workers, and represent a long-awaited recognition of the issue of needlestick safety. OSHA's revised compliance directive and the new state legislation represent an opportunity and a challenge for everyone involved in the delivery of healthcare in the U.S. The conversion to safety will now happen faster than anyone could have imagined ten or even five years ago, and will have a significant impact on healthcare facilities and product suppliers alike.

> ***"One of the biggest advantages of the compliance directive is that it provides a new clarity for manufacturers—it is a signal that it is now safe to invest heavily in safety needle capacity. This sets the stage for safety device prices to become more competitive over the next few years."***
>
> *—Tom Sutton, Bio-Plexus*

Compliance Directive Creates Surge in Demand

The revised compliance directive by itself appears to be providing a big boost to the safety market. According to Tom Sutton, chief operating officer at Bio-Plexus, Inc., "During December 1999 we had about a three-fold increase in sales. We think this was due in large part to the compliance directive, with healthcare employers interpreting the directive as requiring the use of safety needles." Sutton says further, "One of the biggest advantages of the compliance directive is that it provides a new clarity for manufacturers—it is a signal that it is now safe to invest heavily in safety needle capacity. This sets the stage for safety device prices to become more competitive over the next few years."

The medical device industry has not seen a fundamental technology and market shift of this magnitude in decades. Although the eventual conversion to safety-engineered devices was anticipated by many, the momentum created by these legislative and regulatory actions has been stunning. For healthcare administrators and materials managers, "This is more than buying new products," says Kevin Seifert, vice president/general manager of BD Advance Protection Technologies. "We need to help our customers do this right and avoid the complications and costs associated with conversion errors in inventory management, recordkeeping, training, and other areas."

Expanded Opportunities

Medical device companies of all sizes have reoriented their operations—investing in new equipment, retooling manufacturing processes, and expanding facilities—in order to respond to these new market forces and produce at high-volume levels [*see box, right*]. In addition, distribution networks are being broadened and in-service training programs, and the personnel to staff them, are being expanded in order to provide healthcare facilities with the support that such a broadscale and comprehensive product conversion will require. "The recent acceleration in safety adoption in California has shown that clinical support will become an even more critical factor in successful safety transitions," says Steve Fanning, president of Johnson & Johnson Medical. "At Johnson & Johnson, we have a team of over 200 clinical nurse consultants to partner with healthcare facilities in ensuring a smooth and effective transition to our safety I.V. catheters."

Group purchasing organizations (GPOs), such as Columbia HCA, Premier, Tenet, and others, are also making changes in response to the shifting safety climate. "The GPOs are now reexamining the issue of safety needles in terms of existing contracts for needle devices," says Sutton of Bio-Plexus. "For example, Premier is in the process of awarding additional contracts in our product area, safety blood collection needles, similar to what they did for safety syringes and needles. Another major proprietary 300-hospital group has put out a request for information on safety needles in recognition of the new regulatory environment. A definite trend is emerging where individual hospitals, major hospital networks, and GPOs are all recognizing the need to reevaluate the available safety devices on the market and document the evaluation and selection process for their facilities or organizations." According to Sutton, evaluation and documentation are key: "Healthcare facilities are trying to show that they have looked at all the possibilities and have selected the 'safest' device based on objective studies. This provides an expanded opportunity for smaller manufacturers to get hospitals to look at and test their products."

The medical device industry is working hard to supply the products and services its customers require. Manufacturers large and small have voiced their commitment to making the significant investments necessary to increase production in all safety product categories—and to do it as swiftly as possible. As Kevin Seifert of BD says, "So many in this industry have worked long and hard to see this day come, and if customers, manufacturers and distributors work closely together, we can go through this enormous transition in record time without undue expense or any compromise to the delivery of quality healthcare."

Companies Gear Up to Respond to Increased Product Demand

Johnson & Johnson Medical

"Johnson & Johnson Medical has been steadily increasing its manufacturing capacity and efficiencies to meet the rising demand for our line of safety I.V. catheters. We have a new state-of-the-art manufacturing facility that will ensure that we are well positioned to meet the future needs of customers."

—Steve Fanning, President

New Medical Technology

"At New Medical Technology we doubled the size of our marketing and sales staff in late 1999. This year, we will have three additional high-speed automated assembly machines added to our production capacity, which will give the company the ability to produce commercial quantities of our 3cc and 1cc safety syringe with a broad range of needle gauges and lengths."

Rick Field, Executive Vice President

BD Advanced Protection Technologies

"BD started gearing up ahead of recent federal action to meet demand in the U.S., as well as in other countries where the growing conversion movement is also taking shape. Our $300 million production re-tooling program is now well underway and will enable BD to provide advanced products at the necessary volumes embracing all technology platforms in every major clinical category."

—Kevin Seifert, VP/General Manager

Bio-Plexus

"In response to the compliance directive, Bio-Plexus has expanded its sales force, its team of product educators, and its educational tools, including website, videos, and printed materials, as well as increased its manufacturing and overall product capacity."

—Tom Sutton, Chief Operating Officer

CDC Releases National Needlestick Estimates

by Jane Perry, M.A.

Vol. 5, no. 2, 2000

The Centers for Disease Control and Prevention (CDC) presented results of a study estimating the annual number of percutaneous injuries (PI) to U.S. hospital-based healthcare workers at a conference in March 2000. The estimates were based on combined data from two sources: the CDC's National Surveillance System for Hospital healthcare workers (NaSH) database, and the Exposure Prevention Information Network (EPINet) database coordinated by the International Healthcare Worker Safety Center (IHWSC) at the University of Virginia.

The **CDC's estimate of 384,325 annual PIs for healthcare workers in hospitals** is higher than earlier figures developed by the IHWSC. The Center had estimated 295,082 annual PIs to hospital-based healthcare workers based on 1996 EPINet data, using a 39% underreporting rate. The Center doubled this number, since it has been estimated that as many as half of all healthcare workers are employed outside of hospitals, to obtain a figure of 590,164 annual PIs for *both* hospital and non-hospital healthcare workers nationwide. The CDC did not include an estimate for healthcare workers outside of hospital settings. The underreporting rate used in the CDC study, 56.58%, was based on survey data from NaSH hospitals.

The estimates were drawn from 1997 and 1998 data from 15 NaSH hospitals, and 1997 data from 45 EPINet hospitals. Because the EPINet and NaSH networks use similar data collection instruments, it was possible to combine the two data sources. EPINet hospitals tend to be smaller than NaSH hospitals—the average number of beds is 315, compared to 592 for NaSH hospitals—and are concentrated in the southeast and northwest, whereas NaSH hospitals are more scattered regionally, with a number located in the northeast.

The report was presented by Adelisa Panlilio, M.D., and colleagues at the International Conference on Nosocomial and Healthcare-Associated Infections, held in Atlanta, Georgia. Dr. Panlilio, a medical edpidemiologist with the CDC's Hospital Infections Program, commented: "We're hoping that by combining the two data sets and our different hospitals that participate in them, we may get a more representative picture [of needlesticks nationally]."

Using Denominators to Calculate Percutaneous Injury Rates

by Janine Jagger, M.P.H., Ph.D.

Vol. 6, no. 1, 2000

The data that healthcare facilities collect in sharps injury logs can be used to calculate needlestick rates and to make comparisons among different professional groups, device categories, and hospital settings. Rates consist of a numerator and a denominator; in the three methods of rate calculation that follow, the numerator always consists of the number of needlesticks that occurred in a specific time period. The denominator is either the number of occupied beds, the number of full-time equivalent (FTE) employees in a specific job category, or the number of devices in a given device category used in a given time period. The time period corresponding to the numerator and denominator should always be the same.

The rate that is simplest to calculate is the total number of needlestick injuries (NSI) reported during a specific time period (numerator) over the number of occupied hospital beds in an institution for the same time period (denominator). The average daily census of occupied hospital beds for the same year as the reported needlesticks is the relevant number to use in the denominator, since it corrects for unused hospital beds. A realistic example would be:

$$\frac{\text{350 NSI per yr.}}{\text{800 occ'd beds (ADC) per year}} \times 100 = \text{44 NSI per 100 occupied beds per yr.}$$

This gives a rough idea of the institutional needlestick experience, which can then be used to track NSI levels over time. If you are comparing your rates to other institutions, you should be aware that rates are affected by a number of factors, including the level of needlestick underreporting and the types of patients the hospital treats. A regional medical center is likely to treat a higher proportion of patients requiring intensive care than a community hospital, and may therefore have a higher needlestick rate per bed because more needles are used per patient.

Needlestick rates can also be calculated for different professional groups, such as nurses, physicians, laboratory technicians, and housekeepers. With this method, the numerator is the total number of needlesticks reported by the professional group during a given time period. The denominator is the total number of FTE employees for that professional group during the same time period. Using nurses as an example, the rate would be calculated this way:

$$\frac{\text{250 nurse-reported NSI in 1 yr.}}{\text{1,000 FTE nurses employed that yr.}} = \text{.25 NSI per FTE nurse per yr.}$$

A full-time equivalent means that if you have 50 part-time nurses, they count as 25 full-time equivalent nurses. This method gives a more accurate denominator than simply counting the number of nurses employed. For some occupational categories it may be difficult or simply impossible to calculate an accurate rate. For instance, in some hospitals, contract workers who provide services such as phlebotomy are technically not considered employees of the hospital. If those workers report their injuries to their contracting agency, the hospital may not have a record of all or any of their incidents. Furthermore, in many private hospitals physicians are not employed by the hospital they practice in, and their injuries might not show up in hospital records. For other groups such as medical residents, whose working hours may be extremely erratic, it may not be possible to obtain a reasonable full-time equivalent estimate for a denominator. When using this method of calculating rates, it is better to limit the calculations to occupational categories for which a reliable numerator and denominator can be obtained.

Finally, rates can be calculated for specific devices. Rates of needlesticks for specific types of needles are necessary when comparing needlestick risk from different devices and for evaluating the effectiveness of products designed to prevent needlesticks. To calculate device-based rates, a healthcare worker must accurately identify the type of needle involved when reporting his or her needlestick injury. The numerator consists of the number of needlesticks with a particular device during a specific period of time, and the denominator consists of the number of devices used or purchased during the same time period. A device-based needlestick rate, using syringes as an example, would be calculated in the following way:

$$\frac{\text{100 NSI from disposable syringes during a given time period}}{\text{1 million disposable syringes used or purchased during same period}} \times 100{,}000 = \text{10 NSI per 100,000 syringes}$$

If a trial of a safety syringe were carried out, the above rate would be compared to:

$$\frac{\text{\# NSI from safety syringes during given period}}{\text{\# safety syringes used or purchased during same period}}$$

It may be difficult to find out exactly how many devices have been used in direct patient care. Usually, the number of devices purchased is the closest approximation that can be obtained, so the denominator for calculating device-based rates must be obtained from the purchasing department. For devices like I.V. catheters that are used for one purpose, all devices purchased can be included in the denominator. Some devices, such as disposable syringes, are used for many purposes, not all of which are patient-related. If a large number of devices in a given device category are not used for patient care, such as syringes used by pharmacy for mixing drugs, then the devices not used for patient care should be excluded from the denominator when calculating the injury rate.

To determine which rate or rates you want to calculate, decide what questions you want to answer. Do you want to compare your hospital to other hospitals? Compare exposure risk among different professional groups in your facility? Compare injury risk for different devices? Also, the rates you can calculate depend upon the denominator data that are available to you.

Following these steps will help you to calculate meaningful rates and make optimum use of your facility's data.

Healthcare Worker Blood Exposure Risks: Updating the Statistics

by Jane Perry, M.A. and Janine Jagger, M.P.H., Ph.D.

Vol. 6, no. 3, 2003

The availability of data on occupational exposures to bloodborne pathogens has increased dramatically during the last decade. Articles in the medical literature on needlestick injuries and blood exposures have proliferated, as numerous researchers have reported results of single institution or multicenter studies, or focused studies of specific occupational groups and clinical settings.[1-17] There are two ongoing large-scale surveillance programs in the United States (U.S.) that collect data on sharps injuries: the Exposure Prevention Information Network (EPINet) Multi-hospital Needlestick and Sharp-Object Injury database, established in 1993 and maintained by the International Healthcare Worker Safety Center at the University of Virginia; and the National Surveillance System for healthcare workers (NaSH), established in 1995 by the Division of Healthcare Quality Promotion at the Centers for Disease Control and Prevention (CDC). (Other countries, including Italy, Canada, Japan, and Spain, conduct national-level needlestick surveillance as well.)

In the U.S., the most accurate national estimates to date of the number of percutaneous injuries (PIs) sustained by hospital-based healthcare workers (HCWs) have been derived by combining data from the EPINet and NaSH networks. Researchers have also been able to estimate more precise underreporting rates for PIs. And large-scale surveillance of HCWs sustaining blood exposures to hepatitis C-infected source patients has provided a more accurate picture of hepatitis C transmission risk from occupational exposures.

Outdated figures on needlestick injuries and blood exposures continue to circulate in the medical literature, however. One recent article on needlestick injuries that appeared in a nursing journal stated: "It's estimated that up to 96% of all needlestick injuries go unreported. That figure is staggering when you consider that 600,000 to 800,000 percutaneous injuries occur each year."[18] A 2002 report from the Sharps Injury Control Program of California's Department of Health Services cited similar statistics.[19] Such numbers are picked up and cited elsewhere, and eventually regarded as fact. Thus, it is important to review the latest figures on occupational blood exposures, particularly where recent data differ significantly from those previously available. We will review revised estimates for: (1) the annual number of PIs in the U.S. for hospital-based HCWs; (2) underreporting rates for needlestick injuries; and (3) occupational transmission rates for hepatitis C virus (HCV).

Annual Number of Percutaneous Injuries in the U.S.

Estimates of 600,000 to 800,000 needlestick injuries in the U.S. each year (or 800,000 to 1 million) were frequently cited in journal and newspaper articles during the last decade. In 1999, in an effort to develop more precise estimates, the CDC conducted an in-depth analysis in which data from 15 hospitals in the NaSH network and 45 hospitals in the EPINet network were combined. Critical to developing better national estimates for PIs was establishing an accurate underreporting rate (discussed in detail below). It was also important to determine significant variables, such as number of hospital beds, in-patient days, and employees. To account for these variables, researchers weighted and stratified data for each hospital proportionate to its size. NaSH hospitals tend to be large (the average number of hospital beds is 592), while EPINet hospitals tend to be comparatively smaller (average number of hospital beds is 315). EPINet hospitals are mostly located in the southeast and northwest, while a significant proportion of NaSH hospitals are in the northeast, with others scattered across the U.S. Thus the two data sources are complementary and, when combined, provided a balanced statistical sample of U.S. hospitals.

In 2000, based on this study, the CDC published a national estimate for PIs in U.S. hospitals for a one-year period.[20-22] This figure was cited by the Occupational Safety and Health Administration (OSHA) in its preamble to the revised bloodborne pathogens standard: **"The [CDC] has estimated that healthcare workers in hospital settings sustain 384,325 percutaneous injuries involving contaminated sharps annually" (95% CI 311,091 to 463,922).**[23]

Estimates of Missing Components

The CDC estimate did not include PIs occurring outside of hospital settings, nor the number of mucocutaneous blood exposures occurring in any healthcare setting. Estimates for these two missing pieces can be sought from existing sources.

Market data indicating the facilities to which needles were sold in 2001 show that 31% of needles were sold to nonhospital buyers (personal communication, Ned Weller, Health Products Information Services, March 2002). If we accept the assumption that needlestick rates are directly related to needle usage, this suggests that 31% of needlesticks remain unaccounted for in the CDC estimate, and would increase the estimated annual number of PIs from 384,325 to 503,466.

Non-percutaneous exposures should also be taken into account to assess the full spectrum of blood exposure risk—specifically, mucocutaneous contact with blood and at-risk body fluids. EPINet data for 1999 indicated that .29 mucocutaneous blood or body fluid exposures were reported for each PI reported. (There is no generally accepted underreporting estimate to apply as a correction factor for mucocutaneous exposures, and none has been applied here.) On this basis, we assumed that .29 mucocutaneous exposures occurred for each reported PI. This would bring the total estimated number of percutaneous and mucocutaneous exposures occurring annually in hospital and nonhospital settings in the U.S. to 649,471. This is a first attempt to identify the missing pieces of the full exposure spectrum and contains untested assumptions. The most accurate figure available is still the CDC estimate cited above.

Underreporting Rates

The problem of needlestick underreporting was first documented by Hamory in 1983[24]; numerous studies conducted over the last decade cited a wide range of underreporting rates (see Table 1).[25-36] Studies yielding the highest underreporting rates involved physicians and medical students; many were conducted at a single institution and thus had small sample sizes. For example, in 1990 McGeer et al. con-

Table 1. Rates of Underreporting of Percutaneous Injuries in Healthcare Workers

Author/Year? Publication	Country	Study design	Population #responses	Underreporting Rate (%)
Hamory 1983 (24)	U.S.	Survey	1 univ. hospital, 726 respondents	75%
Jagger et al. 1988 (25)	U.S.	Questionnaire	1 univ. hospital, 326 NSIs	39%
McGeer, et al. 1990 (25)	Canada	Survey	1 univ. hospital, 88 med. students residents, interns, 372 PIs	95%+
Mangione et al. 1991 (27)	U.S.	cross sectional survey, 1988-1989	3 teaching hospitals, 86 residents/interns, 103 PIs	70%
Tandburg et al. 1991 (28)	U.S.	Survey	1 univ. hospital, 259 respondents (emergency MDs, RNs, EMTs) 643 PIs	65% overall 87% ED MDs 34% ED RNs 33% EMTs
O'Neill et al. 1991 (29)	U.S.	Survey	1 univ. hospital, 550 med students & residents	91%
Albertoni et al. 1992 (30)	Italy	Validation study	1 teaching hospital	85% (MDs) 69% nurses
Lynch & White 1993 (31)	U.S.	Comparison of blood exposure incident reports with data from OR study	3 hospitals, OR personnel	96% (OR personnel)
Chamberland M et al.1995 (32)	U.S.	1993-94 survey	6 hospitals, HCWs performing phlebotomy	68% residents 60% med. students 35% nurses 11% phlebotomists
Roy & Robillard 1995 (33)	Canada	1-year survey (1991-92), compared to incident reports	5 hospitals 838 exposures (PIs & BBF)	47% (range 29%-61%)
Henry & Campbell 1995 (34)	U.S.	1990 survey	65 hospitals, 100 infection control officers at randomly selected U.S. hospitals	18.5%
EPINet 1997 (survey for NIOSH study, unpublished data)	U.S.	Survey	6 hospitals, 2,544 respondents	38.6%; Range: 25.8%, housekeepers; 72.7%, MDs
Osbom et al. 1999 (35)	U.S.	7-year (1990-96) longitudinal study	1 univ. hospital, 119 med. students 129 PIs	Range: 55%-35% (decreased over study)
Haiduven et al. 1999 (36)	U.S.	Survey (1992-95)	1 teaching hospital, 549 HCWs	46%
CDC study 2000 (A. Panlilio personal communication)	U.S.	2-year survey (1997-98)	12 hospitals, 23,738 HCWs	56.6%

PIs = percutaneous injuries   HCWs = healthcare workers   BBF = blood or body fluid

Table 2. Infection Rates Among HCV-Exposed Healthcare Workers*

Source	Country	# exposed	# infections	Infection Rate % (95% CI)
Hernandez 1992 (39)	Spain	81	0	0 (0–4.4)
Mitsui 1992 (40)	Japan	68	7	10.3 (3.0–17.5)
Sodeyama 1993 (41)	Japan	90	2	2.2 (0.2–7.8)
Lanphear 1994 (42)	U.S.	50	3	6.0 (1.2–16.5)
Zuckerman 1994 (43)	UK	24	0	0 (0–14.2)
Monge 1995 (44)	Spain	603	2	0.3 (0.04–1.2)
Arai 1996 (45)	Japan	56	3	5.4 (1.1–14.9)
Serra 1998 (46)	Spain	443	3	0.7 (0.1–2.0)
Takagi 1998 (47)	Japan	250	4	1.6 (0.4–4.0)
Hasan 1999 (48)	Kuwait	24	0	0 (0–14.2)
Kidouchi 1999 (49)	Japan	4836	15	0.3 (0.1–0.5)
Petrosillo 2001 (50)	Italy	4292	19	0.4 (0.2–0.6)†
Baldo 2002 (51)	Italy	68	0	0 (0–5.3)
Evans 2002 (52)‡	UK	439	1	0.2 (0.006–1.3)
TOTAL		**11,324**	**59**	**0.5 (0.39–0.65)**

**Only the most recent report by the same group of investigators was included. HCV = hepatitis C virus; CI, confidence interval.*

† Percutaneous exposure infection rate, 0.5%; mucocutaneous infection rate, 0.4%.

‡Evans B, Communicable Disease Surveillance Centre. Public Health Laboratory Service, written communication, April 24, 2002.

Table from Jagger J, Puro V, De Carli G. Occupational transmission of hepatits C virus (letter). JAMA 2002 (9/25/02);288:1470.

ducted a study of 88 medical students, residents and interns at a university hospital, and found an underreporting rate of greater than 95%.[26] Similarly, a 1991 study by Tandberg et al. of emergency department workers found an underreporting rate of 87% for ED physicians.[28] A 1992 study by O'Neill et al. of 550 medical students and residents at a university hospital found an underreporting rate of 91%.[29]

The best overall data on PI underreporting for U.S. HCWs comes from a survey conducted by the CDC in 1998 in 12 hospitals participating in the NaSH network.[37] In this survey, 14,215 HCWs indicated if they sustained a percutaneous injury or injuries in the previous year, how many they reported, and their reasons for not reporting. The survey documented an overall underreporting rate of 58%, with a high of 73% for surgeons and 52% for all other HCWs. (Results from two years of CDC survey data, with 23,738 HCWs responding, have not yet been published, but yielded a similar underreporting rate of 56.6%. (A. Panlilio, CDC, personal communication, 2002.) The 1998 survey had a much larger sample size than previous studies, and researchers were able to estimate standard errors and give confidence intervals for the rates, which had not been done before. The CDC study also determined the factors significantly affecting reporting rates, such as hospital size, location, and occupation of the healthcare worker.

HCV Transmission Rate

The average risk of HCV transmission has been most commonly reported as 1.8% in the medical literature, with a range of 0-7% (or sometimes 0-10%), based on studies from the early 1990s.[38] More recent data, including follow-up of more than 11,000 HCV-exposed healthcare workers in six countries, have yielded significantly lower transmission rates (see Table 2).[39-52] The average transmission rate for all reports cited in the table is 0.5%, a rate similar to that for occupational HIV transmission.

Such findings should be taken into account when deciding whether to perform HCV RNA testing after an occupational exposure to an HCV-positive source patient. Current CDC guidelines for HCV postexposure follow-up recommend an ALT activity test at baseline and at four to six months, and state that an HCV RNA test may be performed at four to six weeks if earlier diagnosis of HCV infection is desired.[38] Given a 0.5% average infection rate, if performed routinely, 99.5% of such tests would have negative results. Thus, selective testing of cases with a higher-than-average transmission risk may be an alternative strategy.

Conclusion

The most recent data for the annual number of PIs in the U.S., underreporting of PIs, and occupational transmission of HCV help to define more accurately occupational exposure risk, and will be especially useful for future comparisons as we seek to measure progress in preventing occupational blood exposures. It is also good news for healthcare workers who sustain needlesticks from HCV-positive source patients: those workers can be told that the transmission risk for HCV is substantially lower than once believed.[53]

References

1. Ippolito G, Puro V, Petrosillo N, et al, Studio Italiano Rischio Occupazionale da HIV (SIROH) group. Surveillance of occupational exposure to bloodborne pathogens in healthcare workers: the Italian national programme. *Eurosurveillance.* 2002; 4(3):33–36.
2. Monge V, Mato G, Mariano A, et al. (GERABTAS Working Group). Epidemiology of biological-exposure incidents among Spanish healthcare workers. *Infect Control Hosp Epidemiol.* 2001;22(12):776–780.
3. Whitby RM. McLaws ML. Hollow-bore needlestick injuries in a tertiary teaching hospital: epidemiology, education and engineering. *Med J Aust.* 2002;177(8):418–22.
4. Ng LN. Lim HL. Chan YH. Bin Bachok D. Analysis of sharps injury occurrences at a hospital in Singapore. *Int J Nurs Pract.* 2002;8(5):274–81.
5. Newsom DH, Kiwanuka JP. Needle-stick injuries in an Ugandan teaching hospital. *Ann Trop Med Parasitol.* 2002;96(5):517–522.
6. Memish ZA, Almuneef M, Dillon J. Epidemiology of needlestick and sharps injuries in a tertiary care center in Saudi Arabia. *Am J Infect Control.* 2002;30(4):234-41.
7. Khuri-Bulos NA, Toukan A, Mahafzah A, et al. Epidemiology of needlestick and sharp injuries at a university hospital in a developing country: a 3-year prospective study at the Jordan University Hospital, 1993 through 1995. *Am J Infect Control.* 1997;25(4):322–9.
8. Ippolito G, Petrosillo N, Puro V, et al. The risk of occupational exposure to blood and body fluids for healthcare workers in the dialysis setting. Italian Multicenter Study on Nosocomial and Occupational Risk of Infections in Dialysis. *Nephron.* 1995;70(2):180–184.
9. Phipps W, Honghong W, Min Y, et al. Risk of medical sharps injuries among Chinese nurses. *Am J Infect Control.* 2002;30(5):277–282.
10. Shiao JS, Mclaws ML, Huang KY, et al. Student nurses in Taiwan at high risk for needlestick injuries. *Ann Epidemiol.* 2002;12(3):197–201.
11. Lee CH. Carter WA. Chiang WK. Williams CM. Asimos AW. Goldfrank LR. Occupational exposures to blood among emergency medicine residents. *Acad Emerg Med.* 1999;6(10):1036–1043.
12. Osborn EH, Papadakis MA, Gerberding JL. Occupational exposures to body fluids among medical students: A seven-year longitudinal study. *Ann Intern Med.* 1999;130:45–51.
13. Perry J, Parker G. Percutaneous injuries in home health care settings. *Adv Exposure Prev.* 2000;5(3):32–33.
14. Dale JC, Pruett SK, Maker MD. Accidental needlesticks in the phlebotomy service of the Department of Laboratory Medicine and Pathology at Mayo Clinic Rochester. *Mayo Clin Proc.* 1998; 73(7):611–615.
15. Jagger J, Bentley M, Tereskerz P. A study of patterns and prevention of blood exposures in OR personnel. *AORN J.* 1998; 67(5):979–974, 986.
16. Greene ES, Berry AJ, Jagger J, Hanley E, Arnold WP, III, Bailey MK et al. Multicenter study of contaminated percutaneous injuries in anesthesia personnel. *Anesthesiology.* 1998; 89(6):1362–1372.
17. Folin AC. Nordstrom GM. Accidental blood contact during orthopedic surgical procedures. *Infect Control Hosp Epidemiol.* 1997;18(4):244–246.
18. Metules T. What if you're stuck by a needle? *RN.* 2002;65(11):34ns2-34ns12.
19. California Department of Health Services, Occupational Health Branch. Sharps Injury Control Program Report. Oakland, CA: January 2002, pg. 8. Available on-line at: www.dhs.ca.gov/ohb/SHARPS/sharps.pdf.
20. Panlilio AL, Cardo DM, Campbell S, et al. Estimate of the annual number of percutaneous injuries in U.S. healthcare workers [abstract]. *Infect Control Hosp Epidemiol.* 2000;21:157.
21. Staff report. Latest data show drop in needlestick injuries. *Hosp Employee Health.* 2000;19(6):67.
22. Perry J. CDC releases national needlestick estimates. *Adv Exposure Prev.* 2000;5(2):19.
23. Occupational Safety and Health Administration. Occupational exposure to bloodborne pathogens; needlesticks and other sharps injuries; final rule (29 CFR Part 1910.1030). *Fed Regist.* 2001;66(12):5318–5325.
24. Hamory BH. Underreporting of needlestick injuries in a university hospital. *Am J Infect Control.* 1983;11:174–177.
25. Jagger J, Hunt EH, Brand-Elnaggar J, et al. Rates of needle-stick injury caused by various devices in a university hospital. *N Engl J Med.* 1988;319:284-288.
26. McGeer A, Simor AE, Low DE. Epidemiology of needlestick injuries in house officers. *J Infect Dis.* 1990;162(4):961–964.
27. Mangione CM, Gerberding JL, Cummings SR. Occupational exposure to HIV: Frequency and rates of underreporting of percutaneous and mucocutaneous exposures by medical housestaff. *Am J Med.* 1991;90:85–90.

28. Tandberg D, Stewart KK, Doezema D. Under-reporting of contaminated needlestick injuries in emergency healthcare workers. *Ann Emerg Med.* 1991;20:66–70.
29. O'Neill TM, Abbott AV, Radecki SE. Risk of needlesticks and occupational exposures among residents and medical students. *Arch Intern Med.* 1992;152:1451–1456.
30. Albertoni F, Ippolito G, Petrosillo N, et al. Needlestick injury in hospital personnel: A multicenter survey from central Italy. The Latium Hepatitis B Prevention Group. *Infect Control Hosp Epidemiol.* 1992;13:540–544.
31. Lynch P, White MC. Perioperative blood contact and exposures: A comparison of incident reports and focused studies. *Am J Infect Control.* 1993;21:357–363.
32. Chamberland M, Short L, Srivastava P et al. (Needlestick Surveillance Group, Centers for Disease Control and Prevention). Implementation, impact and compliance with use of safety devices to reduce percutaneous injuries during phlebotomy. Poster presented at "Bloodborne Infections: Occupational Risks and Prevention," Paris, France, June 8–9, 1995. Printed in *Adv Exposure Prev.* 1995;1(4):11.
33. Roy E, Robillard P (Public Health Direction, Montreal, Canada). Underreporting of accidental exposures to blood and other body fluids in health care settings – an alarming situation. Poster presented at "Bloodborne Infections: Occupational Risks and Prevention," Paris, France, June 8-9, 1995. Printed in *Adv Exposure Prev.* 1995;1(4):11.
34. Henry K, Campbell S. Needlestick/sharps injuries and HIV exposure among healthcare workers. National estimates based on a survey of U.S. hospitals. *Minn Med.* 1995;78(11):41–44.
35. Osborn EH, Papadakis MA, Gerberding JL. Occupational exposures to body fluids among medical students. A seven- year longitudinal study. *Ann Intern Med.* 1999;130(1):45–51.
36. Haiduven DJ, Simpkins SM, Phillips ES, Stevens DA. A survey of percutaneous/ mucocutaneous injury reporting in a public teaching hospital. *J Hosp Infect.* 1999;41(2):151–154.
37. Alvarado F, Panlilio A, Cardo D, NaSH Surveillance Group. Percutaneous injury reporting in U.S. hospitals, 1998 [abstract] *Infect Control Hosp Epidemiol.* 2000;21(2):106.
38. Centers for Disease Control and Prevention. Updated U.S. Public Health Service guidelines for the management of occupational exposures to HBV, HCV, and HIV and recommendations for postexposure prophylaxis. *MMWR.* 2001;50(RR11):1–42.
39. Hernandez MF, Bruguera M, Puyuelo T, et al. Risk of needle-stick injuries in the transmission of hepatitis C virus in hospital personnel. *J Hepatol.* 1992; 16(1–2): 56–58.
40. Mitsui T, Iwano K, Masuko K, Yamazaki C, Okamoto H, Tsuda F et al. Hepatitis C virus infection in medical personnel after needlestick accident. *Hepatology.* 1992; 16(5):1109–1114.
41. Sodeyama T, Kiyosawa K, Urushihara A, et al. Detection of hepatitis C virus markers and hepatitis C virus genomic- RNA after needlestick accidents. *Arch Intern Med.* 1993; 153(13):1565–1572.
42. Lanphear BP, Linnemann CC, Cannon CG, et al. Hepatitis C virus infection in healthcare workers: risk of exposure and infection. *Infect Control Hosp Epidemiol.* 1994; 15(12):745–750.
43. Zuckerman J, Clewley G, Griffiths P, et al. Prevalence of hepatitis C antibodies in clinical health-care workers. *Lancet.* 1994; 343(8913):1618–1620.
44. Monge V, Insalud (Grupo Español de Registro de Accidentes Biológicos en Trabajadores de Atención de Salud). *Accidentes Biologicos en Profesionales Sanitarios.* Madrid, Spain: 1995.
45. Arai Y, Noda K, Enomoto N, et al A prospective study of hepatitis C virus infection after needlestick accidents. *Liver.* 1996;16(5):331–334.
46. Serra C, Torres M, Campins M. Riesgo laboral de infeccion por el virus de la hepatitis C despues de una exposicion accidental. *Medicina Clinica.* 1998;111: 645–649.
47. Takagi H, Uehara M, Kakizaki S, et al. *J Gastroenterol Hepatol.* 1998;13(3):238–243.
48. Hasan F, Askar H, Al Khalidi J, et al. Lack of transmission of hepatitis C virus following needlestick accidents. *Hepatogastroenterology.* 1999; 46(27):1678–1681.
49. Kidouchi K, Aoki M, Oka S, et al. Surveillance of blood and body fluid exposures in Japan: international comparisons. 4th International Conference on Occupational Health for healthcare workers. September 30, 1999; Montréal, Canada.
50. Petrosillo N, Puro V, De Carli G, et al. Occupational exposure in healthcare workers: an Italian study of occupational risk of HIV and other blood-borne viral infections. *Br J Infect Control.* 2001; 2(2):15–17.
51. Baldo V, Floreani A, Dal Vecchio L, et al. Occupational risk of blood-borne viruses in healthcare workers: a five-year surveillance program. *Infect Control Hosp Epidemiol.* 2002;23:325–327.
52. Evans B, Communicable Disease Surveillance Centre, Public Health Laboratory Service, United Kingdom, written communication, April 24, 2002.
53. Jagger J, De Carli G, Perry J, Puro V, Ippolito G. Occupational exposure to bloodborne pathogens: epidemiology and prevention. Chapter 28 in: Wenzel R, ed., *Prevention and Control of Nosocomial Infections* (4th edition). Baltimore, MD: Lippincott, Williams and Wilkins. 2003; pp 430–465.

Sample-Size Requirements for Clinical Trials of Safety-Engineered Sharp Devices

by Gina Pugliese, R.N., M.S.; Teresa P. Germanson, Ph.D.; Judene Bartley, M.S., M.P.H.; Judith Luca, R.N.; Lois Lamerato, Ph.D.; Jack Cox, M.D.; and Janine Jagger, M.P.H., Ph.D.

Vol. 6, no. 6, 2003

This article originally appeared in Infection Control and Hospital Epidemiology under the title "Evaluating Sharps Safety Devices: Meeting OSHA's Intent" (2001;22[7]:456–458); it was reprinted with permission in AEP in abridged form.

The Needlestick Safety and Prevention Act of November 2000, which required healthcare employers in the United States to purchase safety-engineered needles and sharp medical devices, greatly accelerated the transition to safety technology. Today, the majority of needles and sharp devices used in clinical settings in the U.S. are of the safety variety, and the number of conventional devices in use has been declining. There remains a continued need, however, for product evaluations in order to assess the performance of safety devices.

Evaluations can be conducted in a number of ways. The most common is the informal device evaluation or product trial, in which subjective feedback from users is elicited; this type of evaluation has no minimum requirement for the number of devices that must be evaluated. Informal product trials are usually brief and involve clinical observations of a relatively small number of devices. They can provide valuable information about user preferences and product characteristics when healthcare institutions are considering the adoption of new devices.

However, such informal evaluations cannot be used to draw objective conclusions about user injury rates or the safety performance of specific devices. This can only be accomplished through a properly designed, large-scale study, with the results subjected to statistical analysis. **Table 1** shows the sample sizes of conventional and safety devices required in order to make statistically valid comparisons of injury rates between the two groups.

Safety Efficacy Studies

Safety efficacy studies are based on comparisons of device-specific injury rates—in this case, rates for conventional devices and their safety counterparts. Because needlesticks are, in statistical terms, rare events, it can be difficult to achieve a sample size large enough to yield statistically valid information. According to published studies, needlesticks occur in the range of 1 to 37 injuries per 100,000 devices used.[1–6] There is an inverse relationship between injury rates and required sample size: the lower the injury rate, the larger the denominator (i.e., number of devices) must be in order to attain an acceptable statistical power of 80% or 90%. Statistical power is the ability of a test to show a statistically significant difference between groups, if a true difference exists.

Sample Size Calculations

To determine the sample size needed to achieve a given level of statistical power, the researcher must first decide the minimum reduction in injuries from the safety device that would be clinically meaningful for that particular study. For instance, should the device prevent at least 75% of injuries, or would it be acceptable to prevent as few as 25%? Reasonable expectations for injury reductions from safety-engineered sharp devices can be derived from the literature; published studies show reductions from safety devices ranging from a low of 25%[3] to a high of 89%.[5] It is up to the researcher to decide what is acceptable under the specific circumstances of the trial. This decision, in turn, determines the number of devices that must be used in the study to achieve statistical significance. It is worth noting that a 100% reduction in injuries is an unrealistic expectation, unless the safety device completely eliminates the sharp.

Table 1 correlates sample-size requirements for the devices being compared (safety and conventional) with the projected injury rates for the conventional device and the desired level of injury reduction for the safety device. The

Table 1. Sample-Size Requirements for Detecting Reductions in Needlestick Rates.†

Injuries per 100,000 conventional needles	% reduction with safety devices	Number of devices required per device type
5	25%	4,600,000
	50%	1,000,000
	75%	375,000
10	25%	2,300,000
	50%	500,000
	75%	186,000
15	25%	1,540,000
	50%	340,000
	75%	125,000
20	25%	1,150,000
	50%	250,000
	75%	94,000

† Fisher's exact test, alpha=0.05, power=80%, two-tailed testing.

sample sizes given in the table are designed to achieve a statistical power of 80%, where alpha is set to 0.05 in two-tailed testing, and are calculated using Fisher's exact test.[7] Since the distribution of rare-event data may not meet the normality assumptions of Pearson's chi-square test, an exact test is the preferred analytic procedure. The selected parameters are derived from current literature and should provide realistic assumptions for determining sample sizes.

For example, if the projected baseline injury rate for the conventional device is 15 injuries per 100,000 devices used, and the researcher hopes to demonstrate a 75% reduction in injuries with the safety device, then 125,000 conventional devices and 125,000 safety devices would need to be used to achieve a statistical power of 80%. On the other hand, if the baseline injury rate for the conventional device is 5 per 100,000 and the researcher wants to show the same 75% reduction in injuries, the sample-size requirement is 375,000 for each kind of device. If the investigator establishes 25% as the minimum reduction worth detecting, then the sample-size requirement would be 1.15 million per group, given a baseline injury rate of 20 per 100,000.

Discussion

The calculations presented in Table 1 show that the number of devices required to compare needlestick rates for conventional and safety devices is enormous even for sample sizes on the lower end of the scale. At the high end, more than four million devices per group would be needed to achieve statistical significance if a conventional device causing 5 injuries per 100,000 devices were compared to a safety device reducing injury rates by 25%. At the low end, 94,000 devices per group would be needed to achieve statistical significance if a conventional device causing 20 injuries per 100,000 devices were compared to a safety device that reduced injury rates by 75%.

For the majority of healthcare facilities, these sample-size requirements would exceed annual device usage in most device categories. In the largest hospitals, an estimated 1.5 to 3 million syringes are used per year; for intravenous (IV) catheters, usage ranges from 75,000 to 300,000 devices per year, with a similar range for winged steel needles and phlebotomy needles. In average-sized or small hospitals, the number of devices used annually in these categories is much lower.

The investigator contemplating a single-center trial must first determine if the annual usage of the device to be evaluated is adequate to meet minimum sample-size requirements. If not, a multi-center study will likely be necessary. Another alternative is to extend the trial for the amount of time necessary to achieve an adequate sample size—if the results would still be worth knowing by the time the trial ended.

Undertaking a study without an adequate sample size increases the chance that statistical significance will not be achieved—even if the safety device is effective. Consider, for example, a study in which 150,000 needles are used in each group, conventional and safety, and the conventional device has a true baseline injury rate of 5 injuries per 100,000 needles. If the true relative effectiveness of the safety device is 50%, then the probability of obtaining statistically significant results in two-tailed testing, where alpha=0.05, is only 27%. In other words, it is most likely that the results from the study will not be statistically significant.

Many challenges face investigators as they embark on efficacy trials of safety-engineered devices, but none is more important than assuring that the sample size is adequate to meet statistical requirements. This will become even more challenging as safety-engineered devices become the predominant technology in the medical device marketplace. Eventually, safety devices with very low baseline injury rates will be tested against other, similar safety devices to determine which is "safest." The result is that sample-size requirements for such studies may reach prohibitive extremes, making multi-center collaborative studies a necessity in order to meet statistical demands.

References

1. Jagger J, Hunt EH, Brand-Elnaggar J, Pearson RD. Rates of needle-stick injury caused by various devices in a university hospital. *N Engl J Med.* 1988; 319(5):284–288.
2. Ippolito G, De Carli G, Puro V, Petrosillo N, Arici C, Bertucci R, Jagger J. et al. Device-specific risk of needlestick injury in Italian healthcare workers. *JAMA.* 1994; 272(8):607–610.
3. Centers for Disease Control and Prevention. Evaluation of safety devices for preventing percutaneous injuries among health-care workers during phlebotomy procedures—Minneapolis-St. Paul, New York City, and San Francisco, 1993-1995. *MMWR.* 1997; 46(2):21–25.
4. Jagger J, Bentley MB. Injuries from vascular access devices: high risk and preventable. Collaborative EPINet Surveillance Group. *J Intraven Nurs.* 1997; 20(6 Suppl):S33–S39.
5. Mendelson MH, Chen LBY, Finkelstein LE, Bailey E, Kogan G. Evaluation of a safety I.V. catheter (Insyte Autoguard, Becton Dickinson) using the Centers for Disease Control and Prevention (CDC) National Surveillance System for Hospital Healthcare Workers database [abstract]. *Infect Control Hosp Epidemiol.* 2000;21:111.
6. Chen LBY, Bailey E, Kogan G, Finkelstein LE, Mendelson MH. Prevention of needlestick injuries in healthcare workers: 27 month experience with a resheathable safety winged steel needle using CDC NaSH Database [abstract]. *Infect Control Hosp Epidemiol.* 2000;21:108.
7. Fisher, RA. *The Design of Experiments.* Edinburgh, Scotland: Oliver and Boyd; 1935.

PART IV

Needle Safety and Needlestick Prevention

U.S. Legislation and Policy

Overview:

Regulations and Legislation in the U.S. for Preventing Occupational Exposures to Bloodborne Pathogens

By Jane Perry, M.A., and Janine Jagger, M.P.H., Ph.D.

The 1991 Bloodborne Pathogens Standard

The bloodborne pathogens standard (BPS) was issued by the U.S. Occupational Safety and Health Administration (OSHA) in December 1991, after a lengthy rulemaking process.[1] It required healthcare facilities to develop an exposure control plan for each area of their institution; use engineering and work practice controls to eliminate or minimize employee exposures to bloodborne pathogens; provide puncture- and leak-resistant sharps disposal containers; train healthcare workers (HCWs) in safe work practices and universal precautions; provide follow-up and treatment, as appropriate, when an employee sustained a blood exposure; and maintain records of reported exposures.

Methods of compliance with the BPS included universal precautions, engineering controls, work practice controls (such as handwashing and not recapping needles), use of personal protective equipment (such as gloves and masks), and appropriate housekeeping and waste handling procedures.

The definition of "engineering controls" in the 1991 BPS included only two examples of what that concept meant in relation to needles and sharps: sharps disposal containers and self-sheathing needles. At that time, safety-engineered sharps technology was still relatively new and untested; little emphasis was placed on safety devices as a means of preventing blood exposures. During the 1990s, however, there was increasing pressure by HCW groups to emphasize safety devices as a primary engineering control to prevent needlesticks. In September 1998, OSHA issued a Request for Information (RFI) specifically on "engineering and work practice controls used to eliminate or minimize the risk of occupational exposure to bloodborne pathogens due to percutaneous injuries from contaminated sharps."[2] OSHA received almost 400 comments from healthcare facilities, including nursing homes, clinics, acute care facilities, and rehabilitation and pediatric hospitals, and published an executive summary in May 1999.[3] One of the key findings was that "safer medical devices are an effective and feasible method of hazard control." According to OSHA, "nearly every healthcare facility responding to the RFI noted that a reduction in injuries had occurred after the introduction of a safer medical device." The responses also indicated that training and education regarding the proper use of safer devices were key to their acceptance, that safer devices in general did not adversely affect patient care, and that the higher cost of these devices was offset by savings from reduced post-exposure testing and treatment.

On the basis of these findings, OSHA issued a revised compliance directive for the BPS in November 1999.[4] Compliance directives provide instruction to OSHA field officers on interpretation and enforcement of OSHA standards. The revised BPS directive placed explicit emphasis on the use of safety devices to prevent occupational blood exposures. The directive noted that since the BPS was issued in 1991, there had been "a substantial increase in the number and assortment of effective engineering controls," and directed employers to continuously evaluate new safety devices as they came on the market.

FDA and NIOSH Safety Alerts

Along with the BPS, there were several other significant federal actions during the 1990s related to the prevention of sharps injuries. In 1992, the Food and Drug Administration (FDA) issued a safety alert advising healthcare facilities to stop using needles to connect intravenous lines or access intravenous ports.[5] EPINet data showed that needles used for this purpose were responsible for a large proportion of needlestick injuries in the U.S., and in 1992 there were already more than a dozen needleless or shielded-needle products available to eliminate this risk. The FDA alert had a significant impact on reducing sharps injuries, as reflected in the EPINet data. In 1993, shortly after the alert was issued, 30% of needlesticks from hollow-bore needles were caused by needles used to access intravenous ports; by 1998, with a significant increase in the adoption of needleless and recessed-needle intravenous (I.V.) systems, the fraction was reduced to 13% in EPINet network hospitals, and continues to decline.

In 1999, the FDA, in conjunction with OSHA and the Centers for Disease Control and Prevention (CDC), issued a second sharps-related warning, this one regarding potential bloodborne pathogen exposures from glass capillary tubes.[6] The advisory urged healthcare facilities to choose safer alternatives (plastic or mylar-wrapped glass capillary tubes), in order to reduce high-risk injuries from blood-

Legislating Safety: U.S. State and Federal Government Actions to Prevent Occupational Exposures to Bloodborne Pathogens

Vol. 6, no. 4, 2003

1987	**August 1987:** Centers for Disease Control and Prevention (CDC) issues Universal Precautions for prevention of HIV transmission in healthcare settings.
1987–1989	Occupational Safety and Health Administration (OSHA) initiates rulemaking process and holds hearings on proposed Bloodborne Pathogens Standard.
1991	**Dec. 1991:** OSHA issues the Bloodborne Pathogens Standard (BPS) to "protect approximately 5.6 million workers in health care and related occupations from the risk of exposure to bloodborne pathogens." Standard requires engineering and work practice controls, personal protective equipment, training, surveillance, hepatitis B vaccination, and other actions to minimize risk of bloodborne disease transmission.
1992	**Feb. 1992:** Congress holds hearings on healthcare worker safety and needlestick injuries; healthcare workers, including a nursing assistant who was occupationally infected with HIV from a needlestick injury, give testimony about the need for safer needle devices. **March 1992:** OSHA issues compliance directive for BPS (directive provides guidance for OSHA field officers on how to enforce the standard when they conduct inspections). **April 1992:** Food and Drug Administration (FDA) issues safety alert on needlestick risk from needles used for piggyback connections and to access I.V. lines.
1997	**Jan. 1997:** CDC publishes two studies evaluating efficacy of safety devices in reducing percutaneous injuries (one study evaluates safety phlebotomy devices, the other a blunt suture needle); finds significant injury reductions from all devices studied. **Oct. 1997:** First federal needle safety bill introduced in U.S. House of Representatives, "Healthcare Worker Protection Act" (sponsored by Rep. Pete Stark [D-CA]; bill did not move forward).
1998	**Sept. 1998:** California passes first state needle safety law, A.B. 1208; this precedent-setting legislation sets the stage for other states to enact similar laws.* **Sept. 1998:** OSHA issues a Request for Information (RFI) on "Occupational Exposure to Bloodborne Pathogens", soliciting input from healthcare facilities on effectiveness of safety devices.
1999	**Feb. 1999:** FDA issues a safety advisory warning about injury/infection risk from use of glass capillary tubes. **May 1999:** A second federal bill, "Healthcare Worker Needlestick Prevention Act," introduced by Rep. Pete Stark in Congress (again, bill did not move forward). **Nov. 1999:** OSHA issues revised compliance directive for the BPS, based on responses to its RFI; for the first time, OSHA makes clear that healthcare facilities are *required* to use safety devices. **Nov. 1999:** National Institute for Occupational Safety and Health of the CDC issues a guideline, "Preventing Needlestick Injuries in Health Care Settings."
2000	**Sept. 2000:** Needlestick Safety and Prevention Act introduced in U.S. House of Representatives by Rep. Cass Ballenger (R-NC); bill passes both houses of Congress unanimously. **Nov. 6, 2000:** President Clinton signs into law the Needlestick Safety and Prevention Act.
2001	**Jan. 2001:** OSHA issues revised BPS, as mandated by the Needlestick Safety and Prevention Act. **April 2001:** Revised BPS becomes effective. **Nov. 2001:** OSHA issues updated compliance directive for BPS, reflecting new requirements in the revised standard.
2002	**April 2001 to May 2002:** OSHA issues 132 citations for failure to use engineering controls (safety devices)—four times more than all citations issued over the previous decade. Fines total over $1 million. OSHA also issues several letters of interpretation on the BPS that make it clear there are no exemptions to the requirement to use safety devices (other than patient safety and market availability).

**By the end of 2001, an additional 20 states had passed laws related to needlestick prevention.*

contaminated glass in the healthcare setting.

In November 1999, a third federal agency—the National Institute of Occupational Safety and Health (NIOSH) of the CDC—took action on sharps safety. NIOSH issued an alert ("Preventing Needlestick Injuries in Health Care Settings") that urged healthcare employers to use safety-engineered sharps devices and establish comprehensive programs to reduce needlestick injuries.[7] While it did not have the force of law, the alert was significant in that the CDC explicitly supported the use of safety devices as a primary means of preventing needlestick injuries.

State Legislation

While the U.S. federal government took significant steps in promoting sharps injury prevention during the 1990s, states were the first to introduce and pass needle safety legislation. In September 1998, California enacted groundbreaking legislation, A.B. 1208, which mandated that needles and other sharp devices have engineered sharps injury protection (built-in protective features). The bill mandated that the state's bloodborne pathogens standard be revised to reflect these new requirements. (California, along with about half of U.S. states, operates its own state OSH program).

From 1999 through 2001, 20 additional states passed legislation related to needle safety. Because the bills varied widely in their requirements and scope, there was increasing pressure for federal needle safety legislation that would bring national uniformity to requirements for safety-engineered sharp devices.

The Needlestick Safety and Prevention Act and 2001 Revised Bloodborne Pathogens Standard

In June 2000, the congressional Subcommittee on Workforce Protections convened a hearing to discuss the impact of the 1999 OSHA compliance directive and determine whether additional federal action was needed. The hearing was chaired by Rep. Cass Ballenger (R-NC), who subsequently introduced the Needlestick Safety and Prevention Act (H.R. 5178) in September 2000. With widespread support from the healthcare industry and medical device manufacturers, the legislation passed both houses of Congress unanimously and was signed into law by President Clinton on November 6, 2000.[8] The first law of its kind in the world, it provided U.S. HCWs with an unprecedented level of protection, and set a global standard for both employee and patient safety.

The Needlestick Safety Act mandated that the bloodborne pathogens standard be revised in several key areas to improve needle safety and strengthen exposure prevention programs; OSHA published the revised standard on January 18, 2001, and it became effective April 18, 2001.[9] The revised standard mandated that employers include non-managerial, frontline healthcare employees who provide direct patient care in the process of evaluating and selecting safer devices. Employers were also required to document in their exposure control plan that they had evaluated and implemented safer medical devices designed to reduce the risk of blood exposures, and update the plans at least annually to reflect changes in sharps prevention technology. Finally, employers were required to maintain a sharps injury log, with information on the type and brand of device causing injury, the department or work area where the injury occurred, and an explanation of how it occurred. The intent of this requirement was to enable healthcare facilities to evaluate exposure risk and device effectiveness. OSHA did not specify the format for the log, as long as it met the minimum requirements for data collection and the confidentiality of employees was protected. Employers with 10 or fewer employees were exempt from this requirement, but not from the other requirements of the revised BPS.

The revised standard included a new term, "sharps with engineered sharps injury protections," defined as "a non-needle sharp or a needle device used for withdrawing body fluids, accessing a vein or artery, or administering medications or other fluids, with a built-in safety feature or mechanism that effectively reduces the risk of an exposure incident." This definition delineates the kinds of procedures for which safety devices must be used. Safety devices are not needed for tasks that do not involve potential exposure to patients' body fluids, such as preparing medications in pharmacies.

With the revised BPS of 2001, OSHA made it clear that use of safety devices was not optional—it was the law.

References

1. Occupational Safety and Health Administration. Occupational exposure to bloodborne pathogens; final rule (29 CFR Part 1910.1030). *Fed Regist.* 1991;56(235):64004-64182.
2. Occupational Safety and Health Administration. Occupational exposure to bloodborne pathogens: Request for information. *Fed Regist.* 1998;63: 48250-48252.
3. Occupational Safety and Health Administration. Record summary of the request for information on occupational exposure to bloodborne pathogens due to percutaneous injury. *Executive Summary,* 1999;www.osha-slc.gov/html/ndlreport052099.html.
4. Occupational Safety and Health Administration. OSHA Instruction: Enforcement procedures for the Occupational Exposure to Bloodborne Pathogens. Directives number CPL 2-2.44D. Washington, DC: U.S. Department of Labor, November 5, 1999.
5. Benson JS, Food and Drug Administration. FDA safety alert: Needlestick and other risks from hypodermic needles on secondary I.V. administration sets - piggyback and intermittent I.V. Rockville, MD: U.S. Department of Health and Human Services; April 16, 1992.

6. Food and Drug Administration, National Institute for Occupational Safety and Health, and Occupational Safety and Health Administration. Glass capillary tubes: Joint safety advisory about potential risks. Rockville, MD: FDA, 1999.
7. National Institute for Occupational Safety and Health. NIOSH Alert: Preventing needlestick injuries in health care settings. DHHS (NIOSH) Publication No. 2000-108. November 1999.
8. Needlestick Safety and Prevention Act of 2000, Pub. L. No. 106-430, 114 Stat. 1901, November 6, 2000.
9. Occupational Safety and Health Administration. Occupational exposure to bloodborne pathogens; needle-sticks and other sharps injuries; final rule (29 CFR Part 1910.1030). *Fed Regist.* 2001;66(12):5318-5325.

California Leads the Way with Healthcare Worker Safety Law

by Jane Perry, M.A.

Vol. 4, no. 1, 1998

On September 30, 1998, California Governor Pete Wilson signed ground-breaking legislation that will require California OSHA (Cal/OSHA) to issue new safety needle guidelines to protect healthcare workers. The legislation, AB 1208, mandates amendments to the state's bloodborne pathogens standard, making California the first state to require healthcare facilities to purchase needles designed to prevent needlesticks.

Emergency regulations will be in effect by January 15, 1999, which give notice to employers of the new requirements. Full compliance will be required by August 1, 1999.

The legislation was introduced in May 1998 by Assemblywoman Carol Migden (D-San Francisco). When the governor signed the bill in September, Migden commented: "This is a long overdue measure for worker safety in California. Safety needles have been on the market for 10 years now. Failure to use this technology to protect workers has been unconscionable. We've finally changed that."

The bill was passed by the California State Assembly on a vote of 46-to-25, and by the State Senate on a vote of 22-to-12, and was sent to the governor on August 28. The governor was expected to veto the bill in response to opposition by the California Healthcare Association (CHA), representing 600 hospitals, health systems and physicians in California. Then, in the final few days, the CHA withdrew its opposition and wrote a letter to the governor supporting the bill.

In signing the bill, Gov. Wilson said, "This legislation recognizes the need to further address the hazard of employee exposure to bloodborne pathogens and calls for reasonable amendments to the Cal/OSHA bloodborne pathogens standard." He added, "In the absence of federal guidance, California finds it necessary to move forward with its own regulatory solution, which will most likely become the model for a national standard." Similarly, John Duncan, director of the State's Department of Industrial Relations, said "California has led the way on worker safety over the years. I challenge [federal regulators] to follow us."

A diverse coalition of groups representing healthcare workers, hospitals, and medical device manufacturers came together in support of the bill. Gov. Wilson called the bill "an important symbol of the consensus among the healthcare industry, labor, and regulatory agencies on the need to move forward with this occupational safety and health issue, and the wisdom of doing so through the public rule-making process."

A driving force behind the effort was the Service Employees International Union (SEIU), which organized a fax campaign to the governor and a candlelight vigil in San Francisco. SEIU president Andrew Stern said that the union had worked for "nearly a decade to get safer needles into the hands of healthcare workers" and was "proud of our role" in getting the legislation passed. He also commented, however, that "California's action begs the question: Why haven't [federal] OSHA and FDA acted to protect healthcare workers nationwide?"

Glenda Canfield, R.N., who coordinates SEIU's Nurses' Alliance in California and who organized the fax campaign to the governor, said that "nurses are the largest group of healthcare workers affected by this bill, and nurses all over California are excited about it." But, she said, until the revised standard comes into full effect in August 1999 and more safety devices are available, healthcare workers are still vulnerable: "Recently I got a call from a public health nurse who had just been stuck with a needle that had been used on an AIDS patient, and she was very upset. Here we are, so close to having the protection this bill will afford, and yet, as far as she's concerned, it's a million miles away. When I say next August to some people, that seems like a long time to wait."

Much of the coalition-building and consensus that made the successful passage of the bill possible took place over the summer during Cal/OSHA advisory committee meetings on the revision of California's bloodborne pathogens standard (BPS). Key participants included representatives from Kaiser Permanente (the nation's largest non-profit healthcare corporation, with 28 hospitals in California), CHA, SEIU, and device manufacturers, as well as nursing and infection control representatives and leading researchers in the prevention of occupational exposures.

Len Welsh, special counsel for Cal/OSHA and co-chair of the advisory committee on the BPS revision, comments that "advisory committee meetings are very powerful tools, if you want to address a technical area [such as bloodborne pathogens] and you can get people with the appropriate expertise to volunteer their time." In this case, he said, "it was particularly helpful that Kaiser and the SEIU were partnering around this issue." Enid Eck, R.N., M.P.H., a senior consultant for HIV and infectious diseases for the California division of Kaiser and a member of the advisory committee, observes that "At Kaiser, on the issue of sharps safety in particular, collaboration between management and labor is not at all unusual. What we brought to the table that facilitated the outcome, if anything, was a history built on trust and experience with our labor representatives. They have been invaluable to us over the years, particularly in the process of evaluating and implementing safety devices."

Roger Richter, a senior vice president at CHA and another member of the advisory committee, agreed that collaboration and compromise were key to getting an agreement on what should be in the revised BPS and how the timetable for its implementation, which is mandated in the bill, should be structured. "CHA opposed [AB 1208], almost until the end, on the basis that it was onerous and extremely costly. But thanks to the good work of the Cal/OSHA staff, a compromise was finally reached on what should be sent to the standards board for adoption. Assemblywoman Migden gathered all interested parties together and got agreement that, while the law said there had to be emergency regs in place by January 15, what would actually be required at that point was that everyone effected by the revised standard would be notified and would start working on implementation, but there wouldn't be any punitive action by Cal/OSHA until August. We felt comfortable that by August 1, 1999, hospitals can have programs in place to convert over to safety devices."

> ***"California is just the starting point for this issue. These are protections all U.S. healthcare workers should have."***
>
> —Len Welsh, Cal/OSHA

How have CHA member hospitals responded to the new requirements? "Since we've supported the bill," Richter says, "we've heard nothing but positive comments from our members. Although hospitals realize that conforming to the new law and regulations will be costly at first, the hope is that the cost will mostly be up front and that once the programs are put in place, the prices on devices will come down." When asked if the CHA had any figures on what the new regulations would cost, he responded, "If a hospital is currently not using any safety devices, implementing the new devices could cost three to seven times what they are currently paying for devices. But it's difficult to come up with numbers because many, if not most hospitals already have at least some safety devices in place."

Enid Eck of Kaiser says that the new legislation and regulations will cost "major money." In general, she notes, safety devices "cost two or three times more than conventional ones—and sometimes even more than that. The impact for those organizations that have not implemented any safety devices will be huge." But, she adds, "For organizations like ours that have been gradually implementing safety devices, while there will be an incremental increase in cost as we standardize, the impact won't be as big. All of the hospitals in the Kaiser system already have at least some safety devices."

Experts in healthcare worker safety and infection control, as well as leaders from industry, have universally praised the bill. Janine Jagger, M.P.H., Ph.D., director of the International Healthcare Worker Safety Center at the University of Virginia, predicts that the legislation "will have an impact far beyond the borders of California. The new law will certainly spur similar legislative initiatives in other states and nationally. Furthermore, the impact on industry will be profound because California represents at least 10% of the medical device market. Supplying California with safety devices will require increases in production and will attract more companies to compete in this new market, hopefully resulting in more choices and lower costs."

Could the legislation have an impact outside the United States? "The United States is the global springboard for new technology," Jagger says. "If we create a thriving safety device market in the U.S., the new technology has a better chance of being adopted in other areas of the world."

Many experts agree that the legislation will likely have a ripple effect. Tom Sutton, vice-president for marketing at Bio-Plexus, thinks the legislation will "cause a wave of similar actions in other states and at the federal level." Murray Cohen, Ph.D., chairman of the Frontline Healthcare Workers Safety Foundation, says "we often look to California to take a leadership role on important issues. I have every hope that this legislation will be seen as a good idea that will quickly spread across the country, and that Congress will take a similar, non-partisan approach to quickly introducing more and better safety devices to protect healthcare workers."

Gina Pugliese, R.N., M.S., former director of infection control and environmental safety at the American Hospital Association and now a consultant with Etna Communications, commended the bill for its thoughtful approach and flexibility: "The new law has some valuable components. First is a requirement for a written exposure control plan that includes a procedure for identifying and selecting existing sharps prevention technology. It often takes one or two years to successfully implement a new safety device, and the plan can be the most important component of that process. The law also allows for exemptions in implementing safer devices—for example, in a case where using safety devices might interfere with patient care. So it gives the employer some flexibility. This will be particularly helpful as devices are evaluated because not all safety devices work in all settings and there are some medical procedures for which effective safety devices are not available. Finally, the bill has a component that requires recording of exposure incidents in a sharps injury log, which will greatly expand the current criteria for reporting of needlesticks that is required by federal OSHA. From my perspective, this law is the most reasonable attempt to implement safer technology that has been proposed thus far."

Like other experts in the field, Patti Tereskerz, J.D., Ph.D., director of health law and policy for the U.Va. Healthcare

Worker Safety Center, hopes that the California legislation will "serve as a catalyst for passage of similar federal legislation" because, she says, "the level of protection a healthcare worker is afforded against contracting occupational infections should not be dependent upon his or her state of employment. Use of safety devices should become the national standard." Len Welsh of Cal/OSHA that belief: "California is just the starting point for this issue. These are protections all U.S. healthcare workers should have." Lynda Arnold, a nurse who contracted HIV from a needlestick injury in 1992 and who has led a national campaign for the past two years to get hospitals to adopt safety devices, says she is "proud of everyone's efforts" in getting the legislation passed. But, she adds, "healthcare workers in California are luckier than those in the rest of the country."

Federal Needlestick Safety and Prevention Act Signed Into Law

Vol.. 5, no. 4, 2000

President Clinton's statement upon signing of H.R. 5178, November 6, 2000:

"Today I am pleased to sign into law H.R. 5178, the Needlestick Safety and Prevention Act. This legislation requires changes in the bloodborne pathogens standard in effect under the Occupational Safety and Health Act of 1970. Supported by healthcare workers and their unions, as well as a bipartisan group of members of Congress, this bill will help to ensure the safety of health-care workers who may be exposed to disease while handling certain medical devices. The Needlestick Safety Act makes clearer the responsibility of employers to lessen the risk of injuries to workers from contaminated sharp devices. It also encourages manufacturers of medical sharps to increase the number of safer devices in the market. This legislation will help to make health care occupations safer."

On the Passage of the Needlestick Safety and Prevention Act

by Janine Jagger, M.P.H.

Vol. 5, no. 4, 2000

November 6, 2000 was an historic day for healthcare workers in the United States. Culminating years of determined advocacy by healthcare workers, unions, researchers and the medical device industry, the Needlestick Safety and Prevention Act (H.R. 5178) was signed into law by President Clinton in an Oval Office ceremony that I was privileged to attend.

I would like to recognize the critical contributions of Congressman Cass Ballenger (R-NC) and his staff, who wrote and introduced H.R. 5178 into the House of Representatives. He skillfully shepherded the bill through the House and the Senate, overcoming numerous obstacles and finally gaining unanimous support in both Houses of Congress.

I would also like to recognize the dedication of the Service Employees International Union (SEIU), which has been a leader for more than a decade in efforts to bring the safest medical devices to the frontlines of healthcare. The American Nurses Association (ANA) and SEIU lobbied tirelessly, with remarkable success, to organize support for needle safety legislation at both the state and federal levels. The Health Industry Manufacturers Association (HIMA) collaborated with other professional associations, also providing important support for state and federal legislation. This unique coalition of forces was an important factor in gaining bipartisan support from legislators.

EPINet and AEP

The International Healthcare Worker Safety Center supports two ongoing initiatives that we believe also made important contributions to the passage of H.B. 5178—our EPINet data-sharing research network and this publication, *Advances in Exposure Prevention*. I would like to say a word about both.

First, I want to acknowledge the dedication of the hospitals that participate in the EPINet network. In every state where needlestick bills have been introduced, EPINet data have been there providing the "power in numbers" to justify and support the need for safer technology for healthcare workers. EPINet was at work again supporting the federal bill at each step in the legislative process, giving voice to the thousands of healthcare workers whose occupational injuries are represented in the data. The effort that these hospitals have put into EPINet surveillance has borne fruit in many ways for the benefit of healthcare workers everywhere.

Second, I would like to thank the readers and supporters of *AEP*—frontline healthcare workers, infection control and employee health professionals, government agency workers, medical device manufacturers, and many others. As the year 2000 draws to a close, we mark the end of our sixth year of publication. From the beginning, our goal for *AEP* was to provide a forum for rapidly communicating the latest findings from our EPINet database and for discussing government policy, legislative developments, and best practices and devices to prevent occupational exposures to bloodborne pathogens. *AEP* has been a unique resource for government policy-makers and legislators seeking support for needle safety initiatives.

One of *AEP*'s most important roles has been providing a voice for healthcare workers occupationally infected with a bloodborne pathogen. The passage of H.B. 5178 honors their dedication and sacrifice.

This landmark legislation without a doubt will save many lives and improve the quality of healthcare. It also sets a world standard and challenges other countries to provide an equal level of protection to their healthcare workers. It is the highest form of recognition of our responsibility to care for those who care for us.

Federal Needlestick Safety and Prevention Act Promises Unprecedented Protection to U.S. Healthcare Workers

by Jane Perry, M.A.

Vol. 5, no. 4, 2000

On November 6, 2000, President Clinton signed groundbreaking legislation designed to protect 8 million U.S. healthcare workers from injuries caused by needles and other sharp medical devices. The Needlestick Safety and Prevention Act passed the House and the Senate quickly, receiving unanimous, bipartisan support in both chambers. The bill requires that healthcare facilities under the federal Occupational Safety and Health Administration (OSHA) use "safer medical devices, such as sharps with engineered sharps injury protections and needleless systems." These include needles that retract, blunt or otherwise shield the sharp point or edge after use.

The legislation, H.R. 5178, was authored by Rep. Cass Ballenger (R-NC), who introduced it in the House with co-sponsor Major Owens (D-NY). The House approved the measure on October 3, 2000. In the Senate, the bill was co-sponsored by Senators Jim Jeffords (R-VT), Mike Enzi (R-WY), Ted Kennedy (D-MA) and Harry Reid (D-NV), and passed on October 26. The bill's success was due to the efforts and support of a broad coalition of healthcare worker and industry groups, including the Service Employees International Union (SEIU), the American Nurses Association and the American Hospital Association. The SEIU, in particular, worked aggressively over the last decade to get needle safety bills introduced and passed, both at the state and federal levels.

"I'm glad that I lived long enough to see this bill pass the Congress and be enacted. My family and I will be celebrating the good news and praying for its speedy and effective implementation so that no other family has to endure what mine already has."

—Lynda Arnold, R.N. *(occupationally infected with HIV)*

The bill's passage was widely acclaimed by healthcare workers, industry representatives and government leaders and lawmakers alike. On the day President Clinton signed the bill into law, Secretary of Labor Alexis Herman commented, "Today, America's healthcare providers have been reassured that their own health is as important as their patients' health. The bill directs an amendment to OSHA's bloodborne pathogen standard to ensure more widespread use of safer medical devices to prevent dangerous needlesticks."

OSHA director Charles Jeffress, who has been supportive of efforts to reduce needlestick injuries through the use of safety-engineered sharp devices, commented, "We never know those whom we have helped, and they will never know us, but this mission is enormously satisfying and critically important. I am glad to have been able to help these past three years in this cause."

Janine Jagger of the International Healthcare Worker Safety Center pointed to the international implications of the bill: "The Needlestick Safety and Prevention Act provides the most significant level of occupational protection afforded to any healthcare workers in the world. It is certain that many countries will follow our lead and step up their own exposure prevention efforts." An article in the *New England Journal of Medicine* that she and her colleagues at the University of Virginia published in 1988 classified medical devices associated with sharps injuries and, for the first time, outlined design criteria for safer devices. This research is reflected in the federal needlestick prevention bill.

Some of the most passionate supporters of the bill have been occupationally infected healthcare workers such as Diane Mawyer, a Virginia nurse who was infected with hepatitis C during her 15 years as an OR nurse and blood bank supervisor.

"I am so pleased that Congress has taken action to protect healthcare workers from potential infection with bloodborne pathogens," Mawyer said. "Had this measure been in place when I was a practicing nurse, perhaps I would have been spared the physical, emotional and financial trauma of years of illness, three organ transplants and the loss of my career."

Who is left out?

Congressman Pete Stark (D-CA), who was an early champion of needlestick prevention legislation (he introduced bills in 1997 and 1999), commented, "We still have work to do to ensure that healthcare workers in public hospitals are also protected. But that does not minimize the importance of what we've accomplished today." Stark referred to the fact that the legislation does not cover public sector (state and municipal) employees in states under federal OSHA. (In states with state OSHA plans, both public and private employees are covered; these states will be required to update their bloodborne pathogens standard to match the revised federal standard.)

Twenty-seven states are under federal OSHA. Of those, seven (Georgia, Maine, Massachusetts, New Hampshire, Ohio, Texas and West Virginia) have passed needle safety laws that cover public employees. That leaves 20 states under federal OSHA where public employees will not have the protection afforded by H.R. 5178: Alabama, Arkansas, Colorado, Delaware, Florida, Idaho, Illinois, Kansas, Louisiana, Mississippi, Missouri, Montana, Nebraska, New Jersey, North Dakota, Oklahoma, Pennsylvania, Rhode Island, South Dakota, and Wisconsin. It is unclear how many of the nation's 8 million healthcare workers fall into the "public sector" category or, more to the point, how many healthcare workers in those 20 states are in the public category. When asked, OSHA was unable to provide such figures.

Rulemaking process preempted

The legislation specifically directs that OSHA bypass its usual rulemaking process, which can take five years or more. In the "Joint Statement of Legislative Intent," written by Senators Jeffords, Enzi, Kennedy and Reid and printed in the Congressional Record along with H.R. 5178, the senators note that "preemption of OSHA rulemaking procedures is not an action to be undertaken lightly," but "the requirements of this bill are driven by the unique circumstances surrounding this narrow and particular public health issue."

Joint statement provides interpretive help

Elsewhere in the Joint Statement, the senators stress that "it is not the intent of this legislation to disturb the underlying flexible, performance-oriented nature of the Bloodborne Pathogens Standard," nor to "alter OSHA's current enforcement of the BBP standard." ("Performance-oriented" means that OSHA does not specify what brands of safety devices a facility should use, only that devices used be safety-engineered. Individual institutions can decide what devices are best for their particular needs.) According to a congressional staffer, the Joint Statement is meant to provide guidance for courts and OSHA administrative law judges on the intent of H.B. 5178 in cases where its "plain meaning" is unclear or in dispute.

Clear authority provided for OSHA

California's needle safety law, A.B. 1208, was both the catalyst and the standard for subsequent needle safety legislation. Using the same mechanism as A.B. 1208, H.R. 5178 requires that federal OSHA's bloodborne pathogens standard be revised to require the use of safety devices. Cal/OSHA, in describing the rationale for revising its bloodborne pathogens standard, said that the "engineering control requirements ... lack a clear assignment to employers of affirmative responsibility to address the use of sharps injury prevention technology. This results in insufficient notice to employers as to what behavior is expected of them, and builds an inherent weakness into any Division enforcement action directed at this issue." Clearly, under both the California and the federal legislation (and the regulatory actions resulting from them), employers now have that "affirmative responsibility" to implement safety devices to protect their employees from percutaneous injuries—and OSHA has a clear mandate to enforce that duty. As Melody Sands, director of federal OSHA's Office of Health Compliance Assistance, says: "This legislation gives us the authority to specifically require [safer] devices."

"When I learned that I had been infected with HIV and hepatitis C as a result of an occupational needlestick injury, I prayed I would see the day that healthcare workers in this country were protected from preventable injuries and potentially deadly illnesses. H.R. 5178 will save countless American healthcare workers from having to tell a story like mine. I thank each of the legislators who supported this bill for demanding a higher standard for the healthcare workers of this country."

— Lisa Black, R.N.

The Needlestick Safety and Prevention Act: What Does It Require?

By Gina Pugliese, R.N., M.S., and Jane Perry, M.A.

Vol. 5, no. 4, 2000

The Needlestick Safety and Prevention Act (H.R. 5178) authorizes federal OSHA to revise the 1991 bloodborne pathogens standard (29 CFR 1910.1030) to require the use of safety-engineered sharp devices. OSHA has up to six months to publish the revised standard in the Federal Register; it will take effect 90 days after publication. According to OSHA, the revised Standard will be in effect by no later than August 6, 2001. Below we review the major provisions of the law and discuss some frequently asked questions about it.

Provisions of the new law:

- Requires healthcare employers to provide safety-engineered sharp devices and needleless systems to employees to reduce the risk of occupational exposure to HIV, hepatitis C and other bloodborne diseases.

- Expands the definition of "engineering controls" to include devices with engineered sharps injury protection. [See box for definitions to be added to the bloodborne pathogens standard.]

- Requires that exposure control plans document consideration and implementation of safer medical devices designed to eliminate or minimize occupational exposure. Plans must be reviewed and updated at least annually.

- Requires each healthcare facility to maintain a sharps injury log with detailed information on percutaneous injuries (including type and brand of device involved in exposure incident, department where exposure occurred and an explanation of how it occurred).

- Requires employers to solicit input from non-managerial (e.g., frontline) healthcare workers when identifying, evaluating and selecting safety-engineered sharp devices, and to document this process in the exposure control plan.

Frequently Asked Questions:

What effect does the new law have on OSHA's November 1999 compliance directive (CPL 202.44D) for the Bloodborne Pathogens Standard?

In drafting the federal Needlestick Safety and Prevention Act, legislators relied on the language and overall content of the compliance directive regarding requirements for the use of safety devices. H.R. 5178 provides legislative authority for OSHA's current enforcement emphasis on the use of safety devices as a primary engineering control to prevent occupational exposures to bloodborne pathogens. In addition to mandating the use of safety devices, OSHA's compliance directive also provides guidance on a number of other issues, such as updated requirements for post-exposure follow-up that include hepatitis C virus. (The revised compliance directive and additional compliance information and training resources are available on OSHA's web site: www.osha-slc.gov/SLTC/needlestick/compliance.html.)

Does OSHA require safety devices now?

Use of sharps with engineered sharps injury protection is required now. OSHA has the authority under the bloodborne pathogens standard to require the use of engineering controls, such as safety devices, to reduce risk to workers. OSHA clarified its position in November 1999 with the revised compliance directive, and outlined the requirements and enforcement procedures for implementation of sharps injury prevention devices. Since November 1999, OSHA has cited healthcare facilities for failure to use safety devices. In determining a facility's compliance with the standard, however, OSHA has considered, among other factors, evidence of adoption of safety devices and whether the exposure control plan includes on-going selection, evaluation, and implementation of such devices, with a timeline for implementation.

The federal Needlestick Safety and Prevention Act does not change the current enforcement activities of OSHA, but rather gives a legislative mandate for OSHA's requirement that healthcare employers provide their employees with safety-engineered sharp devices.

Once the bloodborne pathogens standard has been revised to require facilities to use safer devices, as mandated by H.R. 5178, will OSHA be stepping up its enforcement of the standard?

OSHA has already started conducting more inspections of healthcare facilities. During inspections, compliance with all occupational safety and health requirements is reviewed, including the bloodborne pathogens standard. The increase in inspections is part of a recent initiative that included a letter sent to 2,600 healthcare facilities that had the highest average illness and injury rates, announcing that OSHA would be conducting targeted inspections. The major source for OSHA inspections of healthcare facilities, however, will still be employee complaints. Thus it will be important to adhere to the requirement in H.R. 5178 that frontline workers' input be included in the development and implementation of a sharps injury program, in order to assure that their needs and safety concerns are being met.

What are the requirements for facilities in states with their own OSHA plans?

The 23 states with state OSHA plans must have regulations that are "at least as effective" (that is, at least as protective) as those of federal OSHA. Once the revised bloodborne pathogens standard is published in the Federal Register, state OSHA plans will have six months to revise and publish their corresponding standards so they match federal OSHA. In the interim, state plans will continue to enforce their current requirements. Some states with state OSHA plans, such as California, have already revised their bloodborne pathogens standard to require the use of safer devices.

How does the new federal law apply in states that have passed needle safety laws?

Seventeen states have passed needle safety legislation. The elements of the new federal law that are also part of many state laws include: (1) required use of sharps injury prevention devices; (2) written exposure control plan that is updated annually to reflect consideration and use of safety devices; (3) sharps injury log with detailed information on the type and brand of device causing injury and description of the incident; and (4) involvement of frontline workers in selection, evaluation and implementation of safety devices.

If a state needle safety law has requirements above and beyond what the federal law requires, then the additional state requirements must be followed. (For instance, some states require healthcare facilities to report needlestick injury data to a state agency.) If a state needle safety law is less protective than the federal law, the federal law's requirements must be followed.

Which healthcare facilities are covered by the new law?

The new federal law and (when published) the revised bloodborne pathogens standard apply to any facility where employees may be exposed to blood or other potentially infectious material, such as hospitals, long-term care facilities, and clinical laboratories.

What should our facility do now?

Collect exposure data, using a system such as EPINet. These data are essential to understanding where exposures are occurring in a facility, and what interventions, including safer devices, are necessary to prevent them. If your facility already has a system in place for tracking sharp-object injuries and blood and body fluid exposures, make sure the forms solicit information on the following: type and brand of device causing the injury, department where the exposure occurred, and an explanation of how the incident occurred. The new federal law requires that this information be collected. (The latest version of EPINet, to be released shortly, includes all this information on its data collection forms.)

New and Revised Definitions in the Bloodborne Pathogens Standard

(mandated by H.R. 5178):

(1) The definition of "engineering controls" is expanded to include, as additional examples of controls, "safer medical devices, such as sharps with engineered sharps injury protections and needleless systems."

(2) The term "sharps with engineered sharps injury protections" is added to the definitions (at 29 CFR 1910.1030[b]) and defined as "a nonneedle sharp or a needle device used for withdrawing body fluids, accessing a vein or artery, or administering medications or other fluids, with a built-in safety feature or mechanism that effectively reduces the risk of an exposure incident."

(3) The term "needleless systems" is added to the definitions and defined as "a device that does not use needles for: (a) the collection of bodily fluids or withdrawal of body fluids after initial venous or arterial access is established; (b) the administration of medication or fluids; or (c) any other procedure involving the potential for occupational exposure to bloodborne pathogens due to percutaneous injuries from contaminated sharps."

Establish a sharps safety task force. Each facility should have a multidisciplinary task force and assigned leader to coordinate the sharps injury prevention program. The task force should include both managerial and non-managerial (frontline) workers to assist with the development or revision of a plan for selection, evaluation and implementation of safety devices. Consider safety devices as one component of an overall sharps injury prevention program that includes management commitment to worker safety, education and training on the use of safety devices, strategies to encourage compliance with the use of safety devices, and ongoing evaluation of the effectiveness of safety devices in reducing the risk of injury from contaminated sharps.

Revise exposure control plan. The exposure control plan should be revised to include the plan and timetable for evaluating and implementing safety-engineered devices in all device categories with potential for bloodborne pathogen exposure. The involvement of frontline workers in the device selection process should be documented in the plan.
Select and evaluate devices. The new law does not recommend specific devices, but requires employers to conduct their own evaluations of available safety devices. At present, there are relatively few studies documenting the efficacy of specific safety devices. Therefore, hospitals must select devices to evaluate based on a consideration of their own needs and requirements. If your facility has group purchasing contracts, start by reviewing the safety devices that are currently under contract. Note, however, that healthcare facilities must evaluate any safety device they believe is appropriate and effective for their specific needs, regardless of whether it is covered by a group purchasing contract.

Implement safety devices. When evaluation is complete, devices should be implemented promptly after appropriate education and training on the use of the device.

Excerpts from the "Joint Statement of Legislative Intent on H.R. 5178"

Vol. 5, no. 4, 2000

Editor's note:
The following are excerpts from the "Joint Statement of Legislative Intent on H.R. 5178," published, along with the law itself, in the Congressional Record on 10/26/00. Authored by Senators Jeffords, Kennedy, Enzi and Reid, the statement provides guidance to courts and OSHA administrative law judges in cases where the "plain meaning" of the legislation is unclear or in dispute. It can also help healthcare professionals understand what the law is intended to achieve. Note that the exceptions to use of safety devices found in California's revised bloodborne pathogens standard, but not in H.R. 5178, are discussed in this statement.

"**In modifying the bloodborne pathogens standard (BBP standard) this bill makes narrowly-tailored changes to the standard.** It makes clear ... the direction already provided by OSHA in its Compliance Directive: namely, that employers who have employees with occupational exposure to bloodborne pathogens must consider and, where appropriate, use effective engineering controls, including safer medical devices, in order to reduce the risk of injury from needlesticks and from other sharp medical instruments."

"**[I]t is the intent of this legislation to reflect innovation and evolving technology in the marketplace,** in particular development in safer medical devices such as [sharps with engineered sharps injury protections] and needleless systems."

"**It is ... not the intent of this legislation to disturb the underlying flexible, performance-oriented nature of the Bloodborne Pathogens Standard.** For example, this legislation's reference to the **consideration and implementation of safer medical devices is hinged upon the 'appropriateness' and the 'commercial availability' of such devices.** Finally, while this may be stating the obvious, it is **not the intent of this legislation,** nor for that matter of the current Bloodborne Pathogens Standard, **for employers to implement use of any engineering control,** including a safer medical device, in any situation **where it may jeopardize a patient's safety, an employee's safety or where it may be medically contraindicated.** Moreover, all of the affirmative defenses available to an employer under the current BBP standard remain intact with this legislation. **It is not the intent of this legislation to alter OSHA's current enforcement of the BBP standard in these circumstances.**"

"**The drafters are aware that some of the newer, most effective technologies are more expensive than others and may create higher costs for healthcare facilities. Because some entities largely dependent on Medicare and/or Medicaid, such as long term care providers, will be required to comply with this legislation, we encourage the Health Care Financing Administration to examine the costs of the new technologies and consider these costs when determining Medicare reimbursement rates.** Similarly, we hope that the states will examine these costs and determine whether the costs should be reflected in the Medicaid reimbursement rates."

"**The sharps injury log is to be used as a tool for employers so that they may determine their high risk areas for sharps injuries and use it as a means to evaluate particular devices that may or may not be effective in reducing sharps injuries.** At a House Subcommittee on Workforce Protections hearing in June, representatives of the American Hospital Association testified that many healthcare settings, particularly hospitals, already have in place some type of 'surveillance system' for tracking needlestick and other sharps injuries. The AHA witness noted that hospitals have found this to be an effective tool to provide necessary information to help reduce such injuries."

Cal/OSHA Letter on Reuse of Blood Tube Holders

by Jane Perry, M.A.

Vol. 5, no. 5, 2001

In a previous issue of AEP we published an account of Julie Naunheim Hipps' needlestick and subsequent infection with HCV ("When Home Is Where the Risk Is," vol. 5, no. 3). She was stuck by the tube-puncturing end of a two-ended phlebotomy needle/adapter, which was not exposed until she removed the adapter from the reusable blood tube holder, per instructions from her employer. This practice is risky because it exposes an otherwise shielded, blood-filled needle. The question has been raised as to whether this practice is a violation of the bloodborne pathogens standard, which says that "contaminated needles and other contaminated sharps shall not be bent, recapped, or *removed*" (emphasis added). California OSHA referenced this section of the standard when it issued a citation on 12/10/99 to a healthcare facility that routinely reused blood tube holders. The employer was cited for failing "to ensure that contaminated sharps were not removed from devices."

Below we print a letter, dated 7/13/00, that Cal/OSHA sent in response to an inquiry about the reuse of blood tube holders. It provides helpful interpretation of the bloodborne pathogens standard on this issue.

Cal/OSHA's primary concern with re-use of a needle holder in a venipuncture would be with respect to its involving removal of a needle contaminated with blood or other potentially infectious material (OPIM) from the holder, or the possibility of contamination of the holder, or some other condition, that might cause or contribute to an exposure incident.

Cal/OSHA's Bloodborne Pathogens standard ... states: 'Contaminated sharps shall not be bent, recapped, or removed from devices.' An exception is provided to the three prohibitions in subpart 2: if the otherwise prohibited procedure is performed using a mechanical device or one-handed technique and the employer can demonstrate that no alternative is feasible or that such action is required by a specific medical or dental procedure. [Editor's note: In the federal bloodborne pathogens standard, 1910.1030, this language is found at (d)(2)(vii)(A-B).]

The rationale for the prohibition on needle removal in Section 5193, which is identical in effect to the prohibition found in the bloodborne pathogens standard enforced by federal OSHA, is that any non-essential handling or manipulation of a contaminated sharp object increases the risk of an exposure incident.

Were there to be a situation in which an employer could credibly demonstrate in their particular circumstances that there were no feasible alternatives to re-use of the needle holder in a particular phlebotomy procedure, or that such action was required by a specific medical or dental procedure, then removal of the needle using a mechanical device or one-handed technique would be allowed and conceivably the needle holder could be re-used assuming such re-use was not associated with other occupational safety concerns. Lacking such circumstances satisfying the exception to the prohibition on needle removal, which we presume to be the more usual case, removal of a contaminated needle from a needle holder would be prohibited by Section 5193, in which case the needle holder would need to be disposed of along with the contaminated needle.

Removing a needle adapter from a blood tube holder exposes the "hidden" needle that punctures the blood tube holder. Although the tube-puncturing needle—sometimes referred to as the "back end" of the phlebotomy needle/adapter—is covered by a rubber sleeve, it can still pierce the sleeve and cause a needlestick. On the other hand, because it is covered by the rubber sleeve, healthcare workers may not perceive it as an injury risk.

Therefore, to reduce risk, employers should instruct healthcare workers not to reuse blood tube holders. If the rubber-sleeved needle is not removed from the holder, risk of injury is low. If a double-ended adapter is used, the blood-drawing needle (i.e., the end that is inserted in the patient) should be protected as well. Safety blood tube holders have built-in mechanisms to shield the blood-drawing needle after use, providing protection for both ends of the needle/adapter. Another option is to use a self-blunting blood-drawing needle with a conventional single-use holder.

The trend among device manufacturers is toward phasing out reusable blood tube holders. By instructing employees not to reuse holders, hospital administrators will help reduce the number of preventable sharps injuries and improve safety in the healthcare workplace.

OSHA's 2001 Updated Compliance Directive: Guidance on New Requirements of Bloodborne Pathogens Standard

by Jane Perry, M.A., and Janine Jagger, M.P.H., Ph.D.

Vol. 6, no. 1, 2002

The Occupational Safety and Health Administration (OSHA) released an updated compliance directive for the bloodborne pathogens standard on November 27, 2001—the second time in two years the directive has been revised. (The last update was November 5, 1999.) The directive (CPL 2-2.69) is used by OSHA compliance officers to interpret and enforce the bloodborne pathogens standard (BPS). It was updated to reflect the "major new requirements of the standard," including: (1) documentation in the exposure control plan that safer needle devices are being evaluated and implemented, with the plan updated annually to reflect consideration of new technology; (2) documentation of the involvement of non-managerial, frontline employees in choosing safer devices; and (3) establishment and maintenance of a sharps injury log for recording injuries from contaminated sharps. The revised BPS was published on January 18, 2001, and became effective on April 18, 2001, with enforcement of the new requirements starting on July 18, 2001.

The directive reminds compliance officers that no single safer medical device is appropriate for all situations and that employers "must implement the safer devices that are appropriate, commercially available, and effective." The directive also includes detailed instructions on inspections of multi-employer worksites, including employment agencies, personnel services, home health services, physicians and healthcare professionals in independent practices, and independent contractors.

Following is some of the new language OSHA has added to the directive:

1. **Removal of phlebotomy needles from blood tube holders:** OSHA addresses this issue explicitly for the first time. The directive says: "The practice of removing the needle from a used blood-drawing/phlebotomy device is rarely, if ever, required by a medical procedure. Because such devices involve the use of a double-ended needle, such removal clearly exposes employees to additional risk. Devices with needles must be used and immediately discarded after use, un-recapped, into accessible sharps containers." If an employer claims that no alternative to the removal of contaminated needles is feasible, compliance officers are instructed to "review the exposure control plan for a written justification supported by reliable evidence." This language amounts to a general prohibition on the removal of phlebotomy needles in order to reuse blood tube holders. [See Compl. Dir. XIII(D)(5), which discusses paragraph (d)(2)(vii) of the BPS prohibiting bending, recapping, or removing contaminated needles.]
2. **Use of unwinders in sharps containers:** "The sharps container should not create additional hazards. Some sharps containers have unwinders that are used to separate needles from reusable syringes or from reusable blood tube holders. The use of these are generally prohibited." [Compl. Dir. (XIII)(D)(28)]
3. **Solicitation of employee input:** "Compliance Officers should determine how the devices used in the facility were selected and review the employers' documentation of their employees' input. [...] Employees in various departments and situations should be interviewed to determine the extent to which the employer solicited employee input." [Compl. Dir. (XIII)(C)(6)]
4. **On whether safety needles must be used for non-patient purposes:** "Needles that will not become contaminated by blood during use (such as those used only to draw medication from vials) are not required to have engineering controls under this standard." [Compl. Dir. (XIII)(D)(2)]
5. **On the requirement to maintain a sharps injury log:** "If the nature of the incident is such that determining the type and brand of the device [involved in the injury] would increase the potential for additional exposure (e.g., housekeeper stuck through trash bag), the type/brand may be recorded as 'Unknown.' [...] The purpose of the log is to aid in the evaluation of devices being used in the workplace and to quickly identify problem areas in the facility. Thus, it should be reviewed regularly and during the review and update of the Exposure Control Plan." [Compl. Dir. (XIII)(H)(3)]

If you are responsible for implementing the new requirements of the BPS at your healthcare facility, the updated compliance directive will be an important resource in understanding how OSHA will measure and track compliance. It is available on OSHA's website on its "Bloodborne Pathogens and Needlestick Prevention" page, under "OSHA Directives." Another valuable resource is OSHA's FAQ on the BPS; it offers useful clarification on a number of questions (go to www.osha-slc.gov/needlesticks/needlefaq.html).

Reuse of Blood Tube Holders, Redux

In 2002, OSHA clarified that removing phlebotomy needles to reuse blood tube holders violated the bloodborne pathogens standard—but some lab groups have refused to switch to single-use holders

by Jane Perry, M.A., and Janine Jagger, M.P.H., Ph.D.

Vol. 6, no. 4, 2003

Blood drawing is widely recognized as one of the highest-risk procedures for bloodborne pathogen transmission, because it involves a blood-filled, hollow-bore needle.[1] While many healthcare facilities have made great strides in converting to safety-engineered phlebotomy devices, one component of phlebotomy remains a source of contention.

Vaccum tube phlebotomy sets are comprised of three component parts: a double- or single-ended phlebotomy needle, a blood tube holder, and a blood tube. A double-ended needle has a sharp at both ends: on one end, a needle that is inserted in the patient (the "front end"); on the other (the "back end"), a needle sheathed in rubber that pierces the stopper of the blood tube (the sheath slides back as the needle goes into the tube stopper).[2] A single-ended phlebotomy needle has a tube-puncturing needle at one end, and a luer fitting on the other that can be connected to a luer port on an I.V. line or to the tubing on a butterfly needle.

To perform a blood draw, the phlebotomy needle is first screwed into the blood tube holder (which both holds the needle and helps balance the tube while blood is drawn); the front end of the needle is then inserted into the patient's vein, the blood tube is inserted into the tube holder, the "back end" needle pierces the tube stopper, and blood flows into the blood tube. With a safety phlebotomy needle, after the blood draw is completed the safety feature is activated, shielding the front end of the needle. The filled blood tube is placed in a rack—and then the needle must be disposed of.

It is regarding this step that disagreement arises. Blood tube holders are made in both single-use and multi-use versions. With a single-use holder, the needle remains integrated with the holder and both pieces are disposed of together. Both ends of the phlebotomy needle are thus protected—at the front end by the safety shield and at the back end by the tube holder.

With multi-use blood tube holders, which are made of heavier material than single-use ones, the needle must be detached before the holder can be reused. This exposes the back end of the phlebotomy needle. Although the back end is covered by the rubber sheath, the sheath is designed to easily pull back from the needle. (Of course, if a conventional phlebotomy needle is used with no protective feature, the front end of the needle will be exposed after use and pose a risk as well.)

In 2000, AEP published the account of Julie Naunheim Hipps, a nurse in Missouri who, in 1999, was working in a home-care setting when she was stuck by the back end of a phlebotomy needle that had been used on a patient infected with hepatitis C (HCV). The home health agency for which she worked required employees to reuse blood tube holders; her injury occurred after she had removed the needle from the tube holder and was taking the needle to a disposal container. As a result of her injury, she seroconverted to HCV.[3]

Julie's case provided evidence of the risk of bloodborne pathogen transmission from injuries involving the back end of phlebotomy needles. Moreover, the practice of removing phlebotomy needles from blood tube holders appeared to violate language in the 1991 bloodborne pathogens standard, which said: "Contaminated sharps shall not be bent, recapped, or removed from devices."[4] The standard allowed exceptions only if an employer could demonstrate that no feasible alternative to needle removal existed, or that it was required by a specific medical procedure.

In July 2000, California OSHA cited a healthcare facility for removing needles from blood tube holders; in a letter to the facility, Cal/OSHA stated that this practice increased non-essential handling and manipulation of contaminated needles.[5] Most healthcare facilities in California have since switched to single-use blood tube holders.

In the revised compliance directive for the bloodborne pathogens standard, published in November 2001, federal OSHA specifically addressed the issue of reusing blood tube holders for the first time: "The practice of removing the needle from a used blood-drawing/phlebotomy device is rarely, if ever, required by a medical procedure. Because such devices involve the use of a double-ended needle, such removal clearly exposes employees to additional risk."[6]

This statement was followed, in June 2002, by a Standard Interpretation letter stating in full OSHA's position: "In order to prevent potential worker exposure to the contaminated hollow bore needle at both the front and back ends, blood tube holders, with needle attached, must be immediately discarded into an accessible sharps container after the safety feature has been activated." OSHA added: "Removing contaminated needles and subsequently reusing blood tube holders poses multiple potential hazards. The increased manipulation required to remove a contaminated needle from a blood tube holder is unnecessary and may result in a needlestick from either the front or back end of the needle."

OSHA also warned that a risk to patients from reuse of blood tube holders may exist: "While patient safety is not within the scope of OSHA's mission, we recognize that tube holder reuse poses a potential health hazard to patients. Some clinical studies show that 50-80% of blood tube holders may be contaminated after just one use [Advance/ Laboratory, p. 70, January 2000]. While it may be difficult to document a patient-to-patient cross-contamination exposure, the risk does exist and should be weighed when making decisions regarding overall safety."[7]

OSHA's letter of interpretation made clear that it considered removal of phlebotomy needles from blood tube holders a violation of the bloodborne pathogens standard and, as such, subject to citations and fines. In fact, in September 2002 OSHA cited Laboratory Corporation of America (LabCorp) for this practice.[8] Although most healthcare facilities accepted OSHA's interpretation, as stated in the June 2002 letter, and began converting to single-use holders, LabCorp did not and decided to fight the citation. The company is one of the largest clinical laboratories in the United States, testing more than 300,000 specimens daily and operating a network of 47 primary testing locations, with another 1,000 patient service centers.

The American Clinical Laboratory Association (ACLA), a lobbying group of which LabCorp is a member, submitted a 13 page white paper to OSHA in January 2003 outlining its reasons for opposing OSHA's position on reuse of blood tube holders. The ACLA paper stated that "using only single-use devices ... would increase the cost of doing business approximately fivefold" and "creates up to 10 times as much waste as use of the multiple-use holder."[8] Clearly, the company was worried about the cost of switching to single-use blood tube holders.

In an interview with *CAP Today,* David King, LabCorp's senior vice president and general counsel, stated: "We have looked at this issue very closely, and we don't see evidence of back-end sticks due to the use of the reusable holders." In the same article, JoAnne Glisson, ACLA vice president of government relations, echoed King: "There's absolutely no data to indicate a problem here. There are millions of [blood] draws and virtually no evidence of injuries."[8]

But finding the data on back-end injuries is not a simple matter; sharps injury logs do not typically record which end of the phlebotomy needle injured the worker. And until a few years ago, the maxim "no data, no problem" has applied. The International Healthcare Worker Safety Center decided, therefore, to review EPINet data for evidence that might document the risk; the results are reported in the article on injuries from phlebotomy needles [found in Part I of this book, in the EPINet data articles section]. These data are supportive of OSHA's policy prohibiting reuse of blood tube holders.

Although OSHA dropped its citation of LabCorp in the fall of 2003, it will release a safety and health information bulletin on the disposal of phlebotomy blood-tube holders in the near future, according to Amber Hogan, an industrial hygienist with OSHA's Office of Health Enforcement. This bulletin will further support the single-use policy and reinforce OSHA's position on this contested requirement.

[Editor's note: see reference number 9, below, for publication information on the bulletin.]

References

1. Jagger J. Report on blood drawing: risky procedures, risky devices, risky job. *Adv Exposure Prev.* 1994;1(1):4–9.
2. ECRI. Sharps safety and needlestick prevention; blood collection needles and tube holders. Plymouth Meeting (PA): ECRI. 2001;10(1):47–49.
3. Perry J. When home is where the risk is. *Adv Exposure Prev.* 2000;5(3):25,30–32.
4. Occupational Safety and Health Administration. Occupational exposure to bloodborne pathogens; final rule (29 CFR Part 1910.1030). *Fed Regist.* 1991;56(235):64004–64182.
5. Perry J. Cal/OSHA letter on reuse of blood tube holders. *Adv Exposure Prev.* 2001; 5(5):59.
6. Occupational Safety and Health Administration. OSHA Instruction: Enforcement procedures for the Occupational Exposure to Bloodborne Pathogens. Directives number CPL 2.2.69. Washington, DC: U.S. Department of Labor, November 27, 2001.
7. Occupational Safety and Health Administration, Department of Labor. Standard Interpretations: Re-use of blood tube holders. Washington, DC: U.S. Department of Labor, June 12, 2002. (Available on-line at: www.osha.gov/pls/oshaweb/owadisp.show_document?p_table= INTERPRETATIONS&p_id=24040.)
8. Southwick K. Sticky business with tube holders. *CAP Today.* 2003 (April);17(4):1,12–22.
9. Occupational Safety and Health Administration. Safety and Health Bulletin: Disposal of contaminated needles and blood tube holders used for phlebotomy. Issued October 15, 2003; available on-line at www.osha-slc.gov/dts/shib/shib101503.html.

Risks to Health Care Workers in Developing Countries

by Charles Sagoe-Moses, M.D.†, Janine Jagger, M.P.H., Ph.D.‡, Jane Perry, M.A.‡, Richard D. Pearson, M.D. ††

† Deputy Regional Director of Health Services, Ministry of Health, Accra, Ghana

‡ International Healthcare Worker Safety Center, University of Virginia Health System, Charlottesville, Virginia

††Departments of Internal Medicine and Pathology, Division of Geographic and International Medicine, University of Virginia Health System, Charlottesville, VA

The first report of a health care worker infected with the human immunodeficiency virus (HIV) by a needle stick, published in the medical literature in 1984,[1] launched a new era of concern about the occupational transmission of bloodborne pathogens. In the United States, universal precautions were implemented,[2] regulations such as the Bloodborne Pathogens Standard were issued,[3] and the rate of vaccination against hepatitis B virus (HBV) among health care workers increased dramatically.[4] After a decade of phenomenal technological advances in sharp devices engineered for safety, the federal Needlestick Safety and Prevention Act, requiring the use of safer devices, became law in November 2000.[5–6]

The Risks

Protecting health care workers in developing countries, however—where even the basics of medical care are difficult to provide and where the protection of health care workers does not appear on any list of health care priorities—is a formidable challenge. It is all too easy to ignore a problem about which there are few data. Clearly, health care workers in developing countries are at serious risk of infection from bloodborne pathogens—particularly HBV, hepatitis C virus (HCV), and HIV—because of the high prevalence of such pathogens in many poorer regions of the world.[7,8] HBV and HCV, for example, are endemic in sub-Saharan Africa. A study involving 803 schoolchildren in Ghana found that 61.2 percent had at least one marker of HBV infection and reported a seroprevalence of anti-HCV antibodies of 5.4 percent.[9] Of 303 subjects in three villages in Gabon, 19 percent were carriers of hepatitis B surface antigen, and in one village, 24 percent of subjects had HCV antibodies.[10] By contrast, the prevalence of HBV and HCV in the U.S. population is approximately 4.9 percent and 1.8 percent, respectively.[11,12] Furthermore, other lethal blood-borne pathogens, including Lassa virus, Ebola virus, and other hemorrhagic fever viruses, are endemic in some of the same tropical regions.[7]

Although the prevalence of blood-borne pathogens in many developing countries is high, documentation of infections caused by occupational exposure in these countries is scarce. Seventy percent of the world's HIV-infected population lives in sub-Saharan Africa, but only 4 percent of worldwide cases of occupational HIV infection are reported from this region.[13,14] By contrast, 4 percent of the world's HIV-infected population lives in North America and western Europe, yet 90 percent of documented occupational HIV infections are reported from these areas **(Figure 1)**.[13,14] It is unlikely that surveillance and reporting of occupational exposure to infected blood will be undertaken in places where postexposure prophylaxis, treatment, and workers' compensation are lacking.

In developing countries, the risk of occupational transmission of blood-borne pathogens is increased by the excessive handling of contaminated needles that results from some common, unsafe practices.[7–8,15–22] These include the administration of unnecessary injections on demand, the reuse of nonsterile needles when supplies are low, and the unregulated disposal of hazardous waste. Such practices pose risks of disease transmission to health care workers, patients, and communities at large.

In many developing countries, the high demand for injections derives from the belief that they are more effective than other forms of treatment. In Ghana, 80 to 90 percent of the patients who visited a health center received one or more injections per visit.[23] Similar findings have been reported in Uganda and Indonesia.[23] A correlation has been documented between the frequency of injections and the prevalence of HBV, HCV, and HIV in the population.[7–8]

Figure 1. Distribution of the HIV-Infected Population (Panel A) and of Documented Occupational HIIV Infections (Panel B) Worldwide

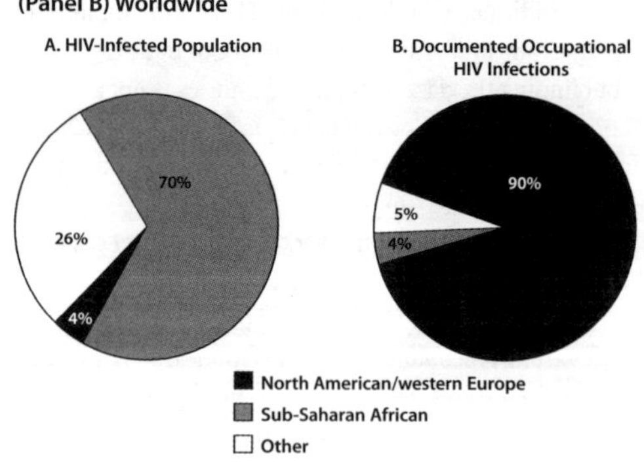

The data in Panel A are from the Joint United Nations Program on HIV/AIDS and the World Health Organization.[13] The data in Panel B are from Ippolito et al.[14] Percentages do not total 100 because of rounding.

Although many developing countries are replacing sterilizable syringes and needles with "auto-disable" and standard disposable syringes, where sterilization is still practiced it is often incomplete.[24] Improperly sterilized injection equipment has been associated with outbreaks of HBV infection, Ebola fever, Lassa fever and tetanus.[7–8]

Unnecessarily hazardous diagnostic equipment, such as nonretracting finger-stick lancets and glass capillary tubes (both of which have been associated with the occupational transmission of HIV[14,25]), is routinely used in developing countries to test for common tropical diseases such as malaria and filariasis. More than 100 million tests for malaria are performed each year.[26] Because much of this testing is done in outreach programs, hazardous equipment endangers not only health care workers but members of the community as well—especially children, who may be tempted to play with blood-contaminated devices that have not been securely contained. Despite the availability of plastic or plastic-wrapped capillary tubes and automatically retracting lancets, which could nearly eliminate this risk,[27–28] no formal recommendations have been made regarding incorporating the safer equipment into programs designed to eradicate tropical disease.

Further exacerbating the risk to health care workers in developing countries is a lack of gloves, gowns, masks, and goggles to protect them from contact with blood. It was recently reported in Tanzania that birth attendants cover their hands with plastic bags to protect themselves from exposure to HIV during deliveries because there are no gloves available.[29] There is also a lack of safe disposal systems for the secure containment and elimination of contaminated waste.[24]

The Costs

Protecting health care workers in developing countries from exposure to blood-borne pathogens will involve some cost. In industrialized countries, the cost of protective devices and equipment that reduce blood exposure may be offset by lower expenditures associated with postexposure testing and prophylaxis, medical treatment of infected workers, institutional insurance premiums, and workers' compensation payments.[30–31] In most developing countries, however, similar economic incentives do not exist; there is little reason for postexposure follow-up in countries that cannot afford prophylaxis, treatment, and compensation benefits.

Nevertheless, there are costs associated with failing to protect health care workers in developing countries. The loss of a wage-earning health care worker can be devastating to the financial security of the worker's family. The loss of health care workers can also have a disproportionate effect on the fragile health care infrastructure of developing countries, where trained health professionals are scarce in relation to the overall populations they serve. Statistics from the World Health Organization (WHO) indicate that there are fewer than 10 physicians per 100,000 population in 15 sub-Saharan countries, as compared with nearly 250 physicians per 100,000 population in the United States.[32] Similar discrepancies exist between the numbers of nurses in these countries and the number of nurses in the United States. Any reduction in the work force further strains understaffed and overextended health care systems. Possibly the largest unrecognized cost of failing to protect health care workers is the loss of the national investment in the training of workers whose careers are cut short by occupationally acquired infections.

A comprehensive cost-benefit analysis of measures to improve the safety of health care workers in developing countries has yet to be undertaken. However, one study, in which the cost of syringe-transmitted disease was factored into the cost-benefit equation, found that switching to a higher-cost syringe with an auto-disable feature to prevent recycling and reuse would result in significant long-term savings in several sub-Saharan countries (Ekwueme DU, Centers for Disease Control and Prevention: personal communication). The total cost per injection with the use of auto-disable, standard disposable, and sterilizable syringes was estimated on the basis of the cost of the equipment, the direct cost of medical treatment for HBV and HIV transmitted by nonsterile, reused syringes, and the indirect cost of lost years of productivity and lost years of life for infected persons. On the basis of this more comprehensive model, the mean cost per injection in Uganda was estimated to be more than five times as high for a standard disposable syringe as for an auto-disable syringe. Similar studies are needed to assess the cost-benefit ratio associated with providing protective equipment, such as safety needles and protective gloves, to health care workers in developing countries. When conducting cost-benefit studies of conventional as compared with safety devices, the cost of treating occupationally transmitted disease and the lost investment in medical and nursing education must be included.

Conclusions

Health care workers are a crucial resource in the health care systems of developing nations. In many countries, including those of sub-Saharan Africa, workers are at high risk for preventable, life-threatening occupational infections. Yet the protection of health care workers in these countries is largely neglected in national priorities for health care and by the international organizations that fund health care initiatives. We must not delay the implementation of effective prevention strategies while we await more data.

International guidelines, endorsed by the appropriate international agencies, are needed to define medically indicated and inappropriate uses of medical needles. Implementing such guidelines should substantially reduce the number of injections that are given and, in turn, the incidence of occupational exposure to blood and cross-infection between patients.

Training and education in injection safety, prevention of

sharps injuries, and universal precautions must be incorporated into the curriculum in medical and nursing schools in developing countries. Health care workers must be instructed to limit the use of needles strictly to medically indicated purposes. In addition, sustained public-education campaigns are needed to dispel much misinformation about injections.

Suppliers of diagnostic test kits that rely on the sampling of capillary blood should be required to bundle noninjurious devices with their kits, including plastic or plastic-wrapped capillary tubes and automatically retracting finger-stick lancets. The safe containment of injection-related waste depends on improvements in the disposal systems for medical waste. International standards must be established for the placement, the puncture resistance, the fluid resistance, the sealing mechanisms, and the transportability of sharps-disposal containers.

HBV-vaccination programs for health care workers are needed in areas where the prevalence of HBV is high. Priority should be placed on vaccinating the health care workers who are at the greatest risk of contact with blood and body fluids. Sentinel surveillance systems are needed in selected developing countries and geographic regions to identify high-risk practices and devices and to help in the planning and testing of effective interventions. Good decisions regarding the allocation of limited financial and material resources require accurate data.

Protective measures for health care workers must be part of every program supported by international organizations for the prevention of the spread of HIV and AIDS and other infectious diseases. The WHO Department of Vaccines and Biologicals has already set a precedent in this area by officially promoting improved standards of injection safety to prevent the transmission of disease among health care workers and patients alike.[33–34] WHO urges immunization programs to make a complete transition to auto-disable syringes by 2004 and promotes the bundling of auto-disable syringes and puncture-proof disposal containers with vaccines.

Other organizations need to follow the example set by WHO. The World Bank has committed $1.7 billion to HIV- and AIDS-related projects in more than 51 countries.[35] One of the primary aims of the prevention initiatives it supports in developing regions is increasing the availability and use of condoms in at-risk populations.[36] However, the use of protective gloves by birth attendants during deliveries is no less important for preventing the transmission of HIV. Therefore, these same prevention programs should also support increased access to procedure gloves and barrier garments for birth attendants and other health care workers who are at risk for contact with blood.

Along with international agencies, national budgets should provide resources to ensure the safety of medical personnel. These expenditures should not be viewed as an increase in the cost of health care in developing nations but, rather, as insurance to protect each nation's investment in its health care work force. The inevitable consequence of continued inattention will be a mounting toll of disease and death among productive health care workers in places where their loss can least be afforded.

Acknowledgement: Dr. Charles Sagoe-Moses was supported as an ITREID (International Training and Research in Emerging Infectious Diseases) Fellow at the University of Virginia, Department of Internal Medicine, Division of Geographic and International Medicine, by NIH Fogarty Center Grant No. 21204. The authors are pleased to acknowledge the support of the ITREID program and Richard L. Guerrant, M.D.

References

1. Needlestick transmission of HTLV-III from a patient infected in Africa. *Lancet.* 1984;2:1376–77.
2. Centers for Disease Control and Prevention. Recommendations for prevention of HIV transmission in health care settings. *MMWR.* 1987;36 (suppl 2S):3s–18s.
3. Occupational Safety and Health Administration. 29 CFR Part 1910.1030: Occupational exposure to bloodborne pathogens. *Fed Regist.* 1991;56:64004–65182.
4. Mahoney FJ, Stewart K, Hu H, Coleman P, Alter MJ. Progress toward elimination of hepatitis B virus transmission among health care workers in the United States. *Arch Intern Med.* 1997;157:2601–5.
5. Perry J. H.R. 5178 promises unprecedented protection to U.S. health care workers. *Adv Exposure Prev.* 2000;5:39–40,44.
6. Needlestick Safety and Prevention Act of 2000, Pub. L. No. 106–430, 114 Stat. 1901 (Nov. 6, 2000).
7. Simonsen L, Kane A, Lloyd J, Zaffran M, Kane M. Unsafe injections in the developing world and transmission of bloodborne pathogens: a review. *Bull WHO.* 1999;77:789–800.
8. Kane A, Lloyd J, Zaffran M, Simonsen L, Kane M. Transmission of hepatitis B, hepatitis C and human immunodeficiency viruses through unsafe injections in the developing world: model-based regional estimates. *Bull WHO.* 1999; 77:801–807.
9. Martinson FE, Weigle KA, Mushahwar IK, Weber DJ, Royce R, Lemon SM. Seroepidemiological survey of hepatitis B and C virus infections in Ghanian children. *J Med Virology.* 1996;48:278–83.
10. Richard-Lenoble D, Traore O, Kombila M, Roingeard P, Dubois F, Goudeau A. Hepatitis B, C, D, and E markers in rural equatorial African villages. *Am J Trop Med Hyg.* 1995;53:338–41.

11. McQuillan GM, Coleman PJ, Kruszon-Moran D, Moyer LA, Lambert SB, Margolis HS. Prevalence of hepatitis B virus infection in the United States: the National Health and Nutrition Examination Surveys, 1976–1994. *Am J Public Health.* 1999; 89:14–8.
12. Alter MJ, Kruszon-Moran D, Nainan OV, et al. The prevalence of hepatitis C virus infection in the United States, 1988 through 1994. *N Engl J Med.* 1999;341:556–62.
13. UNAIDS/WHO. AIDS epidemic update: December 1999. Geneva: Joint United Nations Programme on HIV/AIDS and World Health Organization, 1999. Document No. UNAIDS/99.53E.
14. Ippolito G, Puro V, Heptonstall J, Jagger J, De Carli G, Petrosillo N. Occupational human immunodeficiency virus infection in health care workers: worldwide cases through September 1997. *Clin Infect Dis.* 1999;28:365–83.
15. Jagger J, Hunt EH, Brand-Elnaggar J, Pearson RD. Rates of needle-stick injury caused by various devices in a university hospital. *N Engl J Med.* 1988;319:284–88.
16. Anglim AM, Collmer JE, Loving TJ, et al. An outbreak of needlestick injuries in hospital employees due to needles piercing infectious waste containers. *Infect Control Hosp Epidemiol.* 1995;16:570–76.
17. Ippolito G, De Carli G, Puro V, et al. and the Italian Study Group on Occupational Risk of HIV Infection. Device-specific risk of needlestick injury in Italian health care workers. *JAMA.* 1994;272:607–10.
18. Lymer UB, Schutz AA, Isaksson B. A descriptive study of blood exposure incidents among healthcare workers in a university hospital in Sweden. *J Hosp Infect.* 1997;35:223–35.
19. McCormick RD, Meisch MG, Ircink FG, Maki DG. Epidemiology of hospital sharps injuries: a 14-year prospective study in the pre-AIDS and AIDS eras. *Am J Med.* 1991;91(B):301S–307S.
20. Khuri-Bulos NA, Toukan A, Mahafzah A, et al. Epidemiology of needlestick and sharp injuries at a university hospital in a developing country: a 3-year prospective study at the Jordan University Hospital, 1993 through 1995. *Am J Infect Control.* 1997;25:322–29.
21. Adegboye AA, Moss GB, Soyinka F, Kreiss JK. The epidemiology of needlestick and sharp instrument accidents in a Nigerian hospital. *Infect Control Hosp Epidemiol.* 1994;15:27–31.
22. Miller MA, Pisani E. The cost of unsafe injections. *Bull WHO.* 1999;77:808–811.
23. van Staa AL, Hardon A. Injection practices in the developing world: A comparative review of field studies in Uganda and Indonesia. Geneva: Action Programme on Essential Drugs, World Health Organization; 1996, pp. 15, 43. Report No.:WHO/DAP/96.4.
24. Battersby A, Feilden R, Stilwell B. Vital to health? A briefing document for senior decision makers. Draft. Health Technical Services (HTS) Project, TvT Associates, Inc.; 1998 Sept. Contract No.: HRN-C-00-93-00001-00. Sponsored by the U.S. Agency for International Development (USAID).
25. Aoun H. When a house officer gets AIDS. *N Engl J Med.* 1989;321:693–696.
26. World Health Organization. World malaria situation in 1991–Parts I and II. *Weekly Epidemiological Record.* 1993;68(34, 35):245–242, 253–258.
27. U.S. Department of Health and Human Services, Food and Drug Administration. Glass capillary tubes: joint safety advisory about potential risks. Rockville, MD: February 1999.
28. Jagger J, Deitchman S. Hazards of glass capillary tubes to healthcare workers [letter]. *JAMA.* 1998;280:31.
29. Mfugale, Deodatus. "Health Workers Risk HIV/AIDS with Plastic Bag 'Gloves'". Panafrican News Agency, 5/22/01. Available at: http://allafrica.com/stories/200105030075.html.
30. Jagger J, Hunt EH, Pearson RD. Estimated cost of needlestick injuries for six major needled devices. *Infect Control Hosp Epidemiol.* 1990;11:584–588.
31. Jagger J, Bentley M, Juillet E. Direct cost of follow-up for percutaneous and mucocutaneous exposures to at-risk body fluids: data from two hospitals. *Adv Exposure Prev.* 1998;3:25,34–35.
32. Data from World Health Organization web site, WHO Statistical Information System page (www.who.org/whosis), under "Health Personnel."
33. Safety of injections: WHO-UNICEF-UNFPA joint statement on the use of auto-disable syringes in immunization services. Geneva: World Health Organization,1999. (Report no. WHO/V&B/96.25.)
34. Safety of injections in immunization programmes: WHO recommended policy. Geneva: World Health Organization, 1998. (Report no. WHO/EPI/LHIS/96.05 Rev. 1.)
35. World Bank HIV/AIDS project lending. Washington, D.C.: World Bank Group, 2000. (Accessed July 27, 2001, at: http://www.worldbank.org/html/extdr/pb/pbaids-activities.htm.)
36. National AIDS programmes: a guide to monitoring and evaluation. Geneva: Joint United Nations Programme on HIV/AIDS (UNAIDS), 2000. (Document no. UNAIDS/00.17E.)